The Great Apes

The Great Apes

Edited by David A. Hamburg

and Elizabeth R. McCown

Perspectives on Human Evolution, Volume V
*A Publication of the Society for the Study of
Human Evolution, Inc., Berkeley, California*

The Benjamin/Cummings Publishing Company
Menlo Park, California • Reading, Massachusetts
London • Amsterdam • Don Mills, Ontario • Sydney

This book is in the
Benjamin/Cummings Series in Anthropology

Dedication
Dedicated to the memory of Clarence Ray Carpenter (1905-1975), honoring his many contributions to primatology and particularly his pioneering field studies of the naturalistic behavior of howler monkeys and gibbons.

Editorial/production services by
Phoenix Publishing Services, San Francisco.

Library of Congress Cataloging in Publication Data

Main entry under title:

The Great apes.

  (Perspectives on human evolution ;   v. 5.)
(Benjamin/Cummings series in anthropology)
  "A publication of the Society for the Study of
Human Evolution."
  Bibliography: p.
  Includes index.
  1. Apes—Behavior.   2. Mammals—Behavior.
I. Hamburg, David A.   II. McCown, Elizabeth R.
III. Society for the Study of Human Evolution.
IV. Series.   V. Series: Benjamin/Cummings series
in anthropology.
GN281.P42   vol. 5   [QL737.P96]   573.2'008s
ISBN 0-8053-3669-9   [599'.88'045]   79-11361
ABCDEFGHIJ-MA-782109

The Benjamin/Cummings Publishing Company
2727 Sand Hill Road
Menlo Park, California 94025

The Great Apes

Perspectives on Human Evolution, Volume V

is a book sponsored by:

Society for the Study of Human Evolution, Inc.
Berkeley, California

Organized for the purpose of increasing and disseminating
information in the field of human evolution through a series of
books entitled Perspectives on Human Evolution.

Officers of the Society for 1978:
S. L. Washburn, President; Phyllis Dolhinow,
Glynn Isaac, Alice Davis, Elizabeth McCown

Previous publications of the Society:

Perspectives on Human Evolution, Volume I.
S. L. Washburn and Phyllis C. Jay, Editors.
New York: Holt, Rinehart and Winston, 1968.

Perspectives on Human Evolution, Volume II.
S. L. Washburn and Phyllis Dolhinow, Editors.
New York: Holt, Rinehart and Winston, 1972.

Perspectives on Human Evolution, Volume III.
G. L. Isaac and E. R. McCown, Editors.
Menlo Park, California: W. A. Benjamin, 1975.

Perspectives on Human Evolution, Volume IV.
S. L. Washburn and E. R. McCown, Editors.
Menlo Park, California: Benjamin/Cummings, 1978.

Preface

On May 19, 1975, a heavily armed group of men, for political reasons, crossed Lake Tanganyika in the middle of the night and kidnapped four persons doing primate research at the Gombe Stream Research Centre in Tanzania. Over the ensuing several months, after intensive negotiations, their safe release was obtained. The incident, however, highlights the enormous difficulty of research on great apes in their natural habitat. Such research is possible only in remote locations where human incursions have not yet had devastating effects upon the habitat and upon the animals. Even so, as the Gombe case illustrates, human intervention is an increasingly serious problem, and it may take forms that are menacing not only to the animals and the habitat but also to the scientists. Even without this problem, many difficulties are involved in primate research: geographic access, adequate supplies, attracting scientists and students of high quality in sufficient numbers, coping with disease and isolation, and obtaining the necessary funds. Thus, it has rarely been possible to maintain long-term field stations for great ape research. This situation is particularly unfortunate since the long life span of great apes sharply restricts the value of short-term studies. So it is not surprising that there has been a paucity of reliable information on the behavior and ecology of great apes in their natural habitat. Of the available accounts, much has been descriptive rather than analytic; much has been impressionistic and anecdotal rather than systematic and quantitative.

Since the great apes are more closely related biologically to the human species than any other living organism, their significance for the life sciences is very great. Both as a window on human evolution and as a model for human disorders, great apes can be highly informative. Given the formidable difficulty of studying them in nature, it might be supposed that abundant research in captivity has been undertaken to compensate for the practical limitations of research in the natural habitat, to carry out experimental investigations, and to make kinds of

observations that are impossible in the natural habitat. The fact is, however, that very little research has been done in captivity settings, and
especially in captivity settings that simulate the natural environment.
There are cogent reasons for this deficiency also. Great apes are large,
powerful, and difficult to handle in a laboratory. They are expensive
and difficult to obtain. Since their numbers are dwindling, there is no
desire to decimate their populations further by large-scale importation
even if that were feasible. It is very expensive to maintain adequate
facilities for healthy breeding colonies and good experimental conditions. The net result of these constraints is, again, a regrettable paucity
of data utilizing modern techniques. So, information of great potential
value for behavioral biology that could be obtained by investigation of
great apes in their natural habitats and in diverse captivity settings has
been largely lacking until recent years.

In the 1970s, there has been a modest increase in laboratory investigation and a striking increase in field studies. This book brings a large
portion of that work together in a single volume for the first time.
While attempting to foster an upsurge in such research during the early
1970s, Jane Goodall and I proposed to Lita Osmundsen that the
Wenner-Gren Foundation for Anthropological Research support a
conference that would bring together a number of the young investigators making fresh contributions, leavened with a few experienced
investigators to put the new work in perspective. This led to a highly
stimulating conference at the Wenner-Gren Conference Center in Burg
Wartenstein, Austria, in the summer of 1974. Due to space limitations
at Burg Wartenstein, many promising young investigators could not be
included at the conference. Therefore, I invited an additional group to
prepare papers after the conference. It was my intention to include all
those doing new, careful, systematic research, with special emphasis on
young field workers. In hindsight, I recognize that several thoughtful
workers were left out because we were not aware that they were
actively working in this field. A few of these had made earlier contributions; a few had just entered the field. I regret that this volume could
not have included all of them, but I am gratified that we have been able
to include such a large proportion of those who have new data and
fresh observations to report.

I regret that the book was delayed by several factors: a desire to include a wider range of authors whose papers had to be prepared after the
conference; the kidnapping at Gombe, which seriously disrupted the
lives of some of the people centrally involved in the preparation of the
book; and unforeseeable complications in the publishing world. Indeed,
at times it seemed doubtful that the book would ever be published.

Nevertheless, the book has not only emerged intact but greatly
strengthened by the postconference additions. It contains a great deal

of information, ideas, and insight about three fascinating creatures—chimpanzees, gorillas, and orangutans. Indeed, I believe it contains more dependable information on the great apes than any single volume I have encountered. This is possible only because of the dedicated, thoughtful, and persistent efforts of many people, above all, the authors. I express my deep appreciation to them for their remarkable work and generosity and patience. I am also very grateful to Lita Osmundsen for making the conference possible, to Jane Goodall for help in planning the conference, to Sherry Washburn for long-term stimulation, and to Libby McCown for her fine editing work. All of us who have done research on great apes have been fascinated, and see much potential for understanding, not only the apes, but the human species as well. We only hope that some of this fascination will be contagious for students, scholars, and curious readers everywhere who wonder about human origins and our place in nature.

David A. Hamburg

Contents

Introduction

In 1929 Yerkes and Yerkes published a major review of the behavior of the great apes. In their discussion of the voluminous and widely scattered literature, they stated that "superstitions, surmises, rumors, accidental and unwarranted conclusions have been repeated through the centuries," and that "the hundreds of descriptions, discussions, and rediscussions of anthropoid life and behavior which we have examined make us feel that it is a pity man should write so much, while observing, reflecting, and criticizing himself and his work so little" (p. 2). In their conclusions they commented, "As one reflects on the situation, it appears first incredible, then ludicrous, that as professional scientists we should depend on accident instead of intelligently planned and prearranged conditions for the extension of knowledge and the solution of significant problems" (p. 590). Yerkes not only pioneered in the use of the apes—particularly the chimpanzee—in psychological research, but he also promoted the first field studies devoted entirely to the study of primate behavior. The monographs on the chimpanzee (Nissen 1931), the gorilla (Bingham 1932), and the howler monkey (Carpenter 1934) mark the beginnings of a new era.

The reasons for the "ludicrous" situation described in *The Great Apes* are complex and are worth examining because their consequences are still very much with us. All the classical writings on the behavioral evolution of man predated any reliable account of the behavior of any nonhuman primate. The summary tables in Yerkes and Yerkes (pp. 552-561) offer the best guesses that could be made based on the informa-

The research reported in this article was supported by the Wenner-Gren Foundation. We wish to thank Mrs. Alice Davis for editorial assistance in the preparation of this article.

tion available at that time. According to Yerkes and Yerkes, gibbons, for example, were characterized as nomadic, living in bands or herds, having leaders, but having no family structure. Carpenter's monograph on the gibbon (1940) corrected these erroneous beliefs and his work has to be seen, not only as the first reliable account of the behavior of any ape, but against this earlier background of misinformation.

In the traditional studies of social evolution, the data on human behavior was equally unreliable. Hobbes, Locke, Montesquieu, and Rousseau relied heavily on travelers' tales. Voyagers' accounts were the source of a great deal of the data for many early social thinkers (Penniman 1965). Westermarck *(History of Human Marriage* 1891) believed that the gorilla lived in family groups and that the human family developed from this behavior pattern. At the same time, the accepted view was that these animals were promiscuous and that group marriage preceded monogamy. The purpose here is not to review theories of human evolution, but to emphasize that they rested on data that had been collected in a haphazard manner by untrained people. Primatologists must recognize that, when sociologists or social anthropologists speak of evolution, they are expressing the views of social philosophers who arranged uncertain "facts" according to mistaken biological analogies. The transition between the work of Sir James Frazer and that of Malinowski or Radcliffe-Brown is of the same magnitude as the transition between the work of Yerkes and Yerkes and the recent field studies. Just as in social anthropology, the change has two interrelated features—first, the investigator goes into the field and collects data, and second, the field situation is discovered to be at variance with the prevailing concepts, making a new start necessary. Not only are the new data different, but the questions are of a different order, as can be seen by comparing the articles in this symposium with the summary tables in Yerkes and Yerkes.

Due to geographic, environmental, and, by far the most important, cultural reasons, substantial field studies were slow in being promoted. The gorilla was not even scientifically recognized until the publication of the paper by Savage and Wyman (1847), although there had been many reports of a huge and fierce kind of ape. Much of Huxley's *Evidence as to Man's Place in Nature* (1863) was devoted to describing the discovery of the apes—that there were three great apes and a considerably smaller ape. The existence of the pygmy chimpanzee was not established until the 1930s, and, to keep matters in perspective, the okapi and giant forest hog were discovered[1] in the years 1900-1905 and the mountain gorilla in 1902 (Schaller 1963). The period of discovery really ended about the time of World War I. Until then, zoological interests centered on collection and classification of specimens, and there was great motivation to make new finds.

Until recently, travel was slow and life in the tropics was dangerous for Europeans. In 1972, Kruuk wrote, "I have had a marvelous time in Africa. . ." (p. xv). He is referring to the world of today with the convenience of air travel and the expertise of tropical medicine. But less than a hundred years ago, Africa was a very different place. Alan Moorehead, in *White Nile* (1960), vividly described the European explorations in East Africa—sick explorers being carried on litters, disease so rampant that the question was survival and the wholesale desertion of porters. Lack of time and money and the threat of illness tended to keep scientists at home, although there were infrequent collecting expeditions. These difficulties explain, in part, why scientists did not spend long periods of time studying the behavior of the great apes. A more important reason was that such study was not regarded as necessary, since it was commonly assumed that animal behavior was easy to understand and that reliable information was available from hunters who were reputed to have much knowledge about the animals they encountered. The difference between fact and fancy is nowhere greater than in the lore of the hyena, a well-known animal, which has been the subject of many myths and stories for centuries. It was not until 1972 that Kruuk laid the basis for understanding the behavior of this common carnivore.

Although all great apes live under conditions that make observation difficult, no obstacles can account for the misunderstanding of monkey behavior. Macaques, baboons, and langurs all live close to man, are easy to observe, and are often kept as pets. But whether regarded as gods or nuisances, these animals have never been the subject of a systematic effort to understand their behavior by any earlier culture, including India, the Far East, Egypt, or tribal peoples. This cannot be explained as a lack of interest in animals, since every culture has animal tales and many have animal gods. Hunting, a major human activity until very recent times, did not lead to reliable information on animal behavior. Interest in careful study of behavior is a product of modern science, and many results of such study run counter to the common sense understanding of all cultures. It is no accident that, with the advent of rapid air transportation, the study of primate behavior appeared to start independently in many countries after the suspension caused by World War II. Modern science laid the background for the studies whether they were in Japan, Europe, or the United States, or whether the investigators were anthropologists, psychologists, or zoologists.

The consequences of the new studies of animal behavior in general, and of the primates in particular, are twofold: first, the studies reveal a new and fundamentally different view of life, and second, they suggest that these humanistic and enriching studies should be incorporated into the general education system. Since they are also scientific, such studies are therefore a necessary part of any general behavioral science.

Seeing undisturbed animals, animals unafraid of man, is a remarkable experience. An incident related by James Woodburn, an anthropologist who had been studying the Hadza people, illustrates this point. A Hadza man accompanied him into Nairobi Park and apparently was disinterested in the trip. Then, when he saw the animals, he became very excited. Woodburn explained that this man, a member of a hunting tribe, had never seen so many nor such fearless animals before. This experience vividly illustrates the realization that what western culture has defined as wild animals are really animals afraid of man, and, since our species has hunted for at least three million years, the normal relationship between men and animals is that of hunter and hunted. The antagonistic relationship is so deep that it does not seem obvious to us that when we speak of the buffalo as the most dangerous animal in Africa, we are speaking of a wounded buffalo. Wounded big game may indeed be very dangerous, but the behavior in the parks shows that human behavior creates the flight or fight of so-called wild animals. Misconceptions of the nature of animal behavior were so widespread that only in the last few years could studies such as Schaller's on the gorilla (1963) and the lion (1972) have been undertaken. Many of the differences in accounts of gorilla behavior (as described by Du Chaillu and Fossey) can be attributed to the attitudes of the investigators rather than the behavior of the gorillas. Current behavioral studies, as exemplified by the papers presented in this symposium, are attempting to see the animal world as it is—minus the presence of destructive man—giving us a very different perspective.

Rich descriptions of what animals actually do are essential first steps in understanding, but the history of science shows that description alone is a very weak tool. The contribution of ethology has been the combination of naturalistic observation with ingenious experiments (Eibl-Eibesfeldt 1975). It is necessary that primate field studies, following the example of ethology, begin to employ greater control. The present tendency is to treat the comparative study of behavior as a rather distinct discipline; a whole school of thought—behaviorism—is based on the concept of behavior as an independent science (Skinner 1974).

There should be a constant interplay between the study of behavior and the comprehension of the underlying biology. It is not a question of one or the other, or one first and the other second. For example, in the primates, maturation is increasingly delayed in this order: monkeys, apes, man (Schultz 1969). The phenomenon has been interpreted—particularly in the case of man—as providing more time for learning. Although this may be the case, maturation is also delayed in such forms as elephants and rhinos (Asdell 1964). Since large animals generally mature more slowly than closely related small ones, size seems to be related to length of life. The reasons for the delay are complex, involving, at a minimum, such factors as longer life, larger size, greatly reduced birth potential, longer period of immaturity with consequent longer period of play and learning. It becomes obvious that maturation and learning are not related in any simple way, and, more importantly, neither has been analyzed very satisfactorily.

If an understanding of social behavior is sought, maturation might most usefully be described in terms of maturation of the brain, not in years. For example, if some structures of the brain are particularly important in social behavior, then comparisons might be based on their maturation. Just as sexual maturity has a physiological definition with behavioral consequences, so social maturity could use a physiological definition with behavioral consequences. The general tendency is to stress learning, but is there any evidence that the orang has a capacity to learn more than the highly social ground-living monkeys? The descriptive data show that delay of maturation has been a major factor in primate evolution, and that it has great consequences for social behavior. The effects and costs may be observed, but not the mechanisms or biological causes, and both are necessary for the development of a science of behavior.

Interestingly, although social behavior is usually treated behavioristically with little regard for biology, sexual behavior has been considered from both points of view (Beach 1965; Katchadourian and Lunde 1972). Zuckerman (1932), in his classic study on the *Social Life of Monkeys and Apes,* reviewed the social behavior of the nonhuman primates and concluded that "the factor that determines social grouping in subhuman primates is sexual attraction" (p. 31). He combined what was then known of the social behaviors of primates with that of other mammals and described the basic physiology of the reproductive cycle. It was the combination of the data on social behavior with the underlying physiology that made the theory so attractive and influential. More recent field data show that the behavioral situation was much more complicated than could have been realized at that time, but Zuckerman's use of data from both field and laboratory provided a clear model for behavioral science, which was much more important than his particular theory of society. In *Functional Affinities of Man, Monkeys, and Apes* (1933), Zuckerman reviewed a wide variety of lines of evidence that he considered useful in the study of primates and in the interpretation of their behavior and evolution. It is unfortunate that this volume, which laid a basis for the multidisciplinary study of the primates, exerted so little influence—probably because it was far ahead of its time.

Aggressive behaviors provide useful examples of the importance of considering both field studies and biology. The kinds of situations eliciting aggressive behaviors have been reviewed by Hamburg and van Lawick

(1974). The neural (Mark and Ervin 1970) and hormonal (Hamburg 1971b) bases for such behaviors are remarkably similar in mammals. In spite of the general similarity in basic biology, there are marked differences among the species in the anatomy of such behaviors as bluff, threat, and fighting. The field worker sees these behaviors—and their effects within the groups, between groups of the same species, and between species. In most instances, males appear to be far more aggressive than females (B. Hamburg 1978).

The anatomy of females has been traditionally described in terms of a percentage of the male anatomy (Crook 1972), but another view might be considered: that the basic anatomy of the species is that of the female. In free-ranging primates, the females sleep, sit, move from place to place, and eat in the same way as the males. To these basic biological systems, the males have the added anatomy of bluff and aggression (Washburn and McCown 1973). Although females may also be aggressive, they lack the anatomy for successful fighting against the larger males. Even in gibbons, the least sexually differentiated of the apes, the much higher rate of broken canines in the males (Frisch 1973) shows the results of male aggressive behavior.

Differences between males and females in the anatomy of fighting are much greater than they appear if the usual methods of comparison are used. Crook (1972) states that the female's weight is about 70% of that of the male in *Cercopithecus,* and 50% in the baboon *(Papio)*. But in *Cercopithecus aethiops,* the temporal muscle of the male is more than twice the size of that of the female, and in some baboons, four times. Differences in the size of the canine are much greater, and the difference in the anatomy of fighting is far greater than might be expected on the basis of comparisons of body size. It is grossly misleading to use sex differences in body weight as a measure of the difference in fighting ability, and male chimpanzees probably should be regarded as twice the fighting size of females, rather than considering females as 90% the size of males.

The issues concerning basic anatomy are two. First, if both anatomy and behavior are considered, the sex differences seen in the field appear much greater than if information about the differences is restricted to the external appearance of the animals. Second, if female size is regarded as the basic size of the species, and if the sex differences (aside from the primary ones) are seen as the anatomy of aggression, then a very large and diversified amount of anatomy may be viewed as adaptation to bluff and fighting. Bluff is exceedingly important and takes many forms (Guthrie 1970); the anatomy of fighting consists of gross body size, the canine jaw muscle complex, neck muscle size, orbit size, and many cranial dimensions. The behavior elucidates the anatomy, just as the anatomy helps in understanding the behavior.

If the sex differences of the monkeys and apes are viewed in historical perspective, evolutionary theory demands that they be considered as adaptive. Phenotypically, the differences are numerous and large. Such complex differences cannot be due to chance—although some, at the molecular level, may be. Since adaptation is for the behaviors that lead to reproductive success, the fieldworker's task is to identify these behaviors, one being the morphological pattern of bluff and fighting. It is probable that male-male conflict is by far the commonest and most important kind of agonistic behavior in monkeys and apes. In gorillas, protection of the group appears to be crucial in terms of both bluff and attack. In chimpanzees, there is aggressive behavior within the group. An important task is to determine the various ways that agonistic behaviors affect differential reproduction and group survival. If baboons do not protect against predators, have no territory, and no dominance, how can one account for differences in the anatomy of fighting? In the apes, very substantial species differences in the anatomy of bluff and fighting are evident. If the behaviors that account for them can be specified, then both anatomy and behavior will be understood—but neither can be fully analyzed without the other. The utility of the evolutionary approach is to provide an intellectually powerful perspective. If complex structures have survived, and are maintained, they must be adaptive (Washburn and Harding 1975).

In summary, whether one is concerned with social maturation, sexual behaviors, or fighting, interpretation comes both from observation of behavior and from analysis of the facilitating biology. Traditionally the necessary information and methods have been assigned to different disciplines. Primatology has the great advantage of being a meeting ground; it is neither the property of any one department, nor is it limited to any particular set of techniques. Its great strength can be to capitalize on the diversity of problems and techniques to build toward a synthetic behavioral science.

It is hard to realize how quickly knowledge becomes institutionalized and thus the property of a group of specialists. The studies of primate behavior, because of their very success and the quantity of data this is producing, are in danger of becoming a specialized activity. This is already happening at scientific meetings, at which sessions are reserved for papers on primate behavior. With the advance of knowledge, some specialization is helpful, but a framework of operations needs to be developed that encourages the analysis of behavior from an interdisciplinary approach.

Because of their phylogenetic position, the great apes are essential for comparison of human behavior to that of other primates. At the present time, there is no agreement on how closely man is related to other primates. There are eminent scholars who think that the time of separation may be more than 35 million years ago and that no apelike creature existed in human ancestry (Kurtén 1972). The more common view is that the lineages of man and ape separated in the early Miocene —some 20 million years ago.

Biochemical evidence, however, suggests a much closer relationship. As shown in Table 1, man, chimpanzee, and gorilla form a close group. They are approximately equally related—as close as species that have always been regarded as very similar, such as sheep and goat. Orangutan and gibbon are closely related to each other, but are not close to the African apes and man. Biochemically, the monkeys are much further away from apes and man. These data fit the theory that the apes and man, after the separation from the monkeys, shared a long period of common ancestry. The fundamental similarities in the anatomy of the arms and trunk probably evolved at that time. When the apes divided into eastern and western groups, man shared an additional long period of common ancestry with the chimpanzee and gorilla. The ancestral form was probably a knuckle-walker. These relationships are shown in Table 1. Certainly the biochemical taxonomy justifies the emphasis that this volume places on the behavior of the great apes, and particularly the African apes, because they are—by far—our closest living relatives.

The significance of studies of primate behavior, and particularly that of the great apes, for the social sciences, may be illustrated by the findings reported at this symposium. For example, the incest taboo has been widely regarded by social scientists as a hallmark of mankind, basic to all varieties of kinship organization, and necessary for the human way of life. But recent studies of animal behavior show that mating systems that lead to outbreeding are the rule rather than the exception (Bischof 1975). As shown by the papers in this volume, outbreeding is the norm in apes—and in the chimpanzee and gorilla, the behavior of females is of particular importance. As Parker (1976) has shown, exogamy has genetic,[2] psychological and social consequences,

TABLE 1
Relationships of man to apes and monkeys

MAN TO:	CHIMPANZEE	GORILLA	ORANGUTAN	GIBBON	MONKEY
DNA	.7	1.4	2.9	2.7	5.7
Sequences	.3	.6	2.8	2.4	3.9
Immunology	.8	.8	1.6	2.2	3.5
Albumin	7	5	13	14	37
Transferrin	9	8	23	24	49

Source: Table prepared by John E. Cronin. Full references in: "Molecular Systematics of the Primates," by Sarich, V. M. and J. E. Cronin. In Goodman, M. and R. Tashian, eds., *Molecular Anthropology* New York: Plenum Press. 1976.

and all these are elaborated in the case of man. The human situation may be interpreted, however, as a development from a common biological base that man shares with many other mammals.

The studies reported in this volume show that the great apes have a variety of social organizations. Just as in its biochemistry and anatomy, the orangutan is the most divergent from the African apes and man—but chimpanzee and gorilla are by no means identical. The differences make evolutionary reconstruction more difficult, yet open the way for comparisons among man's closest relatives. For example, the behavior of mothers and infants seems remarkably similar in gibbons, orangutans, chimpanzees, and gorillas—but the social systems of the adults are different. This suggests that the primary purpose of mother-infant behavior is the survival of infants, that variation is strictly limited, and that the social systems of adults are not determined by early experiences. The often cited importance of play in socialization is brought into question by the behavior of both gibbon and orangutan. The same comparisons show that there is no simple relationship between ecology and social organization. It is evident that there are multiple solutions to the same ecological problems. So the very diversity of behaviors that makes evolutionary reconstruction more difficult makes the study of comparative behaviors more rewarding.

The behavior of the apes is of such intense interest to human beings that it has proved very difficult to record and delineate in an unbiased way. Just as the early records stressed violence and bravery and adventures of hunters, so more recent accounts stressed amiability and lack of dominance, aggression, or territorial behaviors. The very recent fieldwork and the subsequent articles in this volume clearly show, however, that the species of apes are, to varying degrees, aggressive, dominance-seeking, and territorial. It is hard to reconcile the characterization, a decade ago, of the mild ape with the chimpanzee or gorilla now known occasionally to kill and eat the young of its own species. Just as the image of the violent ape appealed to the nineteenth-century reader, so the portrayal of the amiable ape appeals to the reader of today (Montagu 1976), and it may be a long time before opinion will be based more on information than on human desire. It should be remembered that the evidence for aggressive behaviors does not depend only on behavioral field studies. The apes have the biological basis for aggressive behaviors, and some years ago, this fact led to a theory that the amount of aggressive behaviors had been greatly underestimated (Washburn and Hamburg 1968).

The importance of considering the biology when making behavioral comparisons is shown by the problem of the origin of human language. Chimpanzees cannot be taught to speak (Kellogg 1968), while, conversely, human beings learn to speak so easily that this kind of learning will take place under almost any circumstances. In man, a large part of the cortex of the brain makes possible the learning of complex sound patterns. It is because the basic language elements (phonemes) may be arranged in an almost infinite variety of forms that human language is such a remarkable means of communication. But in nonhuman primates, sounds are controlled by the limbic system (the primitive parts of the brain), and the cortex is not essential in sound production (Myers 1978). In monkeys, sounds may be produced by electrodes implanted in the primitive parts of the brain, and the removal of large amounts of cortex does not affect the production of sounds. In man, comparable damage to the cortex would destroy the ability to speak. If the problem posed is the origin of the human cognitive-phonetic-cortical system, it is very misleading to use the same word—language—for both human and non-human communication. Human phonetic communication is new in its biological base and in its almost unlimited capacity to convey meaning.

Study of the origin of human language shows the necessity of trying to understand the biological basis for behaviors that are being compared. The importance of communication may be seen in the field studies, but without experimenting, it would not be possible to differentiate between human and nonhuman systems. Observation alone will never solve the problems of primate behavior, nor will laboratory experiments show the adaptive nature of the behaviors under natural conditions.

Looking to the future, primatology should strive toward building a behavioral science that avoids the limitations of the traditional departments. This conference has contributed substantially toward understanding the behavior of the great apes, and, to appreciate the progress that has been made, one need only look back to the descriptions in *The Great Apes* (Yerkes and Yerkes 1929). The conference papers also alert us to the necessity for additional fieldwork, better methods of comparison, and more behavioral experiments in both field and laboratory. A major effort is still necessary to make the study of the apes as useful as possible for our understanding of human beings.

Notes

1. Stefansson has described "discovery" as the time of arrival of the first European, preferably an Englishman.

2. The issue is not that incest may produce genetic defects—because these would be eliminated by selection—but that inbreeding reduces the variability, and so the importance of sexual reproduction. The effectiveness of natural selection depends on variability that would be reduced by inbreeding, especially in very small populations.

PART ONE

Chimpanzees

This volume has been divided into five parts. The first three consider the behavior of the great apes—chimpanzees, gorillas, and orangutans, in that order. These are followed by a section on theoretical and laboratory studies and a section consisting of special topics.

The papers on chimpanzee behavior have been placed first because this species is the best known of the apes, and also because students of chimpanzee behavior were the most numerous at the conference. It was the cooperation between Jane Goodall (Gombe Stream Research Centre) and David Hamburg (then at Stanford University) that led to the planning of the conference.

Many readers will have heard Goodall's lectures or read her papers: fewer will be familiar with Hamburg's contributions on aggression and behavior (Hamburg 1971a, 1971b, 1971c). Because the Gombe camp was located in the middle of a chimpanzee territory much of the information on aggression, territorial boundaries, and group transfers is new.

As Hinde has noted in his article, it has been difficult to make a statement on the relations between ecology and behavior because of the general nature of the ecological descriptions. Itani's article helps to balance this by presenting a detailed description of the Ugalla area. Nishida discusses the spacing of troop and unit groups, and inter-unit-group relations seem to involve both social and ecological factors.

Since the time of the early experiments of Yerkes, the pygmy chimpanzee has been recognized as a fascinating animal. Kano describes his field study of pygmy chimpanzee behavior, but because of the many difficulties he encountered, he characterizes it as a very preliminary study.

Jane Goodall, Adriano Bandora,
Emilie Bergmann, Curt Busse,
Hilali Matama, Esilom Mpongo,
Ann Pierce, and David Riss

Intercommunity Interactions in the Chimpanzee Population of the Gombe National Park

The purpose of this article is to describe the interactions between neighboring communities (unit-groups, Nishida 1968) of chimpanzees in the Gombe National Park. First, however, we should very briefly review the *kinds* of behaviors that occur during intergroup interactions in other higher primates as a background against which to view such behavior in chimpanzees. Space does not permit an exhaustive review, and only a few of the species exemplifying each type of encounter have been selected.

Monkeys and apes sleep, travel, and feed in areas usually referred to as *home ranges* (Burt 1943). Within the home range, certain *core areas*

(Kaufmann 1962) are used extensively, while other more peripheral areas may be visited only occasionally, usually in connection with seasonal availability of food. Many of the higher primates have home ranges that overlap, to a greater or lesser extent, with those of neighbors, and it is in these overlap zones that interactions between groups tend to occur most often.

Intergroup interactions in many primates are sometimes peaceful—at least lacking in overt hostility. Neighboring groups may mingle briefly or even travel and feed together for periods of time (e.g., mountain gorilla, Schaller 1963) or they may remain peacefully in proximity (e.g., rhesus monkeys, Southwick, Beg, and Siddiqi 1965; Nilgiri langur, Poirier 1968). Sometimes, when two groups are near one another, young members from each may approach and initiate friendly (often playful) interactions (e.g., mountain gorilla, Schaller 1963; Fossey, this volume; anubis baboons, Ransom 1972).

In many species, groups tend to avoid one another. Loud carrying calls may serve to indicate the whereabouts of neighboring groups within an area so that they are able to avoid meeting (e.g., howler monkeys, Carpenter 1965; orangutans, Galdikas, MacKinnon, Rodman, this volume). In other species, one or more males may reveal the whereabouts of their troop by sitting high in a tree near the boundary of the home range (e.g., the St. Kitts green monkey, Poirier 1972).

Often there is a good deal of tension when two groups encounter one another—males may show vigilance behavior, threatening and staring intently at the neighbors (e.g., Barbary macaque, Deag 1973), or they may lead or herd their females in the opposite direction (e.g., mountain gorillas, Fossey, this volume; anubis baboons, Ransom 1972). Fossey even records gorilla groups leaving their nests and traveling some distance at night after hearing calls from another group nearby. Primate groups with a high degree of overlap between their ranges, or which meet frequently, are often ranked: a subordinate group usually moves away at the approach of a more dominant one, sometimes precipitously (e.g., rhesus monkeys, Southwick, Beg, and Siddiqi 1965; Morrison and Menzel 1972; Lindburg 1971).

In most species, there will be at least some hostile encounters, particularly when two groups are competing for a food source. Subordinate groups may be actively chased away and sometimes attacked (e.g., rhesus monkeys, Southwick, Beg, and Siddiqi 1965; chacma baboons, Stoltz and Saayman 1970; Saayman 1971a).

Some species actively defend part or all of their home range from intrusion by neighbors—in other words, they defend a *territory* (Burt 1943). The most truly territorial are probably the gibbon (Ellefson 1968) and the little *Callicebus* monkey (Mason 1968). Hanuman, Ceylon Grey, and Nilgiri langurs (Yoshiba 1968; Sugiyama 1967; Ripley

1967; Poirier 1968) are also territorial, as are vervet monkeys (Struhsaker 1967). The gibbons, langurs, and *Callicebus* all have loud carrying calls, which, in this instance, may serve to *attract* a neighboring group to a boundary area where members from both (usually adult males) may engage in aggressive displays that may lead to fighting.

Lone males (common in a number of primate species) may be chased off when they approach bisexual groups; sometimes they may be severely attacked (e.g., Japanese macaques, Kawai 1965; Nishida 1966; patas monkeys, Hall 1965; baboons, Ransom 1972; Packer, in prep.; Hanuman langurs, Jay 1965; Sugiyama 1967). Lone males may also transfer from one group to another (rhesus monkeys, Koford 1966; Lindburg 1969, 1971; Sade 1972; Boelkins and Wilson 1972; anubis and yellow baboons, Ransom 1972; Packer 1975; Altmann 1970; Rowell 1966; Japanese macaques, Nishida 1966). Sometimes a lone male or a group of males may attempt to drive out the leader and take over the females of a bisexual group (e.g., Hanuman langur, Sugiyama 1967; Mohnot 1971), or a lone male may herd or lead away one female (e.g., mountain gorilla, Fossey, this volume). These "raids" may lead to severe fighting.

Field studies of the same species in different areas show that the patterns of intergroup encounter may vary within a species; e.g., Hanuman langurs at Orchis were not aggressive when they encountered another group, whereas fierce territorial encounters were common in some other areas (Jay 1963; Sugiyama 1967). Moreover, encounters between the same two neighboring groups, within a species, may differ from one occasion to another; e.g., rhesus monkey groups, which usually showed extreme hostility to each other, sometimes fed peacefully together (Southwick, Beg, and Siddiqi 1965). It is becoming increasingly clear, as more and more results from long-term field studies become available, that "it is not the single social group which is the basic unit of social organization, but a system of social groups. The social groups interacting in a restricted area would correspond to the deme of the population geneticist" (Sade 1972). Yet the gathering of information that will enable us to understand properly a basic unit of this magnitude is one of the hardest problems that faces field research.

First, before we can comment meaningfully on why interactions between groups within species vary, we must know, among other things, something of the history of the relationship between neighboring groups. Figure 1 is a hypothetical diagram of the division of neighboring primate troops. If an investigator arrived during 1957 he would almost certainly observe differences in the interactions between the neighboring troops in his area. Troops A1 and A2 have been separate for six years: they will probably behave differently, during an encounter, than troops B1 and B2 which have only just split. The contacts

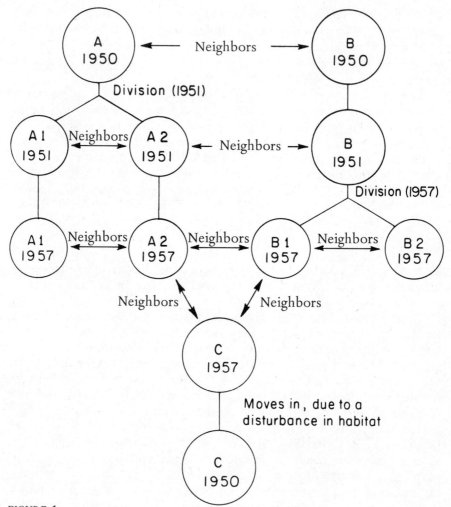

FIGURE 1
Hypothetical diagram representing changes in neighboring primate groups over time

between troops A2 and B1 with troop C will undoubtedly be of a very different nature, since the former have been neighbors for many years and the individuals have some familiarity with each other, while C is an intruder and its members are complete strangers in unfamiliar surroundings. An understanding of this history would explain many differences observed during intertroop interactions that, without such knowledge, could not properly be interpreted.

Second, we must bear in mind that a whole complex of variables are liable to affect the nature of encounters between neighboring groups; here we list merely some of the more obvious ones:

1. Of major importance is the type of habitat and the availability of food since this will affect the density of the population, the spacing between neighboring troops, and the size of the home range.

2. The availability of food at a given time of year.

3. The relative composition of neighboring troops: the male-to-male ratio is particularly important (or the size and disposition of individual males of one-male groups); the presence or absence of females in oestrus may also be significant.

4. The location of an encounter between troops relative to the core area of the home range of each is likely to be important since animals tend to be bolder and more aggressive when on familiar ground and increasingly nervous and apprehensive as they move further into less well-known areas (e.g., Carpenter 1965).

5. The *type* of encounter, and we can label at least six: (a) Chance encounter —when two traveling or foraging groups happen to pass one another, (b) Food competition encounter—when two groups are attracted by a food source in an overlap zone. This will be further affected by the attractiveness of the crop in question and the availability of other staple foods at the time, (c) Water or sleeping site encounter—very similar to the above, when two groups gather at a water hole or sleeping rock in habitats where these commodities are scarce, (d) Boundary patrol encounter —when one or both groups appear to be deliberately seeking out an intergroup encounter with neighbors near the boundary of their home ranges, perhaps after one of the two has called, (e) Raid or invasion—when one or more males approach a bisexual group and attempt either to drive out the leader male of that group and take it over or to herd or lead away one or more females, (f) Transfer encounter— when one individual (occasionally more than one) tries to transfer into a neighboring group.

6. In all of the above situations we must take into account the personalities of the various animals concerned and their "mood" (or state of arousal) at the time of the encounter.

7. Finally it must be emphasized, since we are pointing out the difficulties facing the field investigator, that a major concern is the degree to which animals of two neighboring groups are habituated to human observers. If one of them is relatively shy, while the other is totally unconcerned by the presence of humans, there may be an overriding effect on the outcome and nature of an encounter.

Information regarding interactions between communities (or unit-groups) of chimpanzees has been obtained from the two long-term studies in western Tanzania; in the Gombe National Park and in the Mahali Mountains area. Throughout the range of the species, so far as can be determined by the literature, the social structure of the chimpanzee community is similar (Goodall 1975). All investigators seem agreed that a group of adult males forms the central focus of a community: these males typically have a home range that is larger than that of most individual females and overlaps a number of smaller female ranges (Goodall 1975; Bygott, this volume; 1974; Wrangham, this volume; 1975).

Nishida (this volume) describes interactions between two unit-groups (communities), which have been studied intermittently for 9 years in the Mahali Mountains area. One major difference between chimpanzees of this area and those at Gombe is the fact that the former migrate annually, a distance of some 20 km, when the larger *M* group displaces the *K* group from the overlap portion of their home range.

Information regarding intercommunity interactions in the chimpanzee population at Gombe has been gradually collected over the years by many different field assistants and students. All data regarding such encounters—collected by Tanzanian field assistants, undergraduate science students from Stanford University and the University of Dar es Salaam, and by graduate students working for higher degrees—have been pooled in order that this chapter could be assembled. We should like to emphasize that this truly represents the spirit of collaboration which prevails at Gombe and to thank all concerned for the contributions they have made.

INTERCOMMUNITY INTERACTIONS AT GOMBE

Despite the fact that we have now documented a variety of interactions between chimpanzees of different communities, our understanding of the real relationship between communities is in its infancy. The picture has been confused by a number of factors: (a) in the early years of the study it was not possible to identify accurately all individuals seen in different areas of the park; (b) the setting up of the artificial feeding area served to attract many individuals to a permanent food source over a number of years (see Bygott, this volume; Wrangham 1974; Goodall 1975); (c) nonhabituated individuals of neighboring communities flee the presence of observers, thus minimizing the opportunity for seeing normal encounters between neighbors; (d) more observation is done in the core area, near the research center, than in peripheral areas where intercommunity interactions normally occur. Nevertheless, some extremely interesting observations have been made and it seems worthwhile to discuss these in this volume.

Communities and Ranging Patterns

Figure 2 (a, b, c) is a map of most of Gombe National Park showing the core areas of the two habituated communities, Kasakela and Kahama, during 1974. It was prepared for us by Steve Smith from the maps that are filled in to show the travel routes of chimpanzee groups by all personnel at Gombe: they are the particular responsibility of the Tanzanian field assistants. The map also indicates the location of the other, unhabituated communities, Mitumba, Rift, and Kalande. Unfortunately we do not yet know either the numbers of chimpanzees or the ranges of these communities. There is almost certainly another small community in the extreme south.

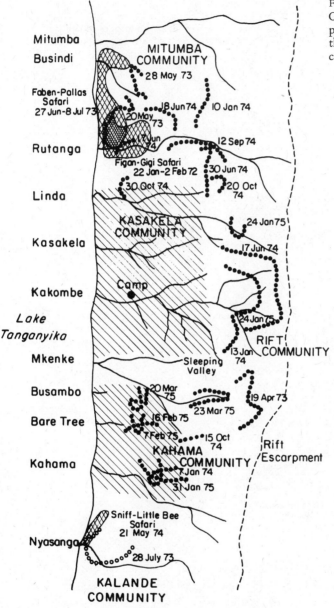

FIGURE 2(a)
Core areas and
peripheral patrols of
the two main study
communities

● ● ● Selected peripheral patrols by Kasakela chimpanzees
○ ○ ○ Selected peripheral patrols by Kahama chimpanzees
\\\\ Approximate core areas
⬭ Safaris in peripheral areas

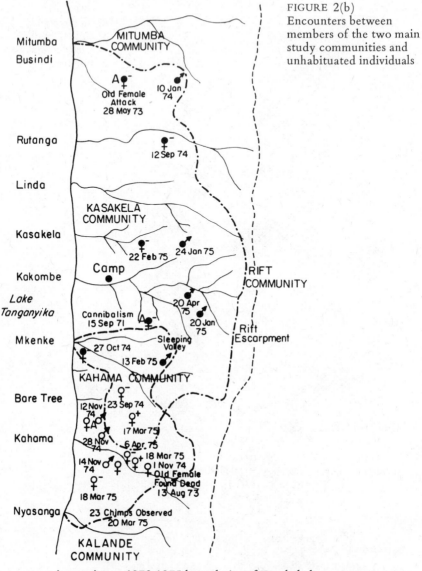

FIGURE 2(b)
Encounters between
members of the two main
study communities and
unhabituated individuals

—·—·— Approximate 1973-1975 boundaries of Kasakela home ranges

—·—·— Approximate 1973-1975 boundaries of Kahama home ranges

- ♀⁻ Visual encounters with anoestrus females by Kasakela chimpanzees
- ♀⁻ Visual encounters with anoestrus females by Kahama chimpanzees
- ♀⁺ Visual encounters with females in oestrus by Kahama chimpanzees
- ♂ Visual encounters with males by Kasakela chimpanzees
- ♂ Visual encounters with males by Kahama chimpanzees
- A Attacks

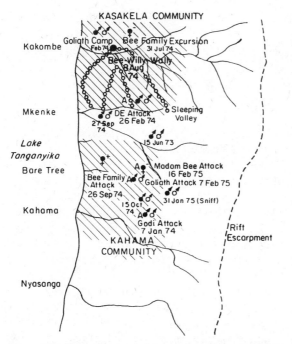

FIGURE 2(c)
Encounters between
individuals of the two
main study communities

♀⁻   Visual encounters with anoestrus females by Kasakela chimpanzees
♀⁺   Visual encounters with females in oestrus by Kasakela chimpanzees
♂♂   Visual encounters between Kasakela and Kahama males
A   Attacks
\\\\\   Approximate core areas
⟡⟡⟡⟡   Bee family excursions: Camp visit and excursion with Willy Wally

The home range of a community may change from year to year. In the early years of the study (1961 to at least 1964), habituated males of the Kasakela community sometimes traveled further north than they are known to do today. At that time, individuals of the Kasakela and Kahama communities associated freely and were considered one community (Bygott, this volume; Wrangham 1974; Goodall 1975). Males who are now members of the Kahama community, which split off to the south, were, in those early years, encountered in and to the north of Mitumba Valley (see Figure 2). The southern boundary of the habituated males at that time was not determined, but they uncertainly ranged as far as Kahama.

In 1971, at the time when the main study community was in the process of splitting (Bygott, this volume), the males of what is now the Kasakela community (the northern study community) were seldom observed to range further north than Linda Valley or further south than

Kahama: the Kahama males traveled most often in Mkenke and Kahama valleys and seldom went north of Kakombe. They occasionally penetrated Nyasanga and, on one occasion, Bygott (1974) followed a mixed group of Kahama males and the Kasakela males and females to Nyasanga. By the end of 1972, the division of the communities was virtually complete: the overlap zone at that time was Sleeping Valley (Figure 2).

During 1973 the Kasakela males began, once more, to extend their range further to the north, while the Kahama individuals frequently traveled as far south as Nyasanga. At the present time, as can be seen from the map, the Kasakela individuals frequently range as far north as Rutanga and they have extended their range to the south. The Kahama community now has a smaller range than during the previous year, and individuals seldom travel south of Kahama, possibly because unhabituated individuals of the large Kalande community (at least 23 individuals have been sighted in one party) are now often encountered in and to the north of Nyasanga Valley.

This oversimplified outline of changes in ranging patterns of males over a 14-year period shows that a chimpanzee can be familiar with a much larger area than he may be using at a given time. The factors that influenced phases of expansion and contraction of home ranges were, without doubt, extremely complex, depending in part on seasonal availability of food, on the extent to which the chimpanzees were provided with bananas, and on a variety of social factors (such as the relationship between the prime males, their personalities, or the availability of females in oestrus).

The ranging patterns of females are more complex and are discussed by Wrangham (this volume), Pusey (this volume), and Nishida (this volume). While young, regularly cycling females frequently accompany parties of males and may travel widely, older mothers with dependent young tend to have smaller ranges than adult males. On the other hand, the life range of a female may be quite large, since it is now an established fact that adolescent or young mature females may, temporarily or permanently, leave their natal communities to associate with neighbors (Nishida, this volume; Pusey, this volume). Each year at Gombe there are a number of sightings of unhabituated females within the community range. In some cases these females gradually become integrated into the social life of the community. Some females, as Wrangham (this volume; 1975) points out, have ranges in the overlap zone between two chimpanzee communities but, to date, the only females known to have peaceful interactions with members of two communities are those ranging between the Kasakela and Kahama communities, which have only divided in the last three years. It seems probable that females may have similar relations with two communities in other areas, but we do not yet have evidence for this.

Visits to Peripheral Areas of the Range

Existing data suggest that visits to peripheral areas of the home range, which are also overlap zones with neighboring communities, may be undertaken for a variety of reasons:

1. Certain preferred foods are not distributed equally through the Gombe habitat and may be more abundant, at certain times, in peripheral areas (Wrangham 1975). On such occasions, chimpanzees tend to travel to these food sources in large parties, often including females. Before moving in to feed, they may spend some time on a ridge, overlooking the valley: often the males will utter pant-hoots (distance calls, Goodall 1968) and may perform charging displays. Then they remain silent as though listening for a response. If all is quiet they continue their journey. At other times a party may travel to the same area with apparent unconcern.

2. Frequently a male chimpanzee travels with a female in oestrus to some peripheral area where it is less likely that the pair will encounter the males of his community (McGinnis 1973; Tutin 1975). One pair (Figan, Gigi) remained in the north of the community range for about a month, during which time they were continuously observed: Gigi was anoestrus for most of the time, but consortships may be formed with anoestrus females (Goodall 1968; McGinnis 1974). This and other peripheral safaris are shown in Figure 2(a). During such safaris the pair tends to be silent (McGinnis 1973).

3. Young females in oestrus may travel toward overlap zones with neighboring communities, often seeming to do so of their own accord (McGinnis 1974; Pusey, this volume).

4. Visits to peripheral areas may be to patrol or monitor their current boundaries. Two or more prime males are invariably present on such patrols. Chimpanzees on a patrol feed very little; they travel silently and give the impression that they are actively looking for neighbors, or indications as to the whereabouts of neighbors. Nishida (this volume) has observed similar excursions in the Mahali mountains' chimpanzees and has termed the behavior scouting.

Boundary Patrols

During a 27-month period, from January 1973 to April 1975, adult Kasakela males (13 years old or more: see Table 1a) were observed on 25 occasions in peripheral areas to the north or east of their home range when they were either patrolling or were observed to respond to the sight and sound of chimpanzee parties presumably from neighboring communities to the north or east of their home range, or to unhabituated females in these areas. (In view of the long history of past association between Kasakela and Kahama individuals, excursions to the south of the Kasakela home range will be discussed in a separate section.) The *total* number of occasions when Kasakela chimpanzees were observed in these peripheral areas (i.e., when they are feeding there, or on safari, but were not patrolling and not observed to encounter neighbors) has not yet been extracted from the records.

TABLE 1a
Kasakela individuals frequently referred to in text: they are ranked in order of the frequency with which they were observed participating in patrols.

		PATROLS			ATTACKS	
INDIVIDUAL	COMMENTS	NORTH & EAST	SOUTH	TOTAL	ON♂	ON♀
Figan (FG)	Prime: alpha since May 1973: excitable: often leads retreat	8	9	17	2	2
Faben (FB)	Elder brother to FG: totally paralyzed right arm: excellent bipedal locomotion	7	7	14	3	2
Jomeo (JJ)	Heaviest male (110 lbs): low ranking: prime	7	7	14	3	1
Sherry (SH)	Younger brother to JJ: not ranked fully socially mature until 1974: extremely active in patrols and attacks	8	5	13	2	—
Satan (ST)	Prime: few years older than Sherry: also very motivated to detect neighbors: active in destroying nests: spectacular waterfall displays	6	7	13	1	2
Humphrey (HM)	Alpha male January 1971– May 1973: just past prime: active on patrols in north and east but apprehensive to travelling south	10	1	11	1	1
Evered (EV)	Prime: just older than Figan but intimidated by FG and FB together: recently became rather peripheral but previously active on patrols	8	2	10	2	2
Goblin (GB)	Early adolescent (born September 1964): often travelling with adult males: does not yet show adult patrol behavior	6	4	10	1	—
Hugo (HG)	Very old: during 1974 he gradually dropped out of strenuous excursions: died January 1975	4	1	5	1	—
Mike (MK)	Very old: alpha 1964 to end 1970: dropped out like Hugo during 1974: disappeared since February 1975	2	—	2	—	—
Gigi (GG)	Sterile: has been cycling regularly since 1965: very often accompanies males: shows much "masculine" behavior (aggressive patterns, displays)	6	2	8	1	1

TABLE 1b
Kahama individuals frequently referred to in text

INDIVIDUAL	COMMENTS
Charlie (CH)	Prime: alpha male: has always been known for his fearless temperament
Godi (GI)	Prime: challenging Charlie for alpha, October 1973: younger than Charlie: similar age as Figan
Dé (DE)	Prime: sick in August 1973: became rather emaciated
Sniff (SF)	Prime young male: Leader in intercommunity interactions
Willy Wally (WW)	Similar age as Dé: semicrippled in one leg, probably polio, since 1966
Goliath (Gol)	Very old: fairly solitary: alpha prior to Mike
Madame Bee (MB)	Old female: mother of Honey Bee and Little Bee: paralyzed arm, probably polio, since 1966
Little Bee (LB)	Young, cycling female: right "clubbed" foot, probably birth defect
Honey Bee (HB)	Adolescent female

An early adolescent male (Goblin, see Table 1a), usually without his mother, traveled with the males on 18 of these occasions, and females were present on 16 of them. Eight of the 13 Kasakela adult females were involved in such excursions but only one, the sterile female Gigi (see Table 1), consistently traveled with the males on such occasions. Thus she was present on 48% of the excursions while the female next most frequently involved (Patti, cycling, nulliparous) was present only on 16%. For the most part, females were in oestrus at such times: 69% of the occasions when they were involved they were half to fully swollen. Gigi, who cycles regularly, was only anoestrus during three excursions. As we have mentioned, females tend to have smaller ranges than do males, and those females involved in excursions to the north or east tended to be those who are normally found in areas to the north or east of the community range.

Chimpanzees taking part in patrols tended to travel in close, compact groups. Travel was silent with frequent pauses to look and listen. Often an individual stood bipedally, to see over tall grass or stare down into a valley or ravine ahead. From time to time the party stopped and sat silently, watching and listening: sometimes they climbed into a tree; at other times they sat, often within arms' reach, on some ridge overlooking a neighboring valley.

During patrols, males (rarely females) sometimes intently smelled the ground and the vegetation. They also smelled discarded food wadges and feces. We do not yet know whether chimpanzees recognize a distinctive odor for different individuals or sexes, or whether they are simply trying to determine the time which has elapsed since other chimpanzees passed by. Of interest in this connection are observations of excessive defecation (diarrhoea) during a boundary patrol when strangers had been heard (Wrangham 1975; A. Pierce) and of excessive urination in a similar situation (A. Pierce). A relevant fact, though it did not occur during a patrol, was observed in a nervous adult male (Evered). He had been absent from the Kasakela core area for 10 weeks and was apprehensive of encountering the alpha male. As he traveled, on his own, he suddenly began to smell the ground very intently, going down on his elbows to do so. As he did, he urinated a number of times (E. Bergmann). Some primates are known to scent mark the boundaries of their range: this is common among the prosimians and some New World monkeys and has also been observed in vervet monkeys. These monkeys rub their cheeks and chins on branches or rocks along territorial boundaries: neighboring vervets may subsequently come to sniff such areas (Jackson and Gartlan 1965; Gartlan and Brain 1968). We are not suggesting that chimpanzees scent mark their boundaries, but it is possible that the proximity of a neighboring community may lead to a state of arousal, which might result in excessive defecation and urination. This can then serve as clues to patrolling neighbors.

When chimpanzees encountered fresh nests in peripheral areas, they often inspected and smelled the nests intently. Wrangham (1975) was the first person to observe this behavior at Gombe. Nest inspection has been seen now on eight occasions: five times one or more of the males present displayed "at" a nest until it was partially or totally destroyed.

Perhaps the most striking characteristic of patrolling chimpanzees is the silence which they may maintain for well over 3 hours. Occasionally, as they travel, a male may perform a charging display, but he will not call out as he does so. Females normally scream during copulation, but in peripheral areas matings are usually silent (McGinnis 1974; Tutin 1975b). Occasionally a party will move stealthily towards some creature rustling in the vegetation: on occasions such sounds were made by bush-pigs or baboons, but the chimpanzees did not break their silence.

Sometimes the patrolling party sat in a tree, overlooking some valley, for as long as an hour without feeding, calling, or even rustling the vegetation. Only younger individuals sometimes broke silence during patrols (see section on individuality, Goblin).

By contrast, when patrolling chimpanzees returned once more to familiar areas, there was often an outburst of loud calling, drumming displays, and even some chasing and mild aggressive contact between individuals. Particularly spectacular were displays along stream beds or around waterfalls. Busse observed a display of this kind, which went on for a full 10 minutes (performed by Satan). It is possible that this kind of noisy and vigorous behavior, first observed by Bygott in 1972, serves as an outlet for the tension and social excitement, engendered by journeying into unsafe areas, which the chimpanzees have suppressed for so long during silent travel.

Females were present on 12 of the 16 patrols to the north and east. Again, they were usually in oestrus when they accompanied males on such occasions: anoestrus females were present on only four patrols. Gigi, who took part in eight patrols, was in oestrus on all but two of them. When females went on patrols, they usually did not take an active part, but tended to trail along in the rear. They were less attentive, for the most part, to the possible whereabouts of unhabituated individuals than were the adult males. Only Gigi was observed to inspect the nest of a presumed stranger, and only she has actively joined the males as they smelled the ground.

Response to the Sight or Sound of Unhabituated Neighbors

On 14 occasions the Kasakela individuals were observed as they responded to the sight or sound of neighbors or unhabituated individuals unknown to the human observers. At the first indication of the proximity of strangers, the chimpanzees often showed reassurance contact behavior (Goodall 1968), reaching out to touch one another and sometimes grinning (teeth exposed as lips are retracted, Goodall 1971) and usually with erect hair. Responses varied and presumably depended on a variety of factors, which, for the most part, observers were unable to record because of the shyness of the neighbors. However, there is evidence that a small party of males may flee from a larger one (Bygott 1974; Wrangham 1975). Two parties of more or less equal size may approach and display at one another until one or both groups retreat.

In response to hearing calls, presumably made by neighbors, the chimpanzees sometimes approached silently and cautiously, or they broke into a bedlam of noise and raced towards the sounds, particularly if they were made by single females or mother-infant pairs. Sometimes they appeared to hunt the perpetrator of the call, traveling slowly, peering around intently, and sniffing the ground and vegetation as they went. When they caught up with unhabituated individuals, the latter often were chased or attacked. On a number of occasions, the habituated party ran so fast that it was lost by human followers.

Table 2 gives some idea of the behavior shown by the Kasakela chimpanzees on those occasions when they were observed responding to unhabituated individuals. It lists (a) the response to the known or presumed presence of neighboring males (when they were either seen or a chorus of pant-hoots indicated they were there) and (b) the response to the sight of an unhabituated female or to single calls, most of which were known to have been made by females. Although we need many more observations, the data do suggest that parties of adult males are

TABLE 2a
Responses of habituated chimpanzees to presumed neighbors—to the sight of unhabituated males or calls from parties when the pant-hoots indicated that males were present

DATE	COMPOSITION OF PARTY		RESPONSE OF HABITUATED PARTY				LOST
	HABITUATED ♂ ♀ A♂	UNHABITUATED	CALL	APPROACH	DISPLAY	RETREAT	
25.5. 1973	9 5 1	Few pant-hoots heard	Screams, waa barks, pant-hoots	Race toward	Many male displays possibly directed at involving strangers		x
10.1. 1974	6 2 —	6 adult males seen	—	Slow, cautious		Run off when charged	
24.1. 1975	5 4 1	Mixed group, bad visibility	—	Slow, cautious	Display: both parties	Presently retreat	
20.4. 1975	4 — 1	Hear drumming, pant-hoots. Later see 3♂♂	Pant-hoots, waa barks	Race toward	Display: both parties	Strangers flee as last Kasakela chimps arrive	
21.5. 1973	7 3 1	Chorus pant-hoots heard	—	Fast			x
30.5. 1973	9 — 1	Chorus pant-hoots heard	—	Race: subsequent slow, cautious		Search, then give up and return	
18.9. 1973*	5 3 —	Chorus pant-hoots heard	♀♀ pant-hoots	Females, Evered wander toward		Other males return	
8.1. 1974	4 1 —	Chorus pant-hoots heard	—	Race toward			x

* Observation by R. Wrangham (1975)

motivated to approach individuals of other communities, even if they subsequently retreat upon finding themselves outnumbered or faced with more formidable opponents. Unhabituated females in peripheral areas are often chased and attacked (as they may be also when they appear in core home range areas, Pusey, this volume).

Responses to the Proximity of Unhabituated Males. A few examples will serve to give a better idea of behavior that may occur when unhabituated males are encountered. On April 20, 1975, a group of four adult males (Figan, Humphrey, Satan, and Sherry), with the adolescent Goblin, were eating meat in Kakombe Valley (Figure 2b). Suddenly there was an outburst of pant-hoots and drumming from high in the valley: the chimpanzees became very excited, gave pant-hoots and waas, and raced off towards the calls. Sherry and Goblin were in the lead with the older males some distance behind. They continued to run until they caught up with a party of three adult males: there was much noise with both parties calling loudly as they displayed towards each other, branch waving and drumming. The strangers displayed at Sherry, coming to within about 10 meters before Humphrey and Figan charged up, both waving branches and throwing rocks. At this the strangers turned and fled. The Kasakela males continued to call and display but did not chase the three males. On this occasion it seems that the strangers may not have fled the observer (P. Leo): he was lying flat behind some bushes and he may not have been seen.

TABLE 2b
Responses of habituated chimpanzees to presumed neighbors—to the sight or sound of unhabituated females, or to a single call probably made by a female

DATE	COMPOSITION OF PARTY				RESPONSE OF HABITUATED PARTY			
	HABIT-UATED ♂ ♀ A♂			UNHABIT-UATED	CALL	APPROACH	DISPLAY/ ATTACK	RE-TREAT
28.5. 1973	9	3	1	Single pant-hoot heard	Scream, waa, pant-hoot	Race toward: then long, silent hunt, smelling	Chase, 3♂♂ attack old female severely	She escapes
11.6. 1973	8	2	1	Single pant-hoot heard. Later *see* mother and infant	Pant-hoots	Just look	She escapes. Chase her	She escapes
12.7. 1973	8	1	1	Single pant-hoot heard	—	All run toward	Satan displays again and again	
21.12. 1973	8	3	1	Infant scream heard	—	All run toward	Fifi (♀) kicks mother 30 meters from tree	
16.7. 1974	2	—	1	See young female (Harmony)	—	Run toward	Figan, Faben attack, display	She escapes
21.6. 1973	4	1	—	One chimpanzee pant-hoots	—	Slow, cautious, hunt	Give up and return	

On a second occasion, six Kasakela males were feeding in the north of the home range with Goblin and a mother (anoestrus) and dependent young. Suddenly Goblin gave pant-hoots and stared to the north: a party of six unhabituated adult males were sitting in a tree some distance away. The Kasakela chimpanzees uttered no further sounds but stared towards the strangers, hair erect. After a short while they began to approach, traveling cautiously. The six males continued to sit, looking at them, and the Kasakela party halted, then moved on again. At this point the six males swung down from the tree and charged despite the presence of humans: the Kasakela chimpanzees turned and fled. Presently they stopped and looked back: there was no sign of the strangers but Humphrey, after uttering a soft pant-hoot, turned southward again and hastily led the party back towards more familiar ground (report by H. Mkono).

Three detailed observations involved the males of the Kahama community and their interactions with males presumed to belong to the Kalande community to the south. Once the young prime male, Sniff, who had been feeding with Charlie and Willy Wally, suddenly stared and began a display. In response there were some calls to the south and the observers (Mpongo and Bandora) noticed two unhabituated chimpanzees, a male and a female. The female ran off as Sniff, followed by the other two Kahama males, ran towards them. The male also tried to escape but was hindered by a semiparalyzed leg. He was caught and briefly attacked by Sniff and Charlie, while Willy Wally displayed nearby. As the stranger escaped, he was chased for a short while.

The second incident was observed by Matama. The same Kahama males were in association. As Charlie sat, feeding on the carcass of an infant bushpig, an unhabituated adult male suddenly appeared on the opposite side of a narrow ravine. Charlie left his meat and fled northward: the other Kahama males also raced off. The observers heard screaming and other calls near Kahama stream and hastily followed: they found Charlie, screaming loudly, but more sounds were coming from just ahead and they moved on. Soon they came across Sniff who was displaying along a track part way down a very steep-sided ravine, about 20 feet from the bottom. Other chimpanzees were below him, hidden in the thick bush. Both Sniff and the unseen party were calling loudly, uttering mainly waa barks. As Sniff displayed, he repeatedly picked up quite large rocks and threw them so that they landed near the strangers below him. Matama counted about 13 missiles thrown in this way. At the same time, it appeared that the others were also hurling things: every so often a rock or a stick flew upward from the undergrowth below but, falling way short of Sniff, rolled back into the ravine. Matama moved round to try to get a better view of the strangers and saw three males: there were almost certainly more. As the strangers retreated, they repeatedly displayed and there was a good deal of drumming and pant-hooting.

The third event involved the same three Kahama males. As they traveled along the southernmost Kahama ridge, Charlie suddenly gave an alarm call (wraaah, Goodall 1968) as a party of seven to nine chimpanzees was seen to the south. The three males charged uphill, seemingly to get a better view of the strangers. Charlie and Willy Wally then fled northward. Sniff, however, remained behind and did not follow his companions until two males of the party charged to within 25 m of him (Observers Pierce and C. Chiwaga). The three observations occurred on November 12, 14, and 28, 1974. Their locations are shown on the map (Figure 2b).

Response to Unhabituated Females. During the past few years, a number of unhabituated females, both young nulliparous individuals and older mothers with infants, have been observed traveling within the range of the Kasakela and Kahama communities. The exact number of such females in any given year has not yet been determined, since it is seldom possible to identify accurately chimpanzees who are rarely seen, who are shy, and who are encountered by a succession of different observers. Some of these females have been positively identified as individuals who gradually associate more and more frequently with habituated chimpanzees until they become habituated themselves and are eventually included among the known members of the community concerned (see Pusey, this volume).

Sometimes these females are encountered, during a follow, in association with habituated individuals: such incidents have not been included in Table 2 which only documents responses to such females encountered on their own or with other unhabituated individuals. When Kasakela males heard calls uttered by these females, or by their infants, they often showed instant excitement, called loudly, displayed, or attacked.

During the 1973-1975 period, only one serious attack was witnessed. The victim was an old female (no offspring visible), and the event was witnessed by C. Tutin, C. Kakuru, and R. Bambaganya. Tutin's detailed report describes the entire sequence of events, and we are grateful for her permission to summarize it. During the morning of May 28, 1973 a large party of Kasakela chimpanzees (nine males, adolescent Goblin, and three females with young) was feeding to the south of Rutanga Valley (see Figure 2b). When a single chimpanzee call was heard on the far northern side of Rutanga Valley, there was instant pandemonium, pant-hoots, waa barks and screaming from the Kasakela party. Humphrey and Evered led off at full speed toward the sound and all the others followed, except the one anoestrus female. When the chimpanzees reached the stream, travel became silent and they moved on, in a fairly compact group, until they reached a place, which, Tutin reckoned, was close to where the single call had been made. There the Kasakela chimpanzees sat, still silent, for 15 minutes. They moved on, then sat staring over another ravine for half an hour. Then, still maintaining silence, they traveled on until they reached the Busindi-Mitumba ridge. There Humphrey, who was still in the lead, began intently smelling the ground and tree trunks: the others followed his example. Soon after this, it looked as though the party had given up, since they spread out a little and began to feed. But then, 3 hours and 20 minutes after the original call had been made, Figan and Evered began to race to the east, retracing their steps, and the rest of the party ran after them. Four minutes later, there was a tremendous outburst of calling—screams, waa barks, and the wraaa of alarm. As the observers arrived they saw Humphrey, Evered, and Satan jointly attacking an old anoestrus female. She was screaming and crouched to the ground as they pounced on her. She finally managed to escape and ran off, pursued by the Kasakela party. The chimpanzees traveled rapidly and noisily and were lost by the observers. The quantity of blood at the site of the attack suggested that the female had been quite badly hurt.

Other serious attacks were seen in 1971, prior to the community division, by Bygott (1972; 1974). In the first instance, five males (Mike, Humphrey, Satan, Jomeo, and Figan) suddenly raced forward, there was an outburst of screaming and barking, and when Bygott caught up he found some males attacking, while others displayed around, an unhabituated female. Finally she escaped; Humphrey next appeared holding an infant chimpanzee, between 2 and 3 years old, which he and Mike killed by eating. They tore at an arm and leg despite the fact that the infant still screamed and kicked. A few days later, Humphrey, Jomeo, Figan, and Faben were again observed as they chased and severely attacked an unhabituated female thought, by Bygott, to be the same individual.

One final observation may be relevant. In 1973 a small group of Kahama males apparently made a detour in order to approach and inspect the dead body of a very old female. There were bad puncture wounds on her back, which could have been inflicted by chimpanzee canines. The only other animals in the area that could have inflicted such wounds are leopards or baboons. The body was not eaten, nor were there any lacerations such as would be expected if the attack had been made by a leopard. Baboons have never been known to inflict serious injury on a chimpanzee, even in severe competition during banana feeding or during a hunting episode when chimpanzees were attacking infant baboons. It seems likely, therefore, that she was also a victim of chimpanzee aggression. (The observations were made by Y. Selemani and reported by Wrangham 1975).

Females of the Kasakela community often took part in chases that preceded male attacks on unhabituated females. Once, after a large party had raced noisily in the direction of a single call, the unhabituated female who was found was attacked by only one member of the Kasakela party: adult female Fifi kicked the stranger some 30 meters to the ground after pursuing her through the branches. When young unhabituated females begin to move into the core area of the Kasakela community, associating more and more frequently with habituated chimpanzees, the Kasakela females very often direct aggression toward them, threatening, and sometimes attacking (Goodall 1971; Pusey, this volume).

History of the Relationship
Between Kasakela and Kahama Communities

We have already mentioned that, until 1972, the Kasakela and Kahama communities were considered one big community. Bygott outlines the gradual decrease in frequency of meetings and associations between males of the two groups (this volume; in more detail, 1974). At the time of the split in 1971, the Kasakela (northern) group comprised eight fully adult males: five prime males (Humphrey, Evered, Figan, Jomeo, and Satan), one prime crippled male (Faben), and two old males (Mike and Hugo). The Kahama (southern) group comprised only five fully adult males: three prime males (Charlie, Godi, and Dé), one slightly older male (Hugh), one prime crippled male (Willy Wally), and one old male (Goliath). Also there was one adolescent male (Sniff). However, although the Kahama group was smaller, the two top-ranking males, Hugh and Charlie (suspected siblings), were able to intimidate the alpha male of the Kasakela group, Humphrey.

The Kahama group, or part of it, had always ranged farther to the south and had always been less frequent visitors to the banana feeding area: after the change in the feeding system in 1968 (Wrangham 1974) when bananas were no longer provided daily, the camp attendance of the southern, Kahama, chimpanzees dropped even more and by 1971 they ranged, for the most part, to the south of Kakombe Valley and camp.

One female who had previously visited camp regularly (Mandy) and another who had appeared fairly often (Madam Bee) also began to stay away for longer and longer periods during 1971. Bygott (1974) writes the following: (In September, 1971)

> A significant event in the southern subcommunity was that Mandy (a middle-aged female) came into oestrus for the first time in six years. She seemed to be very attractive to all the males and when receptive she was commonly accompanied by all or most of the southern males.
>
> Thus on the few occasions when the southern males visited Kakombe Valley, they all came together in a cohesive party, with Mandy, and all returned south together. Similarly, when the northern males went south and met southern males, they often found them in a large party. When a large party of southern males encountered northern males (in *any* part of their range) they tended to display-charge in parallel and cause the northern males to scatter, although after the

initial excitement both sides usually settled down to groom or feed peacefully together. It was perhaps for reasons of security that northern males, when they traveled south, usually went in parties of at least five. Thus there developed a pattern of "expeditions" in which a party of males from one sub-community left their normal range and traveled perhaps 1 km. or more until they met males of the other sub-community. At least 13 such expeditions were recorded between August 25th and November 7th.[1]

These expeditions are reminiscent of boundary patrols and many of the incidents Bygott describes resemble intercommunity avoidance interactions as he points out in this volume.

Kasakela and Kahama males continued to meet occasionally during 1972, with some peaceful interactions, but they showed increasing avoidance of one another and the Kahama chimpanzees had stopped visiting Kakombe Valley, as a group, by September of that year. Only the male, Goliath, and Madam Bee and her two daughters have appeared in camp since that time. After 1972 the only nonaggressive interactions seen between Kasakela and Kahama males involved the three old males (Mike and Hugo with southern males, and Goliath with northern males), but these became increasingly rare (Wrangham 1975; Gombe Stream Research Centre records).

During 1973 the only significant overlap zone between the ranges of the Kasakela and Kahama males was Sleeping Valley, to the north of Mkenke stream (Figure 2a): the two groups were now considered as separate communities (Wrangham 1975). In January 1973, a party of Kasakela males was observed as it traveled to a ridge overlooking this overlap area and sat, silently, until they heard calls from the Kahama males in the distance (Nyasanga/Kahama ridge). Only then did they move down to feed in Sleeping Valley (Wrangham 1975). In May, 4 prime Kasakela males were followed on a patrol to the south: they overlooked Kahama Valley for some time, saw and heard nothing, and returned to their core area. In June, seven Kasakela males heard pant-hoots to the south when they were traveling into Mkenke Valley. They broke into a bedlam of calls and displays and then ran silently down towards the stream. Meanwhile the Kahama males were still calling continuously on the south bank. For the last few hundred meters, only four Kasakela males continued to run forward: they reached the stream and sat silently while the other party continued to call and display, not visible in the undergrowth. Soon the Kasakela males retreated, maintaining silence for some time before breaking out into pant-hoots and drumming displays (D. Riss).

In January and February 1974, parties of Kasakela males chased, caught, and severely attacked two males (Godi, Dé) of the Kahama community. (These attacks will be described in detail in the next section.) In April, Riss was working full time in the Kahama area but, at the same time, the Kahama males were mostly traveling alone. By July, when Pierce began her long-term study in the south, the core group of Kahama males was very small: Hugh had disappeared at the end of 1972, and Godi and Dé were not seen after January and April respectively. Thus only Charlie, Sniff (now adult), cripple Willy Wally, and old Goliath remained. These males associated with the two females and their young already mentioned and an unknown number of unhabituated females.

During 1974, the Kasakela males began to push their range to the south. Occasionally small parties were observed traveling and feeding as far south as Busambo without signs of nervousness (although there were also occasions when they were cautious in the same areas). Between August and December, Kasakela males were followed on five long excursions to the south. Once eight males silently and stealthily hunted Sniff, whom they had seen in a tree only some 50 meters distant before he climbed down and moved away (Pusey 1977). The Kahama males, however, were still able to intimidate the Kasakela males: when Faben, Figan, and Jomeo encountered the four Kahama males by Kahama stream, both parties called and displayed, without contact. The Kahama individuals called longest and loudest, and the other party retreated.

The early part of 1975 was marked by Kasakela excursions and patrols deep into Kahama territory (Figure 2a). During February, the third attack on a Kahama male (Goliath) occurred.

The Three Attacks on Kahama Males

These attacks were all characterized by extreme violence and a brutality shocking to observers. Because of the rarity and intrinsic interest of such incidents, we shall describe each in detail (Figure 2c).

Godi Attack. (Report by H. Matama) On the morning of January 7, 1974, a large mixed group of Kasakela individuals traveled slowly southward, with periods of feeding. At 1415 a party of six adult males (Hugo, Humphrey, Faben, Figan, Jomeo, and Sherry), an adolescent male (Goblin), and the sterile female (Gigi, in oestrus) detached itself from the others and began to travel more purposefully southward. During this journey calls were heard to the south. The party began to travel quickly and silently in that direction. Finally they came upon Godi who was feeding in a tree and clearly caught unaware. He leapt down and fled, screaming. Humphrey, Jomeo, and Figan were close on his heels, running three abreast with the rest of the party following. Humphrey grabbed Godi's leg and pulled him to the ground, then jumped on Godi as he tried to escape. The other chimpanzees began hitting Godi as he sat, pinned to the ground by Humphrey who was leaning forward and holding on to the victim's legs with both hands. During the ensuing attack, Humphrey remained in this position: Godi had no chance to escape or to defend himself.

Figan, Jomeo, Sherry, and Evered beat on Godi's shoulder blades and back with their hands and fists: Hugo bit him several times (though his teeth, which were worn to the gums, could not inflict serious wounds). Gigi, meanwhile, raced round and round, screaming loudly. Matama cannot recall seeing Goblin who must have kept out of the way.

Finally Humphrey got off, the others stopped hitting Godi, and the attack, which lasted 10 minutes, came to an end. Hugo, screaming loudly, stood upright and hurled a large rock at Godi. The rock, weighing well over 5 kg, fell short. The attacking party then left, moving rapidly to the south, uttering pant-hoots and displaying. Throughout the attack they had all been screaming quite loudly. Later, calls were heard to the south and the party hurried on towards them: but they did not persist for long and eventually returned to their core area.

Godi remained motionless on the ground when Humphrey got off his head but, as the attackers moved off, he got up and looked after them, screaming. It was clear that he was badly wounded: he had a great gash, from his lower lip down the left side of his chin, and this was bleeding profusely. Blood was also coming from his nose, his upper lip was swollen, and there were bleeding cuts in the right corner of his mouth. There were puncture marks on his right leg and between his ribs on the right side. He had a few small wounds on his left forearm. Since that moment Godi has not been seen, despite the fact that field assistants and students have been working with the Kahama community since April 1974.

Dé Attack. The second attack, observed by E. Mpongo and A. Bandora, took place on February 26, about one and a half months after the Godi attack. This time, the Kasakela party was smaller, consisting of Evered, Sherry, Jomeo, and Gigi (anoestrus). In the early morning the group fed, slowly moving southwards. At 0845 they began to travel more purposefully to the south. As they arrived in Sleeping Valley, they started to walk slowly, often pausing to stare around and listen intently. Once they tensed with hair erect and stared into a tree: two baboons appeared and the chimpanzees gave a few soft grunts and appeared to relax. They then continued their cautious travel, as if they were hunting prey. Suddenly at 0915 the four raced forward, and a moment later there was an uproar of screaming and barking. The observers thought there had been a predation, but as they approached, the three Kasakela males were found attacking Dé. Calling and displaying nearby were Sniff and Charlie: Little Bee (Kahama female, daughter of Madam Bee) was fully swollen and present with Dé.

After two minutes, Evered and Jomeo left Dé and charged southward, apparently in pursuit of Sniff and Charlie, but the observers stayed to watch the struggle that continued between Sherry and Dé. Sherry stamped on and bit Dé, who was unable to escape as the younger male hit him. Perhaps he had already been badly hurt by the fierce gang attack, or perhaps he was still weak from an illness he had the previous year which left him very thin. At any rate, he soon stopped struggling and sat hunched over, uttering squeaks, as Sherry continued the assault. Another two minutes passed, and then Dé appeared to bite Sherry who ran off. Dé managed to climb a tree, but Sherry followed and renewed the attack. Dé began to scream loudly again.

Soon afterwards Jomeo and Evered charged back: Jomeo immediately climbed up to join his brother in the attack. Dé managed to jump away to another tree, now very quiet, but the brothers leapt after him and continued hitting and biting their victim. Dé, again screaming, managed to jump away but the branch he landed on cracked so that he was left hanging close to the ground. Jomeo instantly leapt down, seized Dé's leg and dragged him to the ground. Evered now joined the attack: Dé lay flat on the ground and no longer tried to escape as the three prime males assaulted him, all screaming. Little Bee continued to watch and Gigi began a display, stamping and slapping the ground, and then joining in the attack. The fight became even more vicious, as all four Kasakela chimpanzees hit and stamped again and again on the still prostrate Dé who was uttering a few squeaks. The observers saw the aggressors tear skin from Dé's legs with their teeth, and at some point he was dragged along the ground.

Once again Evered and Jomeo left Dé to charge after the retreating Charlie and Sniff, uttering barks and screams and waving branches. Sherry continued to attack Dé who began screaming loudly again. Gigi moved off a little way and Little Bee traveled to the south, perhaps trying to follow the Kahama males in their retreat. Evered and Jomeo followed her and only then, after the attack had lasted 20 minutes, did Sherry leave his victim to follow the others. Gigi, however, was left near Dé and was not seen again that day.

One and a half hours later, the Kasakela males returned to the site of the attack. As they approached they called and seemed to listen for a response. They followed the track along which they had dragged their victim and intently smelled the ground and vegetation, particularly where blood was seen. Evered stared into the trees where Dé had been attacked. The party spent about half an hour in that vicinity, but although the observers searched for Dé during that time, they could not find him.

Two months later Dé was found, traveling by himself in Kahama. He was incredibly emaciated with his spine, backbone, pelvic girdle, and

FIGURE 3
Dé, a Kahama male, two months after the 26-2-74 attack. Photo by Jim Moore.

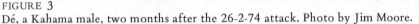

anus protruding (Figure 3). He was lame due to a bad unhealed gash on his inner left thigh that prevented normal locomotion: he had great difficulty in climbing and spent much of his time feeding on the ground. Nails were missing from his fingers, a toe was partially bitten off, and part of one ear was torn away. His once large scrotum had shrunk to a fifth of its former size. He was followed for five consecutive days during which he was always alone. He has not been seen since despite intensive searching (D. Riss).

Goliath Attack. The third attack, on old Goliath, took place on February 7, 1975, and was witnessed by Bergmann and Bandora. During the early afternoon a party of males (Humphrey, Faben, Figan, Satan, Jomeo, and adolescent Goblin) separated itself from a large mixed group of Kasakela chimpanzees and began to travel south. Faben was the leader. Progress was slow and cautious with pauses for intent listening. After 15 minutes of traveling with the group of males, Humphrey turned back. Faben continued to lead as the party crossed Bare Tree Valley. The chimpanzees then climbed into a tree and sat for 48 minutes staring towards Kahama. During this time the only sounds were made by Goblin who fed, rustling the vegetation.

Suddenly the males climbed silently to the ground, sat for a moment, and then Faben gave pant-hoots, displayed, and raced forward some 25 meters to where Goliath was sitting on the ground amidst a tangle of low bushes. Possibly he was hiding. Faben began the attack, leaping at the old male and pushing him to the ground, his hand on Goliath's shoulders. Goliath was now screaming and the other chimpanzees were giving pant-hoots and displaying. Faben continued to hold Goliath to the ground until Satan arrived: both aggressors pulled at the victim who did not try to retaliate, but sat, hunched forward, while Faben and Satan hit him and kicked him all over his body. Jomeo, screaming, ran up and joined in the attack. At one point Goliath seemed to be lifted bodily from the ground and dropped. Goblin then ran up, hit Goliath's head, and ran off. During the attack, which lasted for about 20 minutes, Goblin repeatedly ran in, hit Goliath, and ran off again. As the three males continued to pound and hit their victim, Figan displayed towards them, calling, and hitting out at Goliath as he passed. Unlike the other adult males, Figan never clinched with Goliath, but displayed as he went past, at least eight times, hitting him on each occasion, and once biting him in the thigh.

Meanwhile the other males continued, without pause, to beat up their victim, using their fists and their feet. Goliath tried to protect his head with his arms but soon gave up and, as Dé had done, lay completely still. Faben took one of Goliath's arms and dragged him for about eight meters through the bush. Satan followed and dragged him back again; Faben returned and, jumping onto Goliath's back, stamped on him many times without pause. Goliath still lay, stretched out, his face in the dirt. Faben got off Goliath and began hitting him with his hand again, joined by Satan and Jomeo. Suddenly, with very rapid movements, Jomeo began to drum with his hands on Goliath's shoulder blades. Faben sat on Goliath's back, took one of his legs and, with his hand, seemed to twist the leg round and round.

Eighteen minutes from the start of the attack, Jomeo moved away from Goliath, followed by Satan and Faben. The Kasakela party was incredibly excited as they drummed and pant-hooted. Faben returned towards the victim, displaying, branch-waving, screaming, and charging right over Goliath who lay, motionless, as he had been left. He gave a slight squeak, the only indication that he was still alive. After this, the attacking party set off at a fast run back to the north. There was much vocalization, including loud screaming, and the males drummed on trees as they traveled to rejoin the group they had left earlier in the day.

At the time of this attack, Goliath was looking very old indeed: his shoulders, back, and head were balding, his ribs and spine protruding. After the attack, he was seen to have sustained a very severe wound on his back, low on the spine, one wound behind his left ear which was bleeding profusely, and another on his head. He was clearly in a state of shock after the attack. He tried to sit up when the Kasakela party had left, but fell back and was shivering all over. Several times he tried to sit up, but seemed unable to do so. He was holding one wrist with his hand.

We do not know whether Goliath survived this brutal assault. Intensive searches were made in all likely places, and once he was thought to have been seen, but the small party was with an unhabituated individual and ran off before positive identification could be made. In view of his extreme age and physical condition before the attack, as well as the severity of the fight itself, it seems unlikely that he could have survived.

Interactions Between the Kasakela Community and Kahama Females

As already mentioned, two habituated adult females stopped visiting the banana feeding area at the time of the community split in 1972 and were included in the Kahama community. Mandy, who had previously traveled often with the "northern" Kasakela individuals, moved quite far to the south and was not seen associating with the Kasakela individuals for about two years. Madam Bee has always ranged to the south, like the Kahama males themselves. She was never a frequent visitor at the feeding area, and it seems that the community split affected her ranging patterns very little. Her range includes the zone between the core areas of the two communities, and she and her two daughters (Little Bee, now adult, and adolescent Honey Bee) continue to have some association with Kasakela individuals. The family has been observed feeding with Kasakela females on a number of occasions, though there were few interactions. The Bees have also traveled and fed with the two old Kasakela males, Mike and Hugo. In August 1974, Madam Bee led her daughters to camp for the first visit in two years. During the excursion they encountered Hugo and several females and their young but, again, there was little social interaction. A short while later, Madam Bee again tried to visit camp but was prevented by Kahama male Willy Wally (who was on safari with Little Bee). Each time Madam Bee tried to head down the valley towards camp, which was clearly visible, Willy Wally displayed at her. Finally she gave up and returned southward (Figure 2c).

Little Bee has never severed her relations with the Kasakela community; she has made periodic visits to its core areas during periods of oestrus since 1972 (Pusey, this volume). In February 1974, Little Bee was traveling with the Kahama males at the time of the attack on Dé: she was in oestrus at the time. When the other two Kahama males left the scene (they had been displaying and calling nearby), Little Bee followed: it was at this point that the Kasakela males stopped their assault and hurried in the direction she had taken. It seems that she traveled fast and it took the males some time to catch up with her: the Kahama males were nowhere to be seen. Little Bee tried to continue southward, but the males displayed at her and she stopped. After a short rest the party turned back to the north. As they traveled, Little Bee frequently hesitated but each time one of the males displayed at her, shaking branches, and she typically screamed and presented. She remained with the males until they hunted and caught a monkey: as Evered ran off with it, followed by the others, Little Bee did not follow.

Little Bee was also in oestrus when Kasakela males attacked the entire family on September 26, 1974. She was attacked successively by Evered, Faben, and Gigi. At the same time, Honey Bee was attacked by Satan, and Madam Bee was severely beaten up by Jomeo and then Figan. Only the mother was injured, sustaining a bad gash in one leg, which bled profusely and caused her to limp. After this gang attack, Little Bee remained with her aggressors, this time traveling back with them to their core area and remaining there for 4 weeks. The other two females remained in the south.

The following month, after Little Bee had returned to her family, circumstantial evidence suggests that she was again encountered by Kasakela males in her home range. Pierce was following a group of Kahama males when male pant-hoots were heard in lower Mkenke Valley, to the north. They were followed by female pant-hoots from the area where Pierce had left the Bee family the day before. The Kahama males gave loud calls and displayed, racing over the Kahama stream and then remaining quiet and listening for some time. There were no further calls to the north. Later in the day, it was found that Little Bee was no longer traveling with her family: the following day she was again seen associating with Kasakela individuals in their home range. Again she was in oestrus.

In April 1975, Little Bee was still associating frequently with the Kasakela males, and it is possible that she has transferred permanently into their community. It is interesting that in February 1975 Little Bee (anoestrus) traveled with a party of Kasakela chimpanzees as they patrolled south. After a long period of silent travel, during which the other two anoestrus females dropped out of the patrol, the party paused, staring over a ravine ahead. Then Faben and Gigi raced forward and Faben was seen attacking Madam Bee whom he kicked five times. Madam Bee and Honey Bee fled, screaming. Faben and Gigi returned to Figan, but Little Bee had vanished and did not accompany the males as they returned north. However, she was back with the Kasakela community the following day: possibly she had rejoined her mother for awhile.[2]

Peaceful Associations Between Males of Neighboring Communities?

In 1971, before the community split, unhabituated males were twice seen in association with the "southern" males (i.e., those who subsequently formed the Kahama community): once three prime males were seen traveling with habituated individuals, and a second time one adult male was feeding with them in a tree (Bygott 1974). Did the Kahama males once enjoy a peaceful relationship with males to the south, similar to that which existed until 1972 between the Kasakela and Kahama males? Were these unhabituated chimpanzees part of the same community, but ones who normally ranged even further to the south and who never traveled as far as the feeding station? If so, it suggests that communities at that time were much larger and less compact. Or were intercommunity interactions, in the past, less aggressive in nature? Bygott (1974) has put forward a theory that increasing agriculture outside the boundaries of the park may have driven more chimpanzees into the area, thus increasing their density. This is plausible: certainly it is true that the Kalande community has recently moved further northward and that members of the Rift community are encountered more frequently than before. If the theory is true, it might account for an increased aggressiveness between males of different communities: the primate literature provides much evidence that crowding, particularly in the presence of strangers, leads to increased aggression (e.g., Hamburg 1971).

Certainly at the present time, there is no firm evidence that Kasakela or Kahama males interact peacefully with males of unhabituated neighboring communities. Wrangham (1975) describes an occasion when adult Kasakela male Evered followed three Kasakela females towards a party of chimpanzees that had called from the north. However, neither the sex nor the identity of the calling individuals was established with certainty, nor was it known whether any of the four Kasakela adults did in fact join the party that had called. Evered certainly tends to absent himself from the center of his community range for quite long periods (up to 10 weeks), and he usually seems to go off to the north on such occasions. But, while we have often speculated as to whether he might, perhaps, have associations with members of another community, his absences could be accounted for by safaris with peripheral, or even neighboring, females. More recently, a male chimpanzee who could not be identified by an experienced field assistant (H. Bitura) was seen to make his nest some 100 meters from Figan. When this male was approached, he threatened the observers but did not flee: early the next morning the two nests were empty. Although it is difficult to believe that a neighboring male could wander into the heart of the Kasakela core area and nest so close to its alpha male, we withhold final judgment until more neighboring males have become habituated.

The Individuals

It is quite clear that some individuals play a more prominent role than others in initiating or taking part in boundary patrols and interactions with unhabituated chimpanzees. Table 1 shows the number of times the different males of the Kasakela community were involved in excursions to peripheral areas of the home range when behavior relevant to intercommunity interaction occurred, and the number of times they were involved in boundary patrols. The actual figures, however, do not by any means provide the whole picture. Figan, who became the alpha male during 1973 (Riss, in prep.), has the highest score, yet when it came to actual confrontations with neighboring males the facts show that Figan tended to absent himself. On a number of occasions he hung back while other males in the party charged ahead towards strangers: sometimes he hastily led a retreat. Moreover, although he sometimes led boundary patrols, he frequently traveled toward the rear of the party. Figan was present during two of the attacks on Kahama males but played a relatively minor role in both. He was one of the leaders when the party chased and attacked the old female in the north, but he was not seen to take part in the attack (nor did he play a prominent role in 1971 when the female was attacked and her infant killed). (Figan's record of a great number of patrols might be, in part, due to the fact that he was observed more than any other chimpanzee during the study period.) Perhaps we might mention also that Figan's elder brother, Faben, was present with Figan on all but three of his patrols: Figan had relied heavily on the presence and support of Faben when he took over the alpha position from Humphrey (Riss, in prep.). Often it was one of the two brothers who, after a long period of silence near a boundary, would initiate a series of pant-hoots, the response or absence of response to which would indicate whether or not neighboring males were in the area.

Faben, unlike Figan, and despite his completely useless right arm (Figure 4), seemed motivated not only to take part in patrols, but also to lead travel and take part in the final charge that would rout strangers or lead to a retreat. Faben took part in the attack on Godi and appeared to be the most vicious and brutal assailant during the attack on Goliath which, it appeared, he initiated. After long periods of silent tension, Faben often initiated the outburst of calling, displaying, and drumming which a party engaged in on return to more familiar home ground.

Satan, Evered, and Jomeo were all actively engaged in boundary patrols and intercommunity interactions: all three took part in a number of "final charges" when part of the group remained behind. Jomeo took part in all three of the attacks on Kahama males and was violent and brutal in all of them. Without doubt he was responsible for many of the bad wounds, since he is the largest and heaviest of all Kasakela males (50 kg; average 43 kg). Evered and Satan very often acted as leaders in boundary patrols: they went, together, on one very long patrol in the far eastern part of their range and, a few days later returned to the same area on another long, tense patrol, this time accompanied by a

FIGURE 4
Faben, a partially paralyzed Kasakela male, is an active participant in patrols and attacks. Photo by Larry Goldman.

few other males. The figures shown for Evered in Table 1 are misleading: towards the end of 1974 and during the early months of 1975, he spent long periods away from the Kasakela core area and was simply not around to accompany the other males on their peripheral excursions. During 1973 and most of 1974, he was a very active participant in boundary patrols, being present for nine out of the total of 14 and clearly leading on at least five occasions. He was particularly aggressive when he encountered chimpanzee nests presumed to have been made by chimpanzees of other communities: on four of the five occasions when his response was observed, he not only inspected nests but also displayed "at" them, often destroying them almost entirely (see also Wrangham 1975). (Humphrey and Satan each displayed once at nests, but other individuals merely inspected them carefully.) Evered was an active aggressor in two of the attacks on Kahama males. Satan was also one of the males most eager (subjective impression of observers) to find strangers. Frequently he led patrols and was almost always near the front. A characteristic of Satan was his performance of impressive and spectacular stream-bed and waterfall displays on his return to safe areas (C. Busse).

The behavior of Humphrey is of particular interest. He was alpha male of the Kasakela community in 1971 and 1972, at the time when the Kahama community was gradually splitting off. Even then, Humphrey was intimidated by the presence of Charlie, who almost always traveled around with his presumed elder brother Hugh. Both of these individuals were known for their fearless temperaments: it was mainly the joint performance of Hugh and Charlie with their impressive displays, that made the Kahama males so intimidating when they charged forward in a group. Bygott (1974) describes nine occasions when the Kahama males met and intimidated parties of Kasakela males. On eight of these, Humphrey was involved: seven times he either left his party before the confrontation or soon after the meeting had taken place. A glance at Table 1 shows that Humphrey still tends to avoid confrontation with the Kahama community. He was observed almost twice as often on excursions to the northern areas of the range than to the south, and he was only known to take part in one patrol to the south in two years. Moreover, as recently as March 1975, Humphrey was seen to leave a party with which he had been associating because the other members began traveling purposefully to the south on a patrol. By contrast, Humphrey was present on most of the patrols to the north and east and took part in several chases when it was presumed that strange males were present. Thus, despite the fact that some two years have passed since the community split became finalized, despite the fact that Hugh has been missing for a similar length of time and no longer backs Charlie in his displays, and despite the fact that the little Kahama community has been decimated, it seems that Humphrey still retains his old fears of Charlie and the south.

The youngest of the Kasakela males, Sherry (still a late adolescent during 1973), has become one of the most active of all participants in intercommunity interactions. Recently he has been in the forefront of many border patrols and, several times, has clearly led parties to peripheral areas. It was Sherry who led the charge towards the three unhabituated males on April 20, 1975 and held his ground, even when the strangers displayed within a few meters. There was another occasion in early 1975 while Sherry was patrolling with five other males to the south of Mkenke stream. When calls were heard further to the south, the Kasakela party displayed and called in response: Sherry raced off by himself toward the other party (probably Kahama chimpanzees). After a moment Faben followed him, but the other Kasakela males (led by Humphrey and Figan) hastily retreated northwards. It was another 10 minutes before Sherry and Faben finally rejoined their companions. Sherry took part in the attacks on both Godi and Dé. Twice, when he continued to attack Dé by himself, his aggressive behavior was remarkable for both its intensity and its persistence.

Of interest is the fact that Sniff, the youngest mature male of the Kahama community (about 2 years older than Sherry) shows similar enthusiasm for encounters with neighbors. He was in the forefront of the attack on the male with the paralyzed leg and, on his own, he engaged in a spectacular rock-hurling encounter with a group of at least three adult males, and probably more. Again, when he remained behind by himself to face the charge of two adult males, although he ran to catch up with his companions, he subsequently left Charlie and Willy Wally and returned southward alone. On this expedition his behavior clearly showed that he was searching for the party of neighbors again, as though he had to have "just another look." Perhaps it is significant that Sniff seemed so calm when he spotted the Kasakela patrol (led by Sherry). It seemed almost certain that Sniff must have seen the patrol before climbing down from his tree, yet he made his getaway in a most unflustered manner (Pusey, personal communication).

Early adolescent Goblin has been observed traveling to peripheral areas of the Kasakela community range on 27 occasions. On 19 of these, females in oestrus were also present and Goblin's participation seemed very closely linked with this fact (Tutin 1975; Pusey, in prep.). However, he did accompany the adult males on four *patrols* when no females in oestrus were present (out of a total of ten patrols on which he was seen) and, on his last such excursion (February 7, 1975), he actually left a fully swollen female with whom he had been associating and accompanied a party of males on a long patrol to the south.

Goblin seems to have learned that he should suppress vocal sounds when the adult males are maintaining silence: if he pant-grunts in submission, he does so very softly, and twice he threw temper tantrums, when threatened by superiors, that were remarkable in that they were almost totally silent (Tutin, personal communication). On three occasions when he did scream, he was almost immediately embraced or touched by older males (once by old Hugo who had not caused the sound in the first place). This reassurance contact served to quiet him quickly (Pusey, in prep.; Tutin, personal communication). However, although Goblin will now accompany adult males on patrols even without the attraction of females in oestrus and has begun to show appropriate behaviors, such as smelling the vegetation, inspecting nests and supressing vocal sounds when older males do these things, he has not yet acquired the intent, concentrated silence that typifies the behavior of the adult male on such excursions. Sometimes when his elders are sitting motionless, staring across some valley ahead, Goblin continues to feed, snapping twigs and rustling the vegetation. Once he engaged in a wild play session with an infant and, although neither "laughed," they made a good deal of noise in the branches (E. Bergmann). It will be interesting to see at what age Goblin acquires fully adult patrolling behavior. Goblin was present during two of the attacks on Kahama males. During the first, on Godi, he apparently kept out of the way and his behavior was not noted. The second, on Goliath, occurred a year later, and on that occasion Goblin did indeed take part, running in and hitting the old male before getting out of the way.

The two old Kasakela males, Mike and Hugo, both took part in some peripheral excursions and patrols in 1973 but, as they aged and became increasingly solitary, they dropped out of such community activities. In 1973 Mike actually led some patrols and on a few occasions took part in the final charge toward strangers, but seemed able to keep up for only a short distance (Wrangham 1975). Hugo's last observed patrol was in early 1974 when he took part in the attack on Dé. Both Mike and Hugo sometimes traveled alone quite far to the south where they associated peacefully with Kahama females. Their counterpart in the Kahama community, Goliath, had been the alpha male of the community in the early years of the study, before Mike displaced him in 1964 and before the community split. Goliath had frequently ranged far to the north in those days and was not known to associate regularly with the southern males. It was not clear why he moved to the south when the Kahama males did so in 1972. As mentioned, Goliath continued to pay occasional visits to his old haunts after the other Kahama males had stopped doing so. In February 1974, Goliath encountered two Kasakela chimpanzees on the periphery of camp, including one prime male. The three individuals stared at each other for about 2 minutes, but did not interact further. And he was seen peacefully feeding in a group with Kasakela males (Evered, Faben, Figan and Satan) in Mkenke Valley only 5 months before his brutal, probably fatal, attack.

Only one more individual should be briefly mentioned, the adult sterile female Gigi. Her ranging pattern and her frequent involvement in patrols are not typical for the Gombe female chimpanzees. Gigi has been cycling regularly since 1965; when she is swollen, she is sexually popular with the males and usually associates with them. She shows many other behaviors atypical for females, such as her frequent charging displays and attacks on other individuals. It is our subjective impression that Gigi has become increasingly aggressive and increasingly masculine during the course of the years. However, although Gigi was observed on 13 patrols, joined in sniffing the vegetation and, once, inspected nests, she tended to trail along in the rear like other females, and seldom showed the alert, intense, searching behavior of the adult males.

AN APPRAISAL OF INTERCOMMUNITY RELATIONSHIPS

Intercommunity relationships in the Gombe population are complex and fascinating. The Kasakela community has a large home range, the boundaries of which change from year to year. Parties of up to ten adult males, sometimes accompanied by females and young, may patrol peripheral areas near the current boundary, when they apparently search actively for signs of neighbors. When a fairly large party of strangers, including adult males, is encountered, both parties may approach and engage in visual and vocal displays reminiscent of the territorial boundary displays of gibbons, langurs, and *Callicebus* monkeys. Such an encounter may end when one of the parties charges and chases the other away, or when both give up and return towards their core areas. At other times a party, upon hearing or seeing neighbors, may flee, thus avoiding an encounter. Factors determining exactly what response is made need further investigation and cannot be fully understood until individuals of communities to the north and east have been habituated. When single chimpanzees or very small parties are encountered, or their calls are heard, they may be chased and sometimes attacked, particularly if they are females. The relationship between the Kasakela community and the Kahama community to the south should, for the present, be considered separately in view of the known past history of close association: during 1974 and early 1975 parties of Kasakela males made excursions deep into the heart of Kahama core areas, and three times Kahama males were savagely attacked.

It seems that, for a number of primate species, intergroup encounters may be attractive, at least for some individuals. Gibbons and langurs, for instance, travel swiftly to boundary areas in response to the calls of neighbors. A male gibbon may leave his female and young to watch or actually participate in a dispute between two other groups (Ellefson 1968). A langur troop, after hearing the whooping calls of neighbors, may go far out of its normal range to initiate intertroop displays. These conflicts are often started by young males who may actually leave their troop and travel some 500 meters to engage in whooping and branch-shaking encounters with neighbors (Ripley 1967). On Cayo Santiago, parties of male rhesus sometimes sought contact with another troop. A number of monkeys then became involved in the hostile encounter (Morrison and Menzel 1972).

It seems that intergroup encounters may be attractive to mountain gorillas also. Fossey, at the conference, described how some silverbacks seemed to seek out interactions with neighboring groups: one such prime male was Uncle Bert. Sometimes aggressive encounters occurred: one led to over a minute of screaming and skirmishing during which an infant was bitten and killed, probably by Uncle Bert (Fossey, this volume).

Certainly, as we have pointed out, intercommunity contact seems to be sought by some male chimpanzees. Young prime males, in particular, are strongly motivated to travel to peripheral areas where they appear to look intently for signs of neighbors. Silent, sometimes stealthy travel, visual, auditory, and olfactory scanning, and long periods of time during which parties may maintain almost total silence characterize patrols near the boundaries. Some excursions take the chimpanzees into the core area of neighboring communities. The response to the sight or sound of supposed strangers, particularly the sometimes frenzied race towards the location of calls, suggest that encounters with neighbors may be extremely attractive. Often, after a long silent patrol, the males concerned, on returning to home ground, may explode into loud calling, branch-waving and drumming displays, and occasionally dramatic waterfall or stream-bed displays: this gives some idea of the suppressed excitement and tension to which they have probably been subjected.

In some primate species, males may make aggressive attempts to take over one or all of the females of a neighboring group: for the most part these are lone males or all-male groups without females of their own. On occasions when the leader of a bisexual langur group tries to defend his right to his females against an all-male group, the fighting that results may be prolonged and severe. Mohnot (1971) describes raids and skirmishes that continued, on and off, for periods of 3-10 days. Fossey (this volume) has observed a number of occasions when lone silverbacks entered a group and subsequently left, accompanied by one female. Sometimes this appeared to take place peacefully: but there were also aggressive incidents. One lone silverback charged into a group with violent displays and chest-beating. During the confusion an infant was killed: the following week the silverback went off with the mother of the infant.

Some of the Gombe observations are of interest in relation to the appropriation of females from neighboring groups, such as the incidents described previously involving Kahama community's young female, Little Bee, and the Kasakela males. Twice, after severely attacking Kahama individuals, (once Dé, once all three members of the Bee family, including Little Bee herself) a party of Kasakela males traveled toward their core area, accompanied by Little Bee. Both times she was in oestrus. On a third occasion circumstantial evidence strongly suggests that a party of Kasakela males encountered the Bee family and returned to the north with Little Bee who was again in oestrus. The events after the attack on Dé, when the three Kasakela males pursued Little Bee to the south, displayed at her and, subsequently, branch-waved (in threat) every time she hesitated, clearly indicate that the female was forced to travel north against her inclination.

There is insufficient data to draw any firm conclusions as to the significance of the preceding information, but it certainly seems possible that male excursions to areas that are overlap zones with neighboring community ranges may sometimes serve to recruit females into their own social group. And we should note that excursions may be made by a male traveling alone, such as Evered during his long periods of absence from the core area of his community range.

Young nulliparous females during the one-to-two-year period of adolescent sterility (Asdell 1946; Goodall 1968) cycle regularly and, therefore, travel relatively frequently to peripheral areas in the company of their community males (a) during consortship safaris and (b) during other excursions to feed or to patrol near the current boundaries. Young mothers also tend to cycle for a good many months prior to the birth of the next infant (Goodall 1975), so that they, too, tend to travel with their males on such journeys. At such times there is always the risk that they may encounter neighboring males. Females on safari may try to escape from their male consorts (McGinnis 1974): a female who succeeded in doing so, in an overlap zone, would give an excellent opportunity for an encounter by a neighboring male.

Perhaps we should also review some of the puzzling attacks on unhabituated females in this light. A male chimpanzee, during a consortship with a female (who may even be anoestrus at the time) will sometimes attack her if she tries to escape (Goodall 1968; McGinnis 1974). Such attacks serve a similar function to the neck-bite in the hamadryas baboon (Kummer 1968): the female becomes submissive and follows the male. Possibly brief attacks on females encountered in overlap zones between neighboring communities play a similar role and *attract* rather than *repel* the females concerned. Certainly some young unhabituated females not only remain in the Kasakela home range, but gradually move into the core area despite occasional attacks (Pusey, this volume). When such attacks are observed, the fact that the female normally flees can readily be accounted for by the presence of human observers. When Little Bee was attacked she did not flee, but followed the males north. (The gang attacks on the older females—and note that Madam Bee was much more severely attacked than either of her daughters during the family attack—almost certainly fell into a different category and, at present, are not understood). Only when other communities have been habituated, and the different individuals recognized, can we hope to learn more about the motivation behind attacks on these females and the function they fulfil in the complex chimpanzee social structure.

It is now well documented that Hanuman langurs, after driving a leader male from his females, may systematically kill all infants in the troop (Sugiyama 1967; Mohnot 1971; Hrdy 1974). In this way the new leader ensures that most subsequent infants will carry his genes (Popp and DeVore, this volume). Of interest are Fossey's observations relating to infanticide in the mountain gorillas: three infants were killed as a result of aggressive intergroup interactions. Moreover, on one occasion the lone silverback responsible left with the mother of the dead infant a week later.

Two cases have now been recorded of infant-killing in chimpanzees.[3] One of these occurred at Gombe (Bygott 1972) and the other in the Budongo Forest, Uganda (Suzuki 1971). The Gombe incident occurred when a party of males attacked an unhabituated female: the circumstances were not clear in the other event, but Suzuki describes the kind of loud vocalizing and excited displaying and drumming that might well have resulted from an intercommunity encounter. Although it seems extraordinary that a female, after being so severely attacked, would actually remain in the home range of her aggressors, Bygott saw a female, whom he suspected might be the same one, 10 days later! Possibly she was still around a couple of weeks later when she might have come into oestrus (Goodall 1968). Thus, although the killings were rather different from the langur and the gorilla infant killings (since the chimpanzees, in both instances, partially ate the bodies of their victims), it seems possible that a similar mechanism for increasing the genetic contribution of the aggressors may have been operating.

Some of the most severe intergroup aggression reported in field studies of primates occurs in conditions of extreme overcrowding. Intertroop fighting in langurs tends to become more frequent and more severe as the density increases (Sugiyama 1967). Probably the most fierce intertroop fighting observed in monkeys was that described by Southwick, Beg, and Siddiqi (1965) among rhesus monkeys living in a temple in India where there was 90% overlap between the ranges of three troops. Severe fighting was almost a daily occurrence and all the males bore scars and often sustained broken bones.

We have mentioned the possibility that the chimpanzee density within the Gombe National Park has increased in recent years due to more extensive cultivation outside the boundaries and resultant habitat destruction. This could well account for increased intercommunity hostility, but it does not seem to us that there is enough overcrowding to account for the severity of the three attacks perpetrated by the Kasakela community against Kahama males. These incidents are among the most shocking observations in 14 years of research at Gombe, and all were characterized by unusual brutality and persistence. The attack on Godi lasted for at least 10 minutes and the other two for some 20 minutes each. Descriptions of the incidents contain many similarities: all three victims were held down or pressed to the ground; all were hit, stamped upon, bitten, and dragged; all three were severely injured. After the initial phases of attack, both Dé and Goliath lay passively on the ground, making no further attempts to defend themselves. Godi, with Humphrey sitting on his head, was securely pinned to the ground and rendered harmless throughout the fight. The aggressors each time became intensely excited, screaming loudly and, when they were not actually fighting, displaying nearby.

One of the attacks, that on the old male Goliath, is particularly puzzling, both in view of his extreme old age and his history of long and peaceful associations with the Kasakela males. Goliath could in no way be considered a reproductive competitor. (It seems, also, that Dé was not in prime physical condition at the time of the attack.)

Is it possible that the violence of the attacks was, in fact, due to previous familiarity? Bygott (this volume) found that the most severe attacks between male chimpanzees occurred when they had been apart for some time. Wilson and Wilson (1968), working with captive chimpanzees, found that the most severe aggression occurred when they reintroduced into the colony a young male who had been removed for a period of several weeks. Certainly this may have a bearing on these attacks. The only actual attack recorded on a male from an unhabituated community was fairly brief (that by Sniff and Charlie on a supposed member of the Kalande community). Nishida (this volume) describes an attack by the top-ranking male of K-group on the fourth-ranking male of M-group: it was severe but of relatively short duration. Intuitively we feel that additional factors must be involved: after all, Goliath was seen associating peacefully with three of the males who attacked him only a few months previously. Clearly it is necessary to observe other attacks between Kasakela males and those of unhabituated communities before we can draw any conclusions.

One further point should be brought forward. At some time during all three attacks, observers felt that the aggressors were trying to kill their victims. Indeed, some of the attack patterns were not dissimilar to those sometimes shown by chimpanzees trying to kill an adult monkey or to dismember a carcass, such as the twisting round and round of Goliath's leg or the tearing of flesh from Dé's thighs. Moreover, after the 1972 attack on the female, the captured infant was indeed partially eaten, even when it was still alive and screaming.

The intense excitement shown by the aggressors during the attacks was apparent for some time afterward. The party that attacked Godi hurried off, apparently in pursuit of other Kahama individuals heard calling to the south: the party that attacked Goliath continued calling and displaying as they ran back north. The three males who attacked Dé, after following and retrieving Little Bee, returned to the site of the attack, searched around as though looking for their victim once more, and then began to hunt a troop of red colobus monkeys. They persisted in this even after one monkey had been caught only to escape when a male colobus ran to its aide. After 15 minutes, Evered was successful in the hunt. Again, when the Kasakela party had patrolled for 2 and a half hours, "hunting" for Sniff and maintaining almost total silence throughout, most of the males spent some time hunting a baboon troop. Only when they had given up did they break their long silence and burst into loud calls with drumming and branch-waving displays (Pusey, personal communication).

We lack sufficient data to draw any conclusions from these observations, but we mention them for their intrinsic interest to those who are interested in the evolution of human violence and warfare and its relation to hunting for food. Often it has been suggested that we should consider the behavioral and motivational mechanisms of hunting in a completely separate category from aggressive intraspecific interactions (e.g., Moyer 1968): the above observations suggest that, in some instances, similar motivational factors may be involved.

Clearly the intercommunity interactions of the Gombe chimpanzee population are complex and fascinating. Over the next 5 to 10 years, our major research goal will be to try to habituate other communities, in an endeavor to get a truer picture of chimpanzee behavior. If chimpanzees are to serve as models for the better understanding of human behavior, then we must learn more about factors that lead these apes to seek out contact with neighboring groups, that make aggression attractive to some individuals, and that may result in aggressive incidents of exceptional violence.

Notes

1. Quoted with Bygott's permission.

2. In August 1975, after this paper was written, Madam Bee was the victim of a brutal gang attack by Kasakela males. Four days later she died of her wounds. (Goodall, do Fisoo, Mpongo, and Matama, in preparation)

3. Since this chapter was written, members of the Kasakela community have killed three more infants: two infants of unhabituated females were killed by males, one was the 3-week-old baby of a Kasakela female and was killed by an adult female. These incidents are being prepared for publication.

The Ugalla area

Junichiro Itani

Distribution and Adaptation of Chimpanzees in an Arid Area

THE DISTRIBUTION AREA OF TANZANIAN CHIMPANZEES

In 1972, Takayoshi Kano published a detailed report on the distribution of chimpanzees in western Tanzania, excluding Gombe Stream National Park. He divided the chimpanzee habitat into seven geographical areas. One of these, the Ugalla area in the northeast, will be the focus of this account. This western region has been known as one of the most arid of the chimpanzee habitats in all Africa. The primary purpose of this report is to point out those features of the Ugalla area that substantiate the unusual adaptation of Tanzanian chimpanzees to this dry, sparsely wooded environment. The data presented here about distribution, population density, and ecology of chimpanzees in this environment were obtained mainly during a 7-day survey made in 1966.

Generally speaking, all regions inhabited by chimpanzees in the western Tanzanian distribution area share a common floral feature: a wooded savannah. This type of vegetation is necessary for the survival of the chimpanzees and is known there as the Miombo Forest. With its mixed shades of green, it can be described as mosaic vegetation. The light green deciduous trees constitute the dry, open forest, and the dark evergreen trees compose the riverine forest. Each area within the region has its own unique variations of this wooded savannah vegetation. For instance, in the Masito, Lilanshimba, Ugalla, and Mukuyu areas, *Julbernardia globiflora* is dominant; but along the lakeshore of the Mukuyu area, *Brachystegia spiciformis* prevails. The species of Caesalpiniaceae that is most dominant in the Karobwa and Wansisi areas is *Brachystegia* sp. (Kapepe), and that of the Mahale area is *B. allenii.*

Riverine forests have developed along the numerous rivers, which branch out in a network across the entire western region, forming the savannah woodland, in which most of the chimpanzee population resides. The riverine forest is especially well developed in the Masito, Mukuyu, and Karobwa areas, but it is very poor in the eastern Ugalla region.

On the slope facing the lake in the Mahale area is a particularly moist forest, which has developed because of the moisture from the lake. Moist montane forest is found in areas at altitudes over 1500 m above sea level. These areas include the central part of the Masito area, the western part of the Ugalla area, and the central part of the Mahale area.

This rough description of the general distribution area of Tanzanian chimpanzees provides an overall view of the entire western Tanzanian region. To provide a detailed view of a specific living environment, we investigated the vegetation in the Ugalla area more closely.

VEGETATION OF THE UGALLA AREA

The western border of the Ugalla area is formed by a long stretch of hills with complicated natural features that extends approximately 2800 km², at altitudes of 1100-1800 m above sea level. The Ugalla River swamp, which impedes the movement of chimpanzees, forms the eastern boundary. The area is profusely covered with open forest, dominated by *J. globiflora;* but this riverine forest is exceptionally sparse compared to other districts. Especially in the eastern part, the riverine forest is so thin that it becomes almost nonexistent. Along the riverside, a swampy, treeless Mbuga vegetation is found.

Kano estimated the ratio of riverine forest to open forest to be 4.7% in the entire distribution area and 1.1% in the Ugalla area. Part of the

reason for the extreme dryness in the Ugalla area is that it is not affected by the climate of Lake Tanganyika as are the other districts. Rather,
it has a dry, inland climate.

The dry, open forest consists mainly of *J. globiflora*. Near the Busondo Plateau, *J. globiflora* is often found as a pure stand; but as we move
to the east, other tree species are found in mixed woodland. However,
it is true that no other species is as prevalent as *J. globiflora*. The
species of trees growing with *J. globiflora* are *Brachystegia bohemii, B.
bussei, Isoberlinia angolensis, Pterocarpus angolensis, P. tincterius,
Diplorhynchus condylocarpon, Terminalia mollis, T. angolensis, Combretum molle, C.* sp. (Mukokoti), and *Uapaca kirkiana*. The mixture of
B. bohemii, U. kirkiana, and *C.* sp. (Mukokoti) denotes a highland
region; *B. bussei* a steep slope or an escarpment; and *I. angolensis* and *P.
angolensis* a damp terrain. *T. angolensis* is a species not usually found in
the western Masito area since it is a component of the more eastern
district. The existence of *D. condylocarpon* indicates that the area was
affected by men in the past since the original vegetation never recovered.

On Table 1, the most conspicuous types of vegetation in each woodland are listed as they were recorded, chronologically, day by day. We
can see that the characteristics of highlands are obvious near the high
Busondo Plateau. We can observe that the central part of this area is
moist and has more features of the bordering eastern district. We can
also tell that the vegetation of the eastern part near the Ugalla River,
which is now uninhabited, was once occupied by men because, as in the
Mahale area, no regeneration of the *Oxtenanthera* bush was observed.

The Ugalla area is different from other ones in that there are no
woodlands of *B. longifolia* or savannahs of *Combretum* sp. (Sifunfwe),

TABLE 1
Conspicuous tree species in the *Julbernardia globiflora* woodland [a]

DATE	B. BOHEMII, U. KIRKIANA, C. SP. (MUKOKOTI)	B. BUSSEI	L. ANGO-LENSIS, P. ANGO-LENSIS	T. ANGO-LENSIS	D. CONDYLO-CARPON
September 11	3	2	0	0	0
12	2	2	0	0	0
13	4	1	3	3	1
14	0	1	0	0	1
15	2	3	0	0	1
16	1	3	1	1	0
17	3	3	1	0	0
Total	15	15	5	4	3

[a] Observations of the species are recorded day by day, as they were made on a journey
through this area.

and there are few pure stands of *I. angolensis*. The absence of these species signifies that the vegetation of this area is an arid type. Part of the reason is that the whole area is a fold, owing to the formation of the Western Rift Valley. Consequently, a series of escarpments cover a large portion of the total land mass. On these steep slopes along the terraced bluff, striped stands of *B. bussei* forest can be seen. The existence of a dry riverine forest can also be accounted for by the types of rivers found there. Unlike the rivers in the Masito area, the rivers of the Ugalla area are not intricately meandering but straight and simple. The development of a moist riverine forest was disturbed because these rivers accumulate such a small quantity of water, especially in the middle courses where the water completely dries up during the dry season.

The existing riverine forests can be grouped into three types: highland riverine forest, dry riverine forest, and moist riverine forest. In the Masito area, a highland riverine forest can be found on the plateau. It is narrow, but rich in permanent water from springs and is composed of *Syzygium guineense, Vitex* sp., *Uapaca* sp., and *Parinari* sp. evergreens. The dry riverine forest, consisting mainly of *Cynometra* sp., is the type found at the head of a valley or along a rapid stream and is sometimes seen as pure stand. The moist riverine forest develops on the winding lower river basin, and produces trees 40 m in height and stretching across 200 m of land. This is the semideciduous forest made up of *Entandrophragma utile, Cordia millenii, Pseudospondias microcarpa, Ficus* spp., and *Celtis* sp., revealing the features of a tropical rain forest thickly twined by the vine of *Landolphia* spp.

All three types of riverine forests can also be found in the Ugalla area, but they are more poorly developed than those in the Masito area.

TABLE 2
Types of vegetation along rivers surveyed in the 7-day journey, northern Ugalla area [a]

DATE	DRY RIVERINE FOREST	MOIST RIVERINE FOREST	HIGHLAND RIVERINE FOREST	RIVERINE VEGETATION AFFECTED BY MEN	MBUGA VEGETATION
September 11	2	0	0	1	1
12	2	0	0	1	1
13	1	2	1	0	1
14	0	0	0	0	2
15	0	1	0	2	1
16	5	1	0	2	0
17	0	0	0	0	0
Total	10	4	1	6	6

[a]   Observations are recorded on a day-by-day basis, as they were made during a 7-day journey.

TABLE 3

Seventeen species of mammals: identified by remnants and by information supplied by natives

SCIENTIFIC NAME	ENGLISH NAME	SCIENTIFIC NAME	ENGLISH NAME
Cercopithecus ascanius	red-tailed monkey	*Procavia capensis*	rock hyrax
Galago senegalensis	lesser galago	*Orycteropus afer*	ant bear
Potamochoerus porcus	bushpig	*Hystrix* sp.	porcupine
Taurotragus oryx	eland	*Mellivora capensis*	honey badger
Tragelaphus scriptus	bushbuck	*Canis* sp.	jackal
Tragelaphus spekei	sitatunga	*Lycaon pictus*	lycaon
Hippotragus equinus	roan antelope	*Panthera pardus*	leopard
Kobus defassa	waterbuck	*Viverra civetta*	civet
Equus burchelli	zebra		

*The species marked with an asterisk are the ones considered common in the Ugalla area.

Besides these, there are two other types of riverine vegetation, and these probably resulted from human activities in the area. One is the tall, sparsely growing *Albizia glaberrima,* and the other is a thicket type of vegetation. The former one can be found in the central sections, the latter in the eastern section. In the riverine territories where there are no trees, so-called Mbuga vegetation is found both in a central location and along the respective rivers to the east. Swampy Mbuga vegetation is composed of *Papyrus* sp. and *Phagamites mauritanus.* On the river side of the Ugalla River main stream, no Mbuga vegetation is found; instead, the river faces only dry open forest.

The features of the riverine vegetation in the northern Ugalla area are shown on Table 2, where the route is divided, chronologically, by days, as in Table 1. As clearly shown on Table 2, the dry riverine forest is dominant in this area and only a few moist-type forests are found, those being in the central part. It should be noted that the ratio of dry riverine forests to moist ones in the Ugalla area is entirely different from those in the Kasakati Basin of the Masito area where moist forests predominate.

MAMMAL FAUNA

During our 7-day journey, we encountered or heard 16 species of mammals. The fact that we did not encounter them many times does not mean that the mammal fauna of this area is scarce. It is likely that the mammals had already migrated because our journey was made at the end of the dry season. We did find clear remnants of the presence of various mammals (such as footprints and droppings) throughout the whole course of our survey. The different species whose existence was

TABLE 4
Sixteen species of mammals encountered or heard during the 7-day journey

SCIENTIFIC NAME	ENGLISH NAME	TIMES ENCOUN- TERED	IN DRY OPEN FOREST	IN RIVERINE FOREST
Cercopithecus aethiops	savannah monkey	2	2	
Cercopithecus mitis	blue monkey	2		2
Papio cynocephalus	yellow baboon	2	2	
Pan troglodytes	chimpanzee	1		1
Lepus capensis	cape hare	1	1	
Sciuridae	squirrel	1		1
Loxodonta africana	elephant	1	1	
Phacochoerus aethiopicus	warthog	5	5	
Syncerus caffer	buffalo	2	1	1
Alcelaphus lichtensteini	Lichtenstein hartbeest	1	1	
Oreotragus oreotragus	klipspringer	1	1	
Sylvicarpa grimmia	gray duiker	4	4	
Hippopotamus amphibius	hippopotamus	2	2	
Panthera leo	lion	2		2
Crocuta crocuta	spotted hyena	1	1	
Viverridae	mongoose	1	1	

*The species marked with an asterisk are the ones considered common in the Ugalla area.

confirmed by these remnants are listed on Table 3, and those whose existence was confirmed by seeing the animals directly are on Table 4. Some of the information obtained by inquiries was considered reliable, and was the basis for adding seven species to Table 3. The species marked (*) are the ones that frequent the Ugalla area. Additional unob- served species must surely exist there, even though they are not listed; but the two tables are not meant to be complete checklists of large mammals in the area.

Many of the animals listed in these tables as common in the Ugalla area are also found in the Masito and Mukuyu areas. There are six spe- cies of cercopithecoids that inhabit the entire chimpanzee distribution area of western Tanzania; a comparison of those species in five specific areas is made in Table 5. Only in the Mahale area are all six species found. Three species, *P. cynocephalus, C. ascanius,* and *C. aethiops,* inhabit the whole region with one exception—*C. ascanius* is found only in the western and southern parts of Ugalla (Kano 1971a), not in the middle and eastern parts. The environment in middle and eastern Ugalla is too severe for *C. ascanius* to survive, since it greatly depends on a forest and cannot exist in the sparse riverine forest vegetation.

In Table 6, a comparison is made among ten species of antelope that are found in five areas of western Tanzania. Presumably, the Lichten- stein hartebeest and roan antelope frequent the woodland zone. The

Ugalla area compares quite well with other areas in the kinds of antelope species characteristic of the region, but sitatunga is specific only to the swamp area along the Ugalla River.

THE UGALLA AREA AS A CHIMPANZEE HABITAT

Our investigation team has been making ecological surveys in the Mukuyu, Masito, and Mahale areas since 1961, amassing data concerning the chimpanzees living there; but since the Ugalla area is very distant from our main camps, only four surveys have been made in that district so far. In the first survey of November 1963, S. Azuma and K. Izawa encountered chimpanzees in the riverine forest along the Mahumuwe Valley, to the east of the Busondo Plateau. In 1966, Kano and I walked eastward from the shore of Lake Tanganyika and reached Busondo by crossing the Masito Plateau. Then we walked further eastward to the Ugalla River and returned to the Busondo Plateau by taking the southern route (Figure 1). At the end of 1966, Kano walked around the southern mountainous part of the Ugalla area to complete the third survey of the district. During the dry season of 1968, A. Suzuki moved into the area east of Busondo, but he could not reach the Ugalla River because of his depleted water supply.

Our most valuable expedition, in terms of observing and recording extensive data about chimpanzee movement and habitation of the dry Ugalla area, began on September 11, 1966, when Kano and I started from Busondo for the Ugalla River.

The journey took 4 days, even though the straight-line distance totaled only 50 km, since there was no footpath and our route was very winding. It was at the end of the dry season, and we could not obtain a sufficient water supply until we reached the ruined village located near the Ugalla River, on September 14. For our return route, we walked

TABLE 5
Comparison of the distribution of the genus *Cercopithecus* in the chimpanzee distribution area of western Tanzania

NAME OF SPECIES	UGALLA AREA	MASITO AREA	MUKUYU AREA	KAROBWA AREA	MAHALE AREA
Cercopithecus ascanius	+	+	+	+	+
Cercopithecus mitis	+	−	−	+	+
Cercopithecus aethiops	+	+	+	+	+
Papio cynocephalus	+	+	+	+	+
Colobus badius	−	−	(?)	+	+
Colobus angolensis	−	−	−	−	+
Total number of species	4	3	3+(?)	5	6

southwestward from Igalula for 3 days to a place about 20 km south-east of Busondo on the Uvinza-Mpanda Road. Upon completion of our journey, we had traversed the northern Ugalla area twice. We had five African porters on our safari; but to reduce the equipment to a minimum, we did not bring a gauge for meteorological observation.

The route of our 7-day trek was 130 km. We encountered chimpanzees only once on our safari and that was on the second day of our journey, September 12. For that reason, this report emphasizes the descriptions of flora and fauna of the area, although a careful record was made each day of the various signs of the presence of chimpanzees along the route.

AN ENCOUNTER WITH A GROUP OF CHIMPANZEES

On September 12, we left our campsite at 9:25 A.M. We departed from the right bank of a tributary of the Issa River, which flows eastward, and went down along the valley. The narrow riverine forest on each side curves through the middle of the open, U-shaped valley. At 10:40 A.M., we came upon Mbuga vegetation about 5 km from the campsite and heard chimpanzees in the riverine forest of *Cynometra* and *Albizia* beyond us. Their repeated shrieking was long, intoned, and unfamiliar to us.

At 10:45 A.M., carrying only binoculars, Kano and I crossed the river to the right bank; hiding ourselves in the tall grass, we approached the riverine forest where chimpanzees were gathered. We found a group of five chimpanzees in the trees at 10:57 A.M.—a large adult male, a young

TABLE 6
Comparison of the distribution of antelope species in the chimpanzee distribution area of western Tanzania

NAME OF SPECIES	UGALLA AREA	MASITO AREA	MUKUYU AREA	KAROBWA AREA	MAHALE AREA
Taurotragus oryx	+	+	+	+	−
Tragelaphus scriptus	+	+	+	+	+
Tragelaphus spekei	+	−	−	−	−
Hippotragus equinus	+	+	+	+	+
Hippotragus niger	(?)	−	+	+	−
Alcelaphus lichtensteini	+	+	+	+	+
Kobus defassa	+	+	+	+	−
Sylvicarpa grimmia	+	+	+	+	+
Oreotragus oreotragus	+	+	+	+	+
Nesotragus moschatus	−	−	+	+	+
Total number of species	8+(?)	7	9	9	6

FIGURE 1
We arrived at the riverside of Ugalla after a 10-day walk from the shores of Lake Tanganyika. From left to right: Dr. Kano, Mr. Sushira, Mr. Haruna, Mr. Issa, and Mr. Ramadhani.

female, an estrous female, and two medium-sized chimpanzees. We were 150 m away from them as they continuously screamed to each other. We judged from the volume of noises that there were more than 15 chimpanzees altogether. They gradually moved eastward from tree to tree.

Kano headed east, about 60 m away from me. At 11:05 A.M., both of us began walking forward in the same direction into the riverine forest. Then two or three chimpanzees encountered us, but they only gazed at us without trying to run away. I swiftly moved into the forest and crossed the river. There I observed an estrous female chimpanzee on a somewhat isolated tree at the edge of the river. She was only 15 m away from me and was clearly visible, her body completely exposed to my view, but she did not try to flee. However, a male lion disturbed my observation; he probably preceded me when I was entering the forest. As soon as he sensed my presence, he rushed away eastward where he confronted Kano, who was also walking toward the forest. Kano said that the lion suddenly appeared only 7 m to his left, roared at him, and caused him to rush across the river toward me in quick escape. After that abrupt intrusion, we could not continue our observation.

Our encounter with the lion explained why the chimpanzees were making such helpless and strange cries and why they did not flee in spite of our approach. They must have been frightened by the lion as it stalked among the trees. In the same vicinity where we came upon the chimpanzees, we found two of their beds. We concluded that they had been trapped there by the lion since the night before, because we saw them at 11 A.M., a time when chimpanzees usually move to a new feeding site. It is likely that lions are the chimpanzees' only predators in the woodland, and our observation was a rare case at that.

BEDS OF CHIMPANZEES AND THEIR DISTRIBUTION

Throughout the 7-day safari, we recorded observations of 125 chimpanzee beds. The conditions had been good for finding the beds because the deciduous trees had shed all their leaves during that arid season, and also because the riverine forest vegetation was extremely poor in this area. It should be understood, however, that the number mentioned above is exceedingly low compared to other areas like the Masito area. Listed in Table 7 are both the number of beds we discovered each day and the number of beds in each type of vegetation—riverine forest or dry open forest. Excluding September 14, which we spent traversing the territory near the Ugalla River covered with Mbuga vegetation, we discovered beds almost continuously.

After the 7-day safari, I traversed the woodland that extended about 50 km from the Luegele River in the Karobwa area to the Lugufu River in the eastern part of the Mukuyu area. At that time, in the Lugufu basin, I found surprisingly few signs of chimpanzees; however, this does not indicate that the woodland covering the Lugufu basin is a more

TABLE 7
Number of chimpanzee beds recorded daily

DATE	NUMBER OF BEDS	RECORDED IN DRY OPEN FOREST	RECORDED IN RIVER- INE FOREST
September 11	20	19	1
12	17	14	3
13	5	1	4
14	0	0	0
15	30	30	0
16	45	28	17
17	8	8	0
Total	125	100 (80%)	25 (20%)

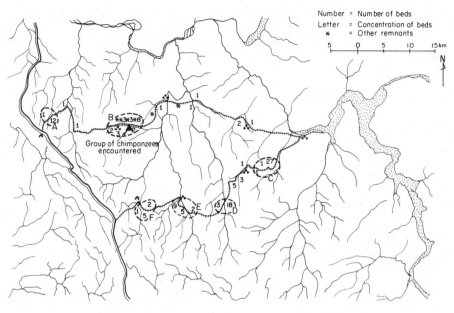

FIGURE 2
Distribution of 125 beds and six concentrations of beds mostly in the Ugalla area

severe environment than the Ugalla. The woodland in the Lugufu is actually better than the Ugalla savannah, but, because the mountains are more suitable for chimpanzees, the population is concentrated in the Mukuyu and Karurunpeta Mountains. On the other hand, it might be correct to say that chimpanzees in the Ugalla area are forced to live in the dry woodland because there is no choice between mountains and woodland as there is in the Lugufu basin.

The fact that 80% of the chimpanzee beds were found in the dry, open forest of the Ugalla is convincing proof that the chimpanzees depend more on woodland vegetation than on the sparse riverine forests. Figure 2 shows the location of the 125 beds that we saw; their concentration into six major groups, or sections, is clearly shown on this distribution map and is as follows:

A. On the escarpment to the east of the Busondo Plateau.

B. Along the tributary of the Issa River, where we came upon a group of chimpanzees.

C. On the escarpment of the table-shaped hills behind the ruined Igalula village, on the left bank of the Ugalla River.

D. Around the narrow, riverine forest area, along the upper stream of the Mogoguwesi River.

E. Around the basin open to the upper stream of the Issa River's great tributary, the Mufwombosi River.

F. Among the hills and escarpment at the upper stream of the Issa River's main stream.

Among the six concentration sections of beds, the dry forest of *B. bussei* develops exclusively in sections of A, C, and F. It can be supposed that chimpanzees gathered in these places to feed on hard nuts of *B. bussei* early in the dry season. Along the Busondo Plateau, stretching between sections A and F, and also on the left bank of the Ugalla River around section C, a tremendous dry forest of *B. bussei* can be found. There are no riverine forests in these areas, so *B. bussei* is the only source of food.

In the vicinity of section B, the escarpment on both sides of the valley is covered with a dry forest of *B. bussei*, as well as a narrow and comparatively moist riverine forest. In this forest, chimpanzees probably find juicy fruits such as *Landolphia, Ficus, Anona,* and *Vitex,* which do not occur so abundantly. My guess is that after a group of chimpanzees passes through this riverine forest, the fruit trees there will not produce new fruit again until about a week or 10 days later. (In contrast, the Masito area has such thick vegetation that an ample fruit supply is available every day.)

The riverine forest, as seen in D, is a very thin, dry forest of *Cynometra,* where chimpanzees can scarcely find any food. The concentration of their beds at D indicates that it is one of the important sleeping places on their nomadic route, which means it is likely that D connects C with F; that is, it is possible that chimpanzees go 20 km away to the Mufwombosi basin after feeding themselves on the left bank of the Ugalla River (C). We will examine this pattern later in more detail.

Mufwombosi (E) is a basin in the vast woodland where a dry, Y-shaped riverine forest is found. No food other than the fruits of *Aframomum* sp. can be found on the edge of the riverine forest, although fruits of *Strychnos* spp., and *Parinari curatellifolia* are available in the woodland. In addition to the six main concentrations of beds, a few more beds were discovered.

I would like to examine the intervening areas between the two neighboring concentrations of beds. In the Mogoguwesi River basin, between B and C, the most moist riverine forest of the Ugalla area is found with its comparatively rich food productivity. It contrasts sharply with the surrounding areas. For instance, between A and B there is only a vast woodland in lowland territory, excluding the dry riverine forest of the Mahumuwe River. Between C and D, D and E, and E and F, just about the same type of vegetation is found. The kinds of riverine vegetation growing there have already been mentioned and are very different from the moist riverine forest.

TABLE 8
The vegetation types and estimated land value of each section[a]

CONCEN-TRATED AREA OF BEDS	VEGETATION TYPE[b]	ESTIMATED LAND VALUE[b]	INTERVENING AREA BE-TWEEN TWO SECTIONS	VEGETATION TYPE	ESTIMATED LAND VALUE
A	BB-JG	3	A-B	JG	1
B	BB-MRF-JG	5	B-C	MRF-JG	3
C	BB-JG	3	C-D	JG	1
D	DRF-JG	2	D-E	JG	1
E	DRF-JG	2	E-F	JG	1
F	BB-JG	3			

[a] Sections refer to areas indicated on Figure 2.
[b] BB: *B. bussei* open forest (estimated land value, 1); MRF: moist riverine forest (estimated land value, 2); DRF: dry riverine forest (estimated land value, 1); and JG: *J. globiflora* open forest (estimated land value, 1).

On Table 8, the vegetation and the land value of the bedding places are shown, based on the chimpanzee's dietary life. However, if we compare the Ugalla area with the Masito, Mukuyu, and Mahale areas, we should note that this area is exceptionally low in its land value on the whole. Downstream of the Mogoguwesi River, there is a relatively large food supply, and it would seem reasonable for the chimpanzees to take advantage of it—but we discovered only two beds around there. They were 8 km and 18 km away from B and C, respectively, as the crow flies. We can presume from the location of the beds that if two separate groups left beds at B or C, either of the two might have moved ahead independently into the moist riverine forest. If the *same* group left beds at B and C, however, this place must have provided a stopping-off point on their route between B and C. Of all these beds—judging from the dryness of the bed materials—those made within the past month included two that we came upon at B, three of five beds found between B and C, fifteen of twenty-nine found at D, and nine of sixteen found at E.

In the latter half of the dry season, chimpanzees seem to move to the central part of this area, not to the east. In other words, the eastern part is not a good place for them to live throughout the entire year. So the unit group, in its seasonal nomadism, moves from the eastern to the central section. Similarly, it might be possible for the group we encountered at section B to make an excursion to the riverside of Ugalla at the beginning of the dry season. (My guess is that they might move from the northern part to the southern part along the narrow riverine forest.)

The analysis of the beds indicates that chimpanzees travel the Ugalla area widely, but they do not remain in any part of it for a very long time. All in all, the Ugalla area has a thin population density of chimpanzees.

OTHER REMNANTS

In addition to the beds, there were other recent remnants of food and droppings. In a week's time, we discovered six remnants, as shown on Table 9. Four of them, 1 through 4, were the remnants that the chimpanzees left on the day before we encountered them. From the location of the remnants, we assumed that this group must have come down the valley from the west. Remnant 5 was 2-3 days old and was found 5 km northeast from the place where we saw the chimpanzees. This evidence proves the existence of some other small population that is related to the nomadic group. Remnant 6 was found on a plateau approximately 2 km east of the mainstream of the Issa River, indicating that chimpanzees can traverse the woodland without necessarily depending on the riverine forest.

Each remnant also shows what kinds of food the chimpanzees eat during the end of the dry season. Five of the six remnants indicate that these animals had eaten *Strychnos* sp. (Kantonga), a woodland tree which bears sour fruit the size of an orange during the dry season. In two of the three droppings, seeds of *Aframomum* sp. were found. This tree grows at the edge of the riverine forest and bears sour-tasting red fruit from the dry season to the beginning of the rainy season, but the fruit is not sufficient in quantity and quality to be a staple food for the chimpanzees. The end of the dry season, which also marks the end of the season for the *B. bussei* nut, is probably a difficult time for the chimpanzees to survive. *J. globiflora* and *Cynometra* sp. produce nuts after the dry season, but we could not confirm that the chimpanzees ate them in this area.

TABLE 9
Chimpanzee remnants other than beds

NUMBER	DATE	KIND OF REMNANT	PLACE	CONTENTS
1	September 12	Dropping	Section B	Seeds of *Aframomum* sp. and *Strychnos* sp.
2	12	Wadge	Section B	Seeds of *Strychnos* sp.
3	12	Dropping	Section B	Seeds of *Aframomum* sp. and *Strychnos* sp.
4	12	Food Remnant	Section B	Fruit of *Strychnos* sp.
5	12	Dropping	About 5 km east from B	Fiber of grasses
6	13	Food Remnant	About 2 km east from Issa camp site	Fruit of *Strychnos* sp.

DISCUSSION

When Kano reported his investigations of the Ugalla area, extending over 2800 km² of inhabited territory, he described the distribution and density of chimpanzee population. In his 1972 report, he estimated the total population to be between 200 and 240, which makes the population density about 0.07 or 0.08 chimpanzees per square kilometer. Furthermore, he estimated that a unit group consisted of 40 chimpanzees on the average, with five or six different groups existing in the Ugalla area.

According to his distribution map, based on the location of beds throughout the area, the chimpanzee population could be divided roughly into north and south Ugalla. If we work from the supposition that three or four of the groups probably live in south Ugalla, where living conditions are more favorable, we can assume that the remaining two or three groups inhabit the north. Since I have been concerned with northern Ugalla, that is where I obtained my data, including the following vital information. Our team confirmed the existence of one group out of two or three possible groups that probably inhabit northern Ugalla. Consequently, if we could confirm the habitation of the other one or two groups, we could support the estimated number that Kano introduced in his report.

We can reconstruct the nomadic range of the chimpanzees by putting together the data we gathered on our 7-day journey. If we assume that the month-old remnant found during our survey is from the only group we came upon, then the region where the new remnant was found will cover approximately 500 km². This range occupies mainly the central part of the northern Ugalla area. As a result, the sections in and around A and C do not support the chimpanzees for a year and they leave. Therefore, as far as this survey is concerned, we should include the sections where we found the only old remnants as belonging within the boundaries of the nomadic range of this group. If we add these sections, however, the nomadic range of this group comes to approximately 1200 km² of the northern Ugalla area, which is too broad an area for one group to cover. It is more reasonable to conclude that the relatively new remnants we found belong to two groups of chimpanzees instead of just one. The new remnants that we found on the way to the Ugalla River were probably left by the group we encountered at B and by the other chimpanzees related to them. The new ones we found on the way back from the Ugalla River must have been left by another group we did not encounter.

The only data that support our supposition is information that we obtained in the village along the upper stream of the Issa River at F on September 11. This village is located below the escarpment, southeastward of the Busondo Plateau, and it was the only populated village

69

encountered along the route of our 7-day journey. The villagers said that they had not heard chimpanzees that day, but they had seen them often lately. From this village to Mufwombosi at E, it is 10 km along the narrow riverine forest. We found nine relatively new beds at E. The natives we met, camping by the upper stream of the Mufwombosi River, had been fishing for catfish for 5 days in the small pools along the dried river bed. They said that they had not seen or heard chimpanzees during that time. Therefore, the group of chimpanzees that had stayed at E until at least 5 days before we found signs of them could possibly have moved on to F; and, as I already pointed out, E is quite possibly connected with C through D.

This information indicates that the nomadic range of a group of chimpanzees extends between the areas of C and F. Consequently, if the information we obtained from the Issa villagers was trustworthy, this group is probably different from the one we encountered.

I would like to comment on our supposition regarding the nomadic range of chimpanzees in the northern Ugalla area. We have found supportive evidence that two groups of chimpanzees inhabited an area of 1200 km². We also obtained some evidence for the existence of a small population that repeatedly joined and parted from those two groups, although we could not obtain positive proof of this. It is my conclusion that the two groups independently inhabit the north and south, and that each group has a wide nomadic range, with the Busondo Plateau as a western boundary and the riverside of the Ugalla River as an eastern boundary. Both the northern and southern ranges encompass about 500 km². According to our current data on the Kabogo, Kasakati, and Mahale areas, a 25-50% overlap of ranges was found between neighboring groups of chimpanzees that sustain close social contacts through periodic exchange of members (Nishida & Kawanaka 1972). It is my belief that these types of relationships also occur in the Ugalla area. Although we did not directly explore the middle streams of the Issa, Mufwombosi, and Mogoguwesi Rivers, it is undoubtedly true that the nomadic ranges of the chimpanzees overlap there, making contacts between the two groups naturally possible. If we add this overlapped area to the regular nomadic area, the nomadic range of a group comes to 700-750 km² —which is much larger than the 470-560 km² that Kano estimated. This nomadic range of a unit group is unusually broad for nonhuman primates; but as I pointed out before, this range allows the chimpanzees to adapt to the extremely severe environment.

As for the group size, we estimated that the group we encountered consists of more than 15 chimpanzees but not more than 40. Although we could not get any information concerning the second group, we can predict that these two major groups (along with some other smaller populations in loose association around them) probably total nearly

100. This figure supports Kano's calculation of population density at 0.07-0.08 per km².

I would like to submit this report of chimpanzees living in the Ugalla area as proof of the adaptation of pongids to an open land. This evidence can serve as the basis for further discussion about chimpanzee adaptation. I am hopeful that future studies of the Ugalla chimpanzees will provide material for consideration of a more detailed investigation of hominization.

Acknowledgment

This field work was supported by the Overseas Scientific Research Fund of the Ministry of Education, Japan. The work was conducted with permission and facilities provided by the Tanzanian Government and the University of Dar-es-Salaam, to both of which the writer is much indebted.

I with to express my deep appreciation to Dr. T. Kano of Ryukyu University and to the five Watongwe friends, who helped me carry out this most difficult but fascinating study.

Anting behavior—two females and two males fishing for arboreal ants with peeled bark

Toshisada Nishida

The Social Structure
of Chimpanzees
of the Mahale Mountains

The field study of chimpanzees of the Mahale Mountains began in October 1965. The study, from the onset, has focused on the elucidation of the social organization of wild chimpanzees and is based on the provisioning and individual identification method.

Although we have used "Mahali" instead of "Mahale" in previous papers, to conform with officially printed documents including maps published in Tanzania, we have decided to change the name to "Mahale" (kitongwe), the name that is used by people locally.

The first stage of the study was to search for the social unit in the society of chimpanzees. Many primate societies are divided into self-supporting social groups, which function as a unit in food-getting, reproduction, defense against natural enemies, and social learning. Such groups are spaced apart and are to some extent antagonistic to one another.

The corresponding group in chimpanzee society proved to be the multi-male bisexual group ranging from 20 to 100 in size, which I have called the *unit-group.* This unit-group, however, is not the usual group with cohesiveness of members. The unit-group is involved in a continual process of splitting into several temporary *subgroups,* which then rejoin, and reseparate into different subgroups. Thus, it is very rare that all members of the unit-group meet together in one group (Nishida 1967; 1968). Moreover, a fair amount of exchange of individuals has been observed between two neighboring unit-groups (Nishida & Kawanaka 1972).

This paper describes the diachronic structure of a unit-group of chimpanzees by analyzing the mechanism and background supporting the two opposite characteristics of the unit-group, i.e., rigidity vs. fluidity, or closedness vs. openness.

FIGURE 1
The Kasoge forest near Nkala Valley

METHOD AND GENERAL BACKGROUND OF THE STUDY

Study Period

This paper is based on the data obtained during a total of 35 months of long-term and intermittent field studies carried out at Kasoge, 138 km south of Kigoma along the east shore line of Lake Tanganyika.[1] My own observation covers less than half of the long-term study, which has lasted for more than 8 years and during which four other researchers contributed to the accumulation of the basic data while working on their own topics of the chimpanzee study.[2]

In addition to our data, two Tanzanian field assistants, who were trained to check attendance records and reproductive cycles in 1968, also kept records of agonistic, grooming, sexual, and feeding behavior mainly near the permanent feeding station since 1969. Their continuous record has filled observational gaps in our intermittent studies.

Study Area

Kasoge is located at the foot of the Mahale Mountains which delineate the largest projection (the Mahale Peninsula) on the east shore of Lake Tanganyika. Running in a straight line from northwest by west to southeast by east, the Mahale Mountains are composed of the main peak Nkungwe (2462 m), and five prominent peaks higher than 2000 m. Moist air blowing from the lake causes considerable cloud and mist development, which supports both extensive montane forests above the altitude of 1300 to 1500 m, and the concentration of gallery forests (the Kasoge forest) at the western side of the mountains. Numerous valleys intersect the mountains, some of which support permanent streams that flow into the lake. The Kasoge forest can be regarded as *forest iselberg* in a midst of Miombo woodland, which extends in a vast area of western Tanzania and eastern Congo (Figure 1).

The year is clearly divided in this area into a dry season (mid-May to mid-October) and a rainy season (mid-October to mid-May). Annual rainfall is presumed to be over 1500 to 2000 mm in the Kasoge forest, but considerably less outside the forest. The study area consists of a narrow mountainous strip stretching for 20 km along the shore of Lake Tanganyika between Katumbi and Msoffwe, which can be divided into two major vegetational zones caused by the marked differences in local rainfall, namely Miombo-gallery forest mosaic and the Kasoge forest. Detailed descriptions of the area's vegetation are being published elsewhere (Nishida, in press).

Study Groups

About five to six unit-groups utilize at least a part of the study area. Among them, the K-group (previously, Kajabala group) was provisioned at the permanent feeding station in 1966 and has been intensively studied to date. Although the study of ranging and feeding behavior of one more habituated unit-group, the M-group (previously, Mimikire group), began in 1965, many chimpanzees were not individually identified until after frequent visits to the permanent feeding station in 1972-1973. The other neighboring unit-groups (B group, N group, etc.) have been tracked and observed several times respectively. None of the chimpanzees of these unit-groups have been individually identified so far, though I have had occasional opportunities of observing them in close proximity.

The home range of the K-group (an area of 10.4 km^2) is covered with a semi-deciduous forest (the northern part of the Kasoge forest) and partly with Miombo-gallery forest mosaic (Myako Valley). The range of the M-group (an area of more than 13.4 km^2) consists mainly of a semi-deciduous forest (the major part of the Kasoge forest), with Miombo woodland in the southernmost part of its range (Nishida 1974). The permanent feeding station was established at the middle stream of Kansyana Valley, in the overlapping area of the ranges of both unit-groups. Since the success in the provisioning of the K-group in 1966, field assistants have continued to supply bait (sugar cane and banana) at the feeding station; however, the K-group usually leaves the overlapping area and migrates to the northern part of its range at the end of the dry season, almost at the same time as the M-group comes to the overlapping area.

Some chimpanzees of the M-group began to take bait at the permanent feeding station in October 1968, but the M-group was moving in its terrain almost independently of our feeding trial until 1973.

Provisioning Process

In October 1965, the secondary bush in the Kansyana Valley was cleared. Seedlings of sugar cane were planted in three acres. The sugar cane plantation was surrounded by a semi-deciduous forest on three sides and by elephant grass bush on one side. In the beginning of November, some full-grown sugar cane was planted in a narrow strip along the edge of the forest. Also, in the forest near the plantation, several hundred ripe sugar cane sticks were set out. In March 1966, some chimpanzees began to take sugar cane here and there in the plantation. The feeding area fixed at the northernmost edge of the plantation in May 1966 made observation possible on a regular basis in a rather narrow area. The feeding area was extended to the forest in September 1966 to accommodate shyer chimpanzees, especially females with infants. It was cleared of grasses and shrubs, but no trees were cut down.

From March to September 1966, as many as 50 sticks of sugar cane were set out daily in the area (30 m x 40 m), so that these might be taken freely by chimpanzees, while I sat on a scaffold built 80 m away. In September 1966, the scaffold was moved to the southernmost corner of the feeding area, where it became possible for me to make observations 10 to 30 m away from chimpanzees. Since 1968, I found it more and more difficult to obtain enough sugar cane. The reduction in the number of sugar cane sticks caused a high frequency of aggressive behavior, however, and sometimes made it impossible for subdominant or shyer animals to get access to bait. Finally, we gave up setting out whole sugar cane sticks. Instead, we cut the sugar cane into very small pieces and scattered them in a wide area so that no chimpanzee could monopolize them and all independent chimpanzees could get something. We began to use bananas as another bait in July 1966. In 1975, we supplied an average of 10 chimpanzees of the K-group with 10-30 bananas and/or 5-10 sugar cane sticks on a daily basis at the permanent feeding station (Figure 2).

FIGURE 2
Permanent feeding area and chimpanzees of M-group

Observational Methods

The observational methods can be divided into the following five categories:

1. Naturalistic Method: The observer simply follows the chimpanzees in the natural habitat without depending upon any artificial feeding method. This method was mainly employed in the initial stages of the study.

2. Mobile Provisioning Method: The observer carries a small quantity of bait while looking for chimpanzee groups in their home range, and gives them food (one to five sticks of sugar cane and/or one bunch of bananas) wherever such groups are encountered. When chimpanzees cannot be located because they do not vocalize, the observer mimics pant-hooting calls, to which chimpanzees of the K-group very often respond. When chimpanzees begin to move after they are provisioned, they are usually followed by the observer for as long as possible. This method was often used for the K-group between September and February in the later periods of the study.

3. Fixed Provisioning Method: Bait (mainly sugar cane) is regularly supplied at the permanent feeding station. Chimpanzees visit there regularly or irregularly while traveling within their home range.

3a. Chimpanzees are observed at the feeding station, but also are followed when they leave the feeding ground. When a group rarely visits the permanent feeding station, this method is similar to the naturalistic approach. Since the M-group neither visits the feeding ground frequently nor stays there for a long time, this is the most important method of observing M-group chimpanzees.

3b. Observers simply wait at the station until chimpanzees arrive, and do not follow groups after they depart. This is the major observational method for the K-group chimpanzees in the earlier periods of the study after provisioning had proved to be successful.

4. Incidental Observation Method: When a unit-group did not utilize the permanent feeding station at all while scientists were absent, Tanzanian assistants generally did not put any food there or make any systematic observations far from the feeding ground. Thus, these periods produced only fragmentary records of behavior due to infrequent encounters.

Analysis of the Provisioning Method

Wrangham (1974) critically stated that: "direct generalizations of the observational results gained at the feeding station to situations outside the feeding area should be avoided in the case of a variety of behavioral and ecological observations." Although this statement is indeed valid, we can point out that provisioning seems to influence not the quality, but the quantity, of some aspects of behavior. All patterns of social behavior observed at the feeding area have also been recorded outside of the area. I do, of course, think that we should refrain from the provisioning method when doing purely ecological studies on a quantitative basis. The sociological study on provisioned primate groups carried out from the holistic viewpoint, however, can produce results that accurately reflect the intrinsic qualities of the animals, as long as we sometimes observe and check relationships among individuals outside of the feeding area.

Criticism of the provisioning method seems to be based partly upon the modern tendency to think that the degree of importance of a phenomenon depends on its frequency. But, it is impossible to say if the frequency of a phenomenon is low, that it will be unimportant and need not be treated seriously. A direct inter-unit-group quarrel, for example, has been observed only once throughout the 7-year study period, but this one case compares favorably with many indirect observations of inter-unit-group antagonism. It is even possible to say that the rarity of the phenomenon implies the deep antagonism between unit-groups and the effective mechanism working to avoid overt clashes.

Artificial feeding does not create any behavioral patterns completely new to the object species or population. It does not change the nature of the animals, but merely the quantity of existing patterns (see also the discussion at the end of this article).

RIGID AND CLOSED ASPECTS OF THE UNIT-GROUP

Consistent Aspects of Membership and Size of the K-Group

The K-group has been intermittently studied since 1966. Although the membership varied every year, 23 of 29 members identified in the initial stage of the study were still found in the K-group after 8 years of observation.

Adult males seem to be stable members. Of six adult males identified in 1966, one was missing in 1969 and another in 1971. It is likely that the former, the oldest and most dominant male (Kasagula), died of old age. The latter, the youngest and lowest-ranking male (Kaguba), was not found in the K-group or any other neighboring groups (M, N, B, etc.). He must have died or moved far from the study area. The remaining four adult males and one young adult male formed a strong social bond with one another and did not move from the home range of the K-group until April 1975, when Kasanga, then the prime-adult male, disappeared. Moreover, Kajabala, then the past-prime male, also disappeared in September 1975. Both of these disappearances occurred very suddenly

FIGURE 3
Record of observation of the chimpanzees identified in the K-group (including only those born before 1964)

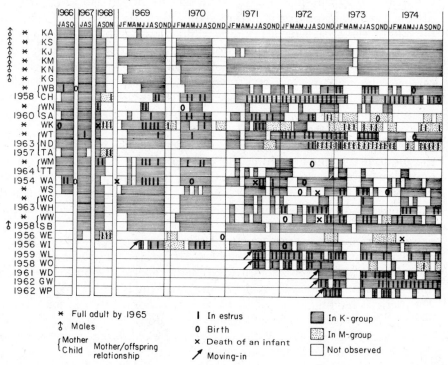

Abbreviation of individual names is as follows: KA=Kasagula, KS=Kasonta, KJ=Kajabala, KM=Kamemanfu, KN=Kasanga, KG=Kaguba, WB=Wabunengwa, CH=Chausiku, WN=Wangala, SA=Sada, WK=Wankungwe, WT=Wantangwa, ND=Ndilo, TA=Tausi, WM=Wamikambi, TT=Tatu, WA=Wakasunga, WS=Wasalamba, WG=Wangulu, WH=Wahanse, WW=Wabwema, SB=Sobongo, WE=Wamkime, WI-Wakasila, WL=Wakiluhya, WO=Wakampompo, WD=Wantendele, GW=Gwekulo, WP=Wakapala.

and for no apparent reason. These two males were not found in any neighboring groups. (The detailed description of moving-out in males and intra-unit-group male aggression and social bond will be published at a later date.)

Three of nine adult females identified initially (Wantangwa, Wabunengwa, Wakasunga) have maintained comparatively strong social bonds with adult males and have never been observed visiting any adjoining group. Although there are many individual differences relative to the extent of association of the remaining six adult females with adult males, they have generally moved within the home range of the K-group and more or less maintained association with males of the K-group throughout the study period, with the exception of Wankungwe (Figure 3).

The size of the K-group has, however, remained remarkably consistent during the study period from 1966 to 1973, probably because factors influencing increase (birth of infants and acceptance of strangers) and decrease (death and moving-out of members) have been well balanced during the study (Table 1). Among ten infants that were born in the K-group since 1967, three died within one year and one within two years after birth. One more infant born in 1966 died before reaching 3 years of age. The birth-interval proved to be 5 years in two females (Wangala and Wabwema) and 6 years in one female (Wasalamba), as long as the previously born offspring did not die. Past-prime females, however, seem to give birth to infants after longer intervals. Two adult males (mentioned above), one adult female, and four subadult females disappeared permanently from the K-group, which, in turn, has accepted six subadult strange females. These will be discussed in relevant sections.

The social composition of the M-group presents a question about the conservativeness of male ranging behavior. Although the age class of adult males of the M-group consists of old, past-prime, prime, and young adult males, I did not see any adolescent males in this group in 1973 (Table 2). Thus, the age gap between the youngest adult male (about 13 years old) and the oldest juvenile male (about 7 years old) amounts to 6 years. The crucial question is whether this age gap is an accidental phenomenon. If it is not accidental, it means that all adolescent males had moved out of the M-group during their progression from the juvenile stage to the adolescent stage, before we provisioned the M-group.

TABLE 1
Annual change in size and composition of the K-group[a]

AGE CLASS	SEX	1966	1967	1968	1969	1970	1971	1972	1973
Adult	♂	6	6	6	6	5	5	5	5
	♀	7	9	8	10	10	10	9	9
Adolescent	♂	0	1	1	1	1	0	0	0
or subadult	♀	2	3	3	2	2	6	9	6
Juvenile	♂	0	0	0	0	0	1	1	1
	♀	1	2	2	4	4	3	4	4
Infant	♂	1	2	2	1	2	0	2	2
	♀	4	5	4	2	2	1	0	0
	?	0	1	1	0	1	2	4	1
Total		21+	29	27	26	27	28	34	28

[a] Any identified individuals who moved with the core-subgroup of the K-group at least once a year are included in the K-group of that year.

TABLE 2

The known age-sex composition of the M-group[a]

AGE CLASS		MALE	SEX UNKNOWN	FEMALE	TOTAL
Adult	III (past-prime)	3		5	8
	II (prime)	6		10	16
	I (young	7		10	17
Adolescent	II (subadult)	0		4	4
	I (adolescent)	0		1	1
Juvenile	II	1		2	3
	I	3	2	2	7
Infant	III	2	2	0	4
	II	3	5	0	8
	I	0	1	1	2
Total		25	10	35	70+

[a] As of February 1974.

Territoriality of the Unit-Group and Social Boundaries

One of the main reasons why I called the study groups *unit-groups* is that they have maintained group ranges with well-delineated and traditional boundaries. All males and most adult females of the K-group have never been seen crossing the boundary line of the group for 7 years. It looks as if invisible barriers exist, which most members of the K-group are prohibited from going beyond. They never fail to turn back on arriving at such points. Similar turning points have also been recognized in chimpanzees of the M-group, although a subgroup of the M-group was once seen to penetrate the exclusive area of the K-group (Case 1). This boundary is neither a geographical boundary as Goodall (1965) mentioned, nor even so much an ecological consequence, as Reynolds (1966) suggested, but a sociological consequence.

Home ranges of the K-group and the M-group, however, do overlap extensively; about half of the home range of the K-group is utilized also by the M-group; conversely, about one-third to one-fourth of the home range of the M-group is utilized by the K-group. Moreover, the southernmost part of the home range of the M-group is sometimes utilized by another different unit-group (the N-group). There are, however, cases where two neighboring unit-groups have scarcely any overlapping of their home ranges; chimpanzees of the B-group ranging to the north of the K-group have not been observed to trespass on the K-group range, and a small subgroup consisting of an adult male and estrous females of the K-group has been observed only once to travel along the margin of the B-group range.

The M-group, dominant over the K-group, has priority of use of the overlapping area. The K-group as a whole generally moves to its northern exclusive area when the M-group comes into the overlapping area, usually at the end of the dry season, which is segregation of the range by time. When two groups cannot avoid utilizing the overlapping area simulataneously, probably because of local differences in production of food resources, they share the same terrain temporarily. The K-group then moves about to avoid any direct encounter with the M-group (Nishida 1968; Nishida and Kawanaka 1972). This coexistence of the two unit-groups in the overlapping area continued as long as 2 months in the study period of 1966.

Displacement Relationship Between Unit-Groups

Unit-groups space apart from one another, and no direct encounter occurs when the food supply is abundant and distributed evenly enough to support each local population. Unit-groups, then, show the simple and discrete pattern of spatial arrangement according to the distribution of staple food. Habitual nomadic routes and seasonal ranges of unit-groups are reorganized when a new staple food with a different pattern of local distribution replaces the older one, although the extent of reorganization of the range is strictly limited by the social factor (Nishida 1968; 1974).

The seasonal northward migration of the M-group usually occurs in September, after several waves of initial exploratory travel by part of the group in June-August. This eventually results in the northward migration of the major part of the K-group, often on the day of arrival of the major part of the M-group in the overlapping area.

The final success in the provisioning of the M-group seems to have influenced the ranging patterns of both the M-group and the K-group in 1972-1973. Since the permanent feeding station was established in the overlapping area of the home ranges of both unit-groups, number of days of occupation of the area by the M-group increased remarkably, while those of the K-group decreased accordingly. The seasonal migration of both groups is evidence of the integrative nature of the unit-groups and also of the dominance of the entire M-group over the entire K-group in the priority of use of the nomadic range.

The K-group is aware of the approach of the M-group by its vocal outburst. The K-group chimpanzees seem to recognize the unit-group by the direction from which vocal outbursts come. In acoustic and (very rarely) visual inter-unit-group contacts, a wide spectrum of behavior is observed, which can be divided into five categories; reassurance, scout, aggressive advance, withdrawal, and friendly advance (Table 3). The kinds of behavioral categories expressed at each encounter depend upon the relative size and composition of subgroups (of different unit-groups), the absolute distance between subgroups, and the location of encounter (the relative distance from the exclusive area of the home

TABLE 3
Behavioral repertoire of chimpanzees observed in inter-unit-group encounters[a]

AGGRESSIVE ADVANCE	1. Vocal outburst
	2. Wraah call
	3. Pant hoot
	4. Drumming (slap or stamp on the buttress of tree)
	5. Branching (shake branch)
	6. Slapping (slap the ground)
	7. Charging display (rush about, break branch, slap the ground, drag the stick, etc.)
	8. Bipedal hunch (shake arms bipedally and stamp on the ground)
	9. Approaching towards enemy
	10. Chasing (run after the enemy)
	11. Grappling (wrestle with and bite the enemy)
	12. Cannibalism (kill and eat the enemy)[b]
REASSURANCE (TOWARDS INTRA-GROUP MEMBERS)	1. Grin (often silent, but sometimes accompanied by scream and penile erection)
	2. Flee to one's companion
	3. Half-embrace (touch one's companion on the shoulder, abdomen, scrotum, etc.)
	4. Gather together
	5. Dorso-ventral embrace or mounting
	6. Ventro-ventral embrace
WITHDRAWAL	1. Run away
	2. Slip off silently
	3. Complete inhibition of calls
	4. Turn back calmly towards the center of the home range
SCOUT	1. Bipedal gaze (stand upright with hair erect and gaze towards the enemy)
	2. Sniff on the ground
	3. Patrol the area
	4. Sit to listen on the spot
	5. Ascend the tree to listen
	6. Ascend the hill to listen
FRIENDLY ADVANCE	1. Inspect the genitalia of strange females
	2. Yawn[c]
	3. Approach the enemy group with pant-grunt, ascend a tree and gaze at the enemy group from the periphery[c]
	4. Follow chimpanzees of the enemy group[c]

[a] In response to strange calls or to approach of strangers.

[b] For a mother-infant pair, males of the neighboring unit-group are often dangerous. An infant of a strange female was killed and eaten by chimpanzees of the habituated community at Gombe (Bygott 1972). At Mahale, an infant of the M-group was killed and eaten by three males of the K-group and on another occasion (although unconfirmed) an infant of a K-group female was possibly killed and consumed by M-group males (Nishida et al., in press). On the other hand, females without infants, or females accompanied only by juveniles, may transfer safely to the neighboring unit-group since they can be estrous. Infanticide may function in changing *mothers* into receptive females as was seen in Hanuman langurs (Sugiyama 1967).

[c] These three categories are observed only in females who subsequently transfer to the neighboring (enemy) group.

range of each unit-group). Reassurance behavior however, is observed whenever chimpanzees suddenly hear strange calls (often, calls of chimpanzees of different unit-groups) or detect the approach of strangers. Among them, the size factor is most important in determining how the K-group chimpanzees react.

Generally, chimpanzees of the K-group retreat rapidly and avoid making any calls, when they hear calls from the M-group, while both groups are in the overlapping area. If they hear only a few calls, or distant calls barely audible, however, they merely stop feeding to listen and remain where they are, although often they retreat calmly to the center of the home range. When K-group chimpanzees roam in their exclusive area of the home range and hear many calls from M-group chimpanzees, who are moving in the overlapping area, they very often make repeated vocalizations and exhibit display behavior accompanied by branching, drumming, charging, slapping, etc., although this behavior is not always confined to this context. K-group chimpanzees may, however, assume aggressive and threatening attitudes when they detect only one individual or a relatively small subgroup of the M-group, even if they are situated in the overlapping area near the exclusive area of the M-group. Although calls of members of the same unit group often attract animals to each other, those of different unit groups are, in principle, repellent.

We have assumed that the antagonism between the two unit-groups is based on the antagonism between adult males of different unit-groups. The reason is that adolescent or adult females often move into and are accepted by the neighboring unit-group, but adult males never transfer to the neighboring group (Nishida 1973). Moreover, adult males have been observed to show scouting behavior, or to engage in charging displays toward enemy groups, and to assume initiative and substantial roles in driving away enemy chimpanzees.

It is mainly the size and composition of the subgroup (especially the number of adult males) that decides the issue in each case of quarrels between subgroups of different unit-groups. Thus, the dominant status of the entire M-group over the entire K-group, observed usually at the end of the dry season, is the logical consequence of the larger size (80 vs. 27) and the larger number of adult males (16 vs. 5, as of 1973-1974) of the M-group.

Examples of Inter-Unit-Group Encounter

In this section, examples of inter-unit-group encounters observed in 1974 are described. The preceding and succeeding discussions are based on these examples.

On 37 days in 1974, acoustic or visual inter-unit-group contacts were confirmed. Visual contacts were observed on 8 days and acoustic contacts on 29 days of 37 total contact days. (By visual contact, I mean that chimpanzees of the different unit-groups *really witness* one another.) On 5 of 8 days of visual contacts, however, K-group chimpanzees encountered not total strangers, but ex-K-group members; namely, those who were originally born in, or belonged to, the K-group and had transferred to the M-group.

Of the remaining 3 days of stranger-to-stranger visual contacts, there was only one true inter-unit-group visual contact (Case 1), when a large subgroup of the K-group encountered a large subgroup of the M-group. All the episodes except Case 1 were observed in the vicinity of the permanent feeding station.

Case 1. January 21, 1974. During the morning, a large subgroup of the M-group was situated at Hasala Valley (0.4 km south of the temporary feeding area), and a large subgroup of the K-group was at Myako Valley (0.4 km north of the area). Although both groups called out several times, they could not hear each other because of the intervening ridge where the feeding area was located.

1232, K-group chimpanzees appeared at the feeding area (Kasonta, Kasanga, Kamemanfu, Sobongo, Wabunengwa, Chausiku, Wasalamba, with a newborn infant, and Madina). 1234, the four males and Chausiku called out. 1239, Gwekulo joined the group from the north. 1245, Kasonta slapped the tree trunk, pant-hooting. Kamemanfu, Kasanga, Sobongo, and Chausiku pant-hooted. This resulted in vocal outbursts from the M-group chimpanzees, who were south of the temporary feeding area. (This area was never used by the M-group for the 7 years. During the period, the northern boundary of the M-group was Nkala Valley. The reason why the M-group penetrated the boundary and approached secretly far northward in this particular observation period is not certain, although it is reasonable to speculate that the M-group chimpanzees were very curious about what was happening around the temporary feeding area.) Chausiku proceeded at once to the southern edge of the ridge as if to inspect what had occurred. Kasonta, his hair erect, rushed out northeastward, slapping the ground violently. Kasanga, Kamemanfu, Sobongo, Wasalamba, Madina, Wabunengwa, Mwese, and Gwekulo slapped the ground concertedly, emitting barking calls and then followed Kasonta. After Chausiku observed some chimpanzees of the M-group, she also followed Kasonta. 1246, Kidole (a middle-ranking male of the M-group) appeared at the southwestern fringe of the feeding ground and rushed about breaking twigs and branches and slapping the ground while emitting calls. Ntologi (a young adult male of the M-group) then dashed along the fringe of the feeding ground, breaking tree twigs and slapping the ground, while emitting calls. 1247, the remaining members of the M-group burst into calls and slapped the tree trunks and the ground to the south of the feeding ground. At the same time, the K-group chimpanzees slapped the tree trunks and the ground while emitting calls. 1249, M-group chimpanzees repeated the same actions near the southern periphery of the ridge. The K-group chimpanzees simultaneously slapped the tree trunks and the ground in the gallery forest. 1250, the K-group chimpanzees retired to an area farther upstream. 1253, the M-group divided into two subgroups, one returned to where it came from and the other turned eastward. Each group called out and slapped the ground. 1254, the M-group chimpanzees called out at Hasala Valley. K-group chimpanzees responded in outbursts. 1256 and 1300, the M-group chimpanzees called out at Hasala Valley. 1303, K-group chimpanzees called out. 1305 and 1310, M-group called out at the southern side of Hasala Valley. 1320, the M-group called out at Nkala Valley. 1330, the M-group called out at Nkala Valley again. 1407, the M-group called out.

1414, Chausiku appeared at the feeding area, emitting pant-grunts. 1418, Kasanga appeared there also emitting pant-grunts. 1423, Kasonta, Sobongo, Kamemanfu, Kasanga, and Chausiku moved to the southern fringe of the ridge to inspect and gaze at where M-group chimpanzees had gone and then they sat down on the ground. 1430, Chausiku, Sobongo, Kasanga, and Kasonta moved into the bush toward Hasala Valley. 1455, they returned to the feeding area and calmed down.

Case 2. Toward the end of January 1974, the majority of M-group chimpanzees had just migrated southward away from the overlapping area of the two groups, although about ten M-group chimpanzees were still ranging in scattered fashion around the vicinity of the feeding station.

On January 30, the core-subgroup of the K-group left the exclusive area of its home range and moved southward. At 1653 of the next day, the core-subgroup (five adult males, two estrous females, one adult female and one mother-infant) visited the permanent feeding station for the first time in 1974, after a 5-month absence.

On January 31, unusually frequent interactions between the core-subgroup of the K-group and individual chimpanzees of the M-group (including ex-K-group chimpanzees) were observed.

1649, Lubulungu, an adult male of the M-group, appeared at the feeding station from the south. 1653, a subgroup of the K-group (Kasonta, Sobongo, Kamemanfu, Kasanga, Chausiku, Wakasila with Milembe and Wantangwa) approached the feeding ground from the west. 1655, on hearing the foliage rustling, Lubulungu ran away eastward and then southward. Meanwhile, K-group chimpanzees arrived at the feeding station. 1657, Kasonta moved and sniffed eagerly the ground where Lubulungu was sitting. Kasanga and Wantangwa followed Kasonta and sniffed the ground.

1725, Gwekulo of the K-group joined her companions at the feeding ground. 1727, Chausiku, Kasonta, Kajabala, and Kamemanfu who were eating calmly at the feeding ground suddenly stood bipedally and moved westward. Panting calls were heard from the west and Wankungwe (ex-K-group chimpanzee) appeared. 1728, Kasonta returned to the center of the feeding ground. Wankungwe climbed a tree near the periphery of the feeding ground and remained for a while, seeming to appraise the situation. 1729, Kajabala, Chausiku, and Kamemanfu returned to the feeding ground, and were joined by Wankungwe.

1730, SB suddenly stood bipedally with full grin, touching Kasonta on his shoulder. All the chimpanzees looked westward. 1733, Kasanga, with erect hair, stood bipedally. Facing westward, Kasanga shook both arms from side to side, while stamping on the ground. 1734, Ndilo (ex-K-group chimpanzee) was seen in the woods. Sobongo rushed toward and attacked Ndilo. Kasonta followed him, shook a branch violently, and ran westward. Wantangwa (the mother of Ndilo) showed no reaction at all. 1735, Kasonta rushed about in the feeding ground, while slapping the ground. 1737, Sobongo returned to the feeding ground. 1738, Wankungwe began to take pieces of sugar cane, while Wantangwa sniffed Wankungwe's buttocks. 1744, Ndilo appeared in the vines at the periphery of the feeding ground. 1748, Kasanga groomed Wankungwe, poked his index finger into her vagina, and licked the finger tip. 1754, Ndilo entered the feeding ground. 1755, Ndilo and Gwekulo (with maximum genital swelling) stood face to face with each other. Gwekulo tried to sniff Ndilo on her buttocks.

1800, Kasonta, Kajabala, and Kasanga, listening intently, faced westward, without eating bait. 1801, Kasonta reached his right hand around the back of Kajabala and them mounted him. All the males with hair erect descended westwards. A strange chimpanzee (probably Musa, an adult male of the M-group) ran away northward, followed by all the chimpanzees except Ndilo, with Kasonta in the lead. 1805, Chausiku returned to the feeding ground. 1806, Kasonta and others slowly returned to the feeding ground.

1808, Fatuma of the M-group who had silently come near the feeding ground suddenly began to pant-bark at the eastern periphery of the feeding ground. Ndilo also began to pant-bark. Kasonta and Kajabala slapped on the ground. Kajabala rushed about and kicked the buttress of a tree. Fatuma climbed a tree at the edge of the feeding ground and calmed down. Kasanga, Wantangwa, and Wakasila returned from the bush. 1812, Fatuma continued to pant-bark, shaking branches and then came down from the tree. Standing on all fours, she slightly lowered her waist and straightened herself up. 1813, Fatuma yawned. Kajabala gazed at Fatuma. 1814, Fatuma climbed the tree again and yawned. 1816, Fatuma yawned again. 1829, Fatuma sat in the tree.

1845, Fatuma finally came down from the tree and joined K-group chimpanzees in feeding on elephant grasses.

Case 3. On February 1, 1974, the core-subgroup of the K-group visited the feeding station, where they peacefully spent the morning.

1241, Mimikire, the fourth-ranking male of the M-group stealthily arrived, was observed to be alone in the banana plantation near the feeding ground. Wantangwa, who found him first, stood bipedally. Kasonta and Kajabala barked together. Kasonta, Kasanga, and Kajabala with erect hair dashed forward and chased the opponent male violently until Kasonta seized the rival after chasing for 200 m. The other two males were left behind on the way. 1247, Kasonta and Mimikire wrestled together in upright posture in a cleared part of the elephant grass bush, until Kasonta forced Mimikire to the ground and bit him on the right thigh. Mimikire retreated little by little on his buttocks, continuing to scream weakly, when Kasonta stamped on his back. Mimikire finally managed to flee and ran away southward. 1250, Kasonta still ran after him, but gave up suddenly.

1251, Kasonta returned, swaggering, from the clash point to the southernmost edge of the banana plantation where he met Kajabala, Kasanga, and Wantangwa. He slapped the ground violently. Kasonta, Kajabala, Kasanga, and Wantangwa, in that order, returned rapidly to the feeding station. 1253, on the way to the feeding station, these three males and Kamemanfu emitted pant-hoots, and slapped the ground. When the adult males reached the feeding station, they ran about dragging fallen branches and slapping the ground. Then they left the feeding ground, went rapidly northward, and climbed toward the left bank of Kansyana Valley. 1327, the K-group chimpanzees called out at the upper stream of Kansyana Valley. 1332, the K-group chimpanzees called out again.

Case 4. April 27, 1974. 1510-1521, five adult males and two adult females of the K-group appeared at the feeding ground by twos and threes. 1527, to the south of the feeding ground, a few M-group chimpanzees called out. Kajabala and Kasanga approached and embraced each other and pant-hooted together. 1529, four adult males moved southward, where vocal outbursts of the M-group had occurred. 1532, Kamemanfu returned alone to the feeding ground. 1534, Wantendele went southward. 1540, Kasonta, Kajabala, and Kasanga pant-hooted and slapped the ground. 1550, the three males and Wantendele returned from the south. 1630, Chausiku and four males pant-hooted. 1730, Kasanga, Kajabala, Kasonta, and Kamemanfu suddenly stood bipedally and embraced one another (possibly they heard calls of M-group chimpanzees). 1740, Kasanga and Kamemanfu moved southward and sat and listened. 1902, all K-group chimpanzees retreated northwestward to sleep.

Case 5. July 7, 1974. 0910, K-group chimpanzees called out to the northwest of the feeding station. 1000-1005, five adult males, Wabwema, and Masisa of the K-group appeared at the feeding ground. 1012, no sooner had a few M-group chimpanzees called out at Kasiha Valley then Kajabala grasped Kasanga on the abdomen. 1020, Kajabala, Kasanga, Kasonta, and Chausiku pant-hooted. Then Kajabala and Kasonta slapped the ground. 1025, a few M-group chimpanzees called out on the left bank of Kasiha Valley (exclusive area of the M-group home range). 1034, Kasonta, Kajabala, and Kamemanfu engaged in grooming one another. 1110, M-group chimpanzees called out at the same place. 1125, five males and Chausiku began to move northward and then ascended the ridge.

Case 6. September 15, 1974. Three adult males (Rashidi, Kajugi, and Ntologi), one estrous female (Fatuma) and a juvenile of the M-group arrived at the feeding ground. 1400, they slept in the vicinity of the station. 1530, they woke up. 1623, as soon as the K-group chimpanzees burst into calls to the northwest of the feeding station, the seven M-group chimpanzees slipped off silently and moved eastward via the banana forest. 1624, Kajabala appeared alone from the west and pant-hooted. 1625, Kasonta, Kasanga, and Sobongo appeared from the northwest and then moved northward. 1626, the K-group chimpanzees arrived at the feeding ground, passed there, advanced slowly into the gallery forest, and sat around a huge buttressed tree *(Pseudospondias)*. 1627, K-group chimpanzees burst into calls and violently slapped the buttress. Then, M-group chimpanzees slapped the ground and called out, about 300 m southeast from the K-group chimpanzees. 1641, Kasonta, Kajabala, Kasanga, and Sobongo proceeded again to the feeding ground. 1644, Kamemanfu followed them. 1645, Kasonta and Kamemanfu hunted about to the south of the feeding ground as if to scout. Sobongo and Kasanga remained sitting side by side at the feeding ground. 1656-1810, K-group chimpanzees again took sugar cane at the feeding ground, which they took to the bush to consume. They finally retreated northward to sleep.

ADULT MALES AS THE CORE OF THE UNIT-GROUP

Adult males are stable members of the unit-group and have the strongest tie with the home range. The position that adult males occupy in a unit-group is investigated in terms of the social relationships among members of a unit-group.

Male Bond or Male Propensity to Gather

Analysis of the composition of subgroups of the K-group indicates that adult males have the highest propensity to gather among chimpanzees of any age-sex class. Of 218 subgroups observed at the feeding ground in the 1966-1967 period, for example, 113 groups (51.8%) were of mixed type, and the average size was 13.1. The average number of adult males and females contributing to the total average size was 4.5 and 3.7 respectively. Furthermore, since the K-group contained 6 adult males and 10 adult females during that period, the expected banding ability would be calculated as 4.5/3.7 x 10/6 = 2; that is, male cohesiveness was roughly twice that of females (Nishida 1968). This also applies to the M-group, (16 adult males and more than 26 adult females), but that almost always visited the feeding ground in subgroups that consisted of more adult males than adult females.

Analysis by inter-individual familiarity index indicates that the strongest social bonds, except mother-offspring groups, are formed among adult males, and that adult males and adult females or adolescents are more strongly attached to one another than is the case among adult females (Nishida 1968). If we draw a dendrogram by hierarchical cluster analysis on a basis of familiarity index, we find a tightly connected male cluster emerging in the dendrogram in any year of the study.

Male Aggressiveness, Dominance, Rank, and Coalition

The bonding ability of an adult male with the other chimpanzees within a unit-group has a given relationship with the position he occupies in the dominance rank of the adult male cluster. There is a tendency for higher-ranking males to be found more often in larger subgroups

TABLE 4
Average size of subgroups in which each adult male was found[a]

RANK ORDER	INDIVIDUAL NAME	1966	1967	1968	NO. OF GROUPS OBSERVED	AVERAGE SIZE DURING 3 OBSERVATION PERIODS: STUDY 1975-1976
1	Kasagula	13.3	16.8	14.0	81	15.1
2	Kasonta	13.6	16.3	13.1	101	14.7
3	Kajabala	9.0	15.9	10.3	145	11.7
4	Kasanga	11.3	16.1	11.5	119	13.3
5	Kamemanfu	8.8	14.6	11.1	145	11.6
6	Kaguba	9.1	15.1	9.5	147	11.2

[a] Observations were made at the feeding ground during three periods.

(Table 4). This implies that high-ranking males are more preferably followed by the other group members than low-ranking ones, since chimpanzees do not necessarily exhibit true leadership behavior.

Data collected intermittently for 7 years indicate that dominance rank order was very stable among full adult males, but that young adult males managed to raise their rank as they matured. The dominance hierarchy recognized among four full adult males in August 1966 (Kasagula > Kasonta > Kajabala > Kamemanfu) did not change until 1975 except that the oldest and top-ranking Kasagula was missing, possibly because of death in 1969.[3]

On the other hand, Kasanga who was a young adult and fifth-ranking male in 1966 surpassed Kamemanfu in 1969. Moreover, he surpassed Kajabala in October 1972, but he was defeated by Kajabala in January 1973. Although he enjoyed a brief victory over Kajabala in November 1973, he was dominated again in January 1974. Kajabala won by cooperating with Kasonta in attacking Kasanga; when Kasonta attacked and bit Kasanga on his back severely, Kajabala rushed out towards Kasanga and also bit him on the great toe. Sobongo, who was an adolescent male and situated at the periphery of the unit-group in 1967, managed to join the cluster of adult males in 1970. He associated very closely with Kamemanfu, then the lowest-ranking male, in 1971, but began to show the threat-display behavior against Kamemanfu in January 1974 when he became about 16 years old.

There seem to be two ways for an adult male to raise his status in the dominance hierarchy—by frequently showing display behavior peculiar to adult males (Nishida 1970) and by keeping close association with a top-ranking male. In reality, display behavior has very often been observed among high-ranking (but not old) males and young adult males. Those who raised their positions (Kasanga and Sobongo) exhibited display behavior more frequently year by year.

Analysis of the annual change of a familiarity index indicates that the index between declining Kajabala and top-ranking Kasonta, which was very high in 1967, became lower than the index between Kasonta and Kasanga after 1968, although Kajabala himself continually chose Kasonta as his favorite subgroup partner more often than any other male. On the other hand, Kasanga has maintained the highest index with Kasonta since 1969. In almost every year between 1971 and 1974, two dyads of stable coalition emerge among five adult males, namely that of Kasonta and Kasanga and that of Kamemanfu and Sobongo.

It is recognized that, with the exception of old males (Kasagula and Kamemanfu), the higher the rank in the hierarchy an adult male occupies, the more often aggressive episodes occur around him. If a subdominant male tries to associate more closely with high-ranking males, he is more likely to be attacked by them. The subdominant male, however, seems to seek association with a top-ranking male. From the study of a captive rhesus group, Chance (1961) points out that the potential threat from the high-ranking males acquires an ambivalent quality for subordinate males—"it attracts as well as repels them, just as the female's display of modified aggressiveness can attract a male." It is likely that complex sequences of threat-submission-reassurance interaction may strengthen the male bond among chimpanzees. Although Goodall (1973) showed that sibling relationships were at least one factor establishing male coalition, it seems that ambivalent psychology on the part of subordinant males towards dominant males may be responsible for the basic background in the male bonding pattern.

Role of a Top-Ranking Male

Little attention has been paid to the social role of a top-ranking male of the unit-group. The long-term observation at Mahale revealed, however, that a top-ranking male must be treated as something more than an alpha male that excels all other males of the unit-group.

Kasonta, the biggest as well as top-ranking male of the K-group since 1969, has been most prominent in the frequency of both aggressive

FIGURE 4
Kasonta (left) displays and Kasanga emits panting hoots.

TABLE 5

The most and the second most frequent male companions chosen by each adult male in each study period

	1967	1968	1969	1970	1971	1972	1973
Kasagula (KA)	KS, KJ	KS, KN	—	—	—	—	—
Kasonta (KS)	KJ, KM	KA, KM	KN, KJ	KM, KG	KN, KM	KN, KJ	SB, KM
Kajabala (KJ)	KS, KN	KS, KN	KS, KG	KG, KS	KS, KN	KS, KN	SB, KS
Kasanga (KN)	KJ, KM	KA, KS	KS, KJ	KM, SB	KS, KM	KS, KJ	KS, SB
Kamemanfu (KM)	KS, KN	KS, KG	KS, KN	KN, KS	SB, KS	SB, KN	SB, KS
Kaguba (KG)	KS, KM	KM, KS	KJ, KS	KM, KS	—	—	—
Sobongo (SB)	KN, KM	KG, KA	KN, KJ	KN, KS	KM, KS	KM, KS	KM, KS

behavior and threat-display (Figure 4). Among adult males, he groomed and was groomed most frequently. The conspicuous nature characterizing top-ranking status, however, is the frequency of being chosen as the object of following by the other adult males (Table 5).

Aggressive and display behavior may operate as an attracting force through the psychological mechanism mentioned above. As a result, a top-ranking male can generally be found in the largest subgroup among members of the unit-group. I pointed out that the presence of a comparatively large subgroup was always recognized throughout a year, though a unit-group may generally split up into several subgroups (Nishida 1968). This core-subgroup may be better defined as the subgroup containing a top-ranking male rather than as the subgroup composed of "all of the adult males and a few adult females as regular members" (Nishida 1968), although the content may be substantially the same in either direction.

At the stage of the study when the function of top-ranking status among adult males was not clearly defined, I conjectured that "seasonal mass migration would not be well understood unless we assumed the presence of a so-called leader who regulates, on specific occasions, the movement of the entire K-group" (Nishida 1970). Now that more data are available, we can state that the seasonal group movement of the K-group is due to the nonintentional leadership of a top-ranking male.

Another quality characterizing top-ranking Kasonta concerns sexual behavior. Kasonta was responsible for 177 (46.2%) of the 383 examples of copulation observed in the K-group. Although the frequency of copulation among adult males appeared to be in direct proportion to status in the dominance hierarchy, the top-ranking male overshadowed the others very conspicuously and had more opportunities to copulate with young newcomer females (Table 6). He was responsible for 40 (72.7%) of the total 55 cases of sexual intercourse involving three newcomer females that had joined the K-group in 1972. We may well presume that he monopolized newcomer females, although it is still unknown whether he has the greatest selective effect upon the gene pool of the K-group.

The part Kasonta played in the direct inter-unit-group contact must be recalled. He often not only assumed leadership in scouting and chasing, but also was the only male to defeat the male of the neighboring group without help from others. On the other hand, Kasonta was the shyest and most careful male towards human observers and/or artificial open environment. He was the last male to take sugar cane at the feeding ground.

We must, however, remember that this description is based on only one top-ranking male, Kasonta, who is now in the prime of life. Kasagula, who was possibly a top-ranking male of the K-group during the initial periods of the study, and Kalindimya, the present top-ranking male of the M-group, were not seen to copulate or to show aggressive behavior very often. The two features that both Kalindimya and Kasagula share with Kasonta is the high frequency with which they were followed by fellow chimpanzees and their carefulness with human observers. They were almost always found in the largest subgroups. Both Kasagula and Kalindimya were very shy animals. Kasagula was next to Kasonta in being slowly habituated. Kalindimya is still reluctant to feed on sugar cane at the open feeding ground and prefers to feed in the bush. Although shyness may be attributable to their old age, Kasonta is almost as old as Kamemanfu and Kajabala.

Both Kasagula and Kalindimya were very large, aged males with remarkable white backs. Although personality difference must be taken into consideration, it is reasonable to presume that both of them showed the same behavioral tendency as Kasonta when they were in their prime. Their top-ranking status may have been the inheritance of

TABLE 6

Comparative frequency of copulation of adult males of the K-group with resident females (members since 1966) and newcomer females (immigrants since 1969)

YEAR	A NO. OF RECEPTIVE NEWCOMERS	B NO. OF RECEPTIVE RESIDENTS	A/(A+B)	C FREQUENCY OF COPULATION WITH NEWCOMERS	D FREQUENCY OF COPULATION WITH RESIDENTS	C/(C+D)
1969	2	6	0.25	5	56	0.08
1970	1	5	0.17	14	40	0.26
1971	3	6	0.33	15	66	0.19
1972	4	4	0.50	42	12	0.78
1973	4	5	0.44	49	50	0.50
1974[a]	2	1	0.67	10	8	0.56

[a] Data obtained up to February 1974.

their past prime of life. Kortlandt (1962) was the first to record such a type of an old male and named him "Grandad". It is interesting that the second-ranking male of the M-group shows the same behavioral tendency as Kasonta.

OPEN AND FLUID STRUCTURE OF A UNIT-GROUP

Dispersion of Adult Females

Adult or adolescent females generally have a tendency to disperse and to form small subgroups in the range, in remarkable contrast to adult males. Moreover, some of them ignore the boundary line of the home range of the unit-group.

As stated earlier, among nine full adult females of the K-group identified individually in the initial stage of the study, three females have always been within the home range of the K-group (Figure 3) and have often been observed moving in large subgroups. However, a period lasting more than 3 months was recorded in which the other females were not seen in the core-subgroups, at least once throughout the 7-year intermittent study period. An older female, Wankungwe, kept going back and forth between the K-group and the M-group, after the death of her 2.5-year-old male infant in 1968. According to her "residence pattern," she spent about 7 months in the K-group and the remaining 5 months in the M-group. Namely, she was traveling with the chimpanzees of either of the two unit-groups, as they occupied the overlapping area of the ranges. This particular female, therefore, seemed not to transfer positively between two unit-groups as much as to remain in *her own individual range.* This assumption seemed to prove valid after the provisioning of the M-group was successful at the permanent feeding station in 1972-1973. Since the M-group occupied the overlapping area most of the year, she continued to stay with the M-group without returning to the K-group (Figure 3). From the onset of the study, however, she was never observed in the exclusive area of the K-group range, and, moreover, she had been seen moving with members of the M-group in the exclusive area of the M-group range since 1972. We can therefore speculate that Wankungwe was originally a member of the M-group and transferred to the K-group every season when the overlapping area began to be utilized by the K-group. After 1973, however, she ceased to move with the K-group completely, although she sometimes happened to meet and mingle with K-group chimpanzees at the permanent feeding station (see Case 2, section 2).

The remaining five females began to keep their distance from the adult males when their dependent young attained the developmental stage of early juvenility to early adolescence, an average of 5.6 years of age (namely, Wangala and Wasalamba with 4-year-old juveniles, Wab-wema with a 5-year-old juvenile, Wamikambi with a 7-year-old adolescent, and Wangulu with an 8-year-old adolescent).

Mothers have not been observed showing estrous swellings until their dependent young become 3 years old. If an infant of any age dies, its mother shows the maximal genital swelling very soon: in six cases of infant death recorded in the K-group, three females resumed estrus within one month of their infants' death and three females within, at most, 6 months. It is possible that the latter three females recovered receptivity earlier, but that it was overlooked because of their low attendance rate for 6 months after the death of their infants (0-3 days a month). Thus, the period of female nonreceptivity continues at least for 3 years after delivery, provided the infant does not die. In many cases female receptivity is not seen until the offspring reaches the age of 5 or more; that is, much individual difference is recognized in the length of nonreceptivity after the delivery, and nonreceptive females rarely copulate. This difference is considered a result of the hormonal mechanism that determines the estrous condition. It may be, for instance, that continuous suckling behavior of infants stimulates the secretion of a pituitary hormone (e.g., oxytocine), which inhibits estrus in mothers. Without such a mechanism, it is very difficult to understand why a mother can show maximal genital swelling as soon as her infant of any age dies, and why there is so much individual difference in the length of the nonestrous period of mothers. Goodall (1973) mentioned that youngsters are not finally weaned, in most cases, until their fifth or sixth year. Therefore, sexual receptivity seems to have something to do with the temporary roaming tendency of the adult females. Although they were not seen in the core-subgroup of the K-group, they may have moved on their own in the K-group range or may have joined one of the neighboring unit-groups. For example, Wabwema with her 7-year-old son, Masisa, was discovered moving in the M-group in 1973 (Figure 3).

The bond of each adult female with adult males changed drastically year by year. Generally speaking, young nulliparous females very often accompany adult males, and thus become regular members of the core-subgroup, although they frequently follow particular pairs of mother-infant (infant of 0.5 to 2 years) during nonestrous periods. When such females give birth, or become heavily pregnant, they usually maintain distance from males and roam on their own or with other nonestrous females or mothers. As the offspring grow, mothers often join the core-subgroup, although they prefer to be on their own or in small groups. After they resume estrus, some become regular members of the core-subgroup again, and others seek adult males of a different unit-group. Thus, the physiological cycle has an important influence on male-female bonds, and on the great individual variations in bonding ability.

Regardless of the female physiological cycle, however, it is possible to descriminate between *central females* and *peripheral females.* Some females (Wamikambi, Wangala, and Wangulu) and their offspring have rarely been seen through the long-term study period. They are presumed to have roamed within the K-group range, because, often, when they were encountered, they were near the center of the K-group range. Some females (Wantangwa, Chausiku, etc) have maintained comparatively strong associations with the core-subgroup even when they carried infants. Most females are between these two extremes in the extent of their association with adult males.

Transfer of Females

1. Females moving into the K-group.

Seven strange females were seen to become gradually incorporated into the K-group and to acquire stable positions in it. Wakasila joined for the first time in 1969, Wakiluhya and Wakampompo in 1971, Wantendele, Wakapala, and Gwekulo in 1972, and Gwamwami in 1975. All were adolescent or young adult females estimated to be 10 to 14 years of age at the time of transfer. Wakasila moved out temporarily to join the M-group, and after a 3-month stay, she returned to the K-group in early 1970. She gave birth to a male infant in 1972. In 1975 after she spent 5.5 years in the K-group, she left when her son became 3.5 years old.

Of the two strange females that joined the K-group in July 1971, Wakiluhya remained, but Wakampompo moved to the M-group in late 1973 after more than a 2-year stay in the K-group. Wantendele, Wakapala, and Gwekulo have always been observed moving with the K-group. The entrance of these seven females into the K-group constitutes some of the most striking evidence indicating the unstable membership of the K-group, although the origin of the females is unknown. It is certain, however, that most of them were not born in the K-group.

Two adolescent females of the M-group temporarily transferred to the K-group in 1974. Faruma, with adolescent swelling, first approached the core-subgroup of the K-group at the permanent feeding station early in 1974 (see Case 2, section 5). After she moved with the K-group chimpanzees for 2 weeks, she returned to the M-group. She again joined K-group chimpanzees for 3 weeks in May-June and for 1 week in July 1974. She was in estrus for 18 days of the total 34 days of observation while in the K-group. Gwakakumo temporarily joined the K-group in May 1974 and stayed for 2 weeks; for 10 days of this period, she was at the height of estrus and copulated with males of the K-group. In both cases, the transfer occurred while

K-group chimpanzees occupied the overlapping area of the two unit-groups. Neither Fatuma nor Gwakakumo have joined the K-group for more than 2 years since then.

2. Females moving out of the K-group.

As mentioned previously, the full adult females of the K-group, except Wankungwe, have never completely severed their connections with core-subgroups (and range) of the K-group, although many of them have become temporarily estranged.

On the other hand, many young females growing up during the period of the long-term study have shown the tendency to leave their natal group. Of six young females who have been in the developmental stage from late infancy to early adolescence at the initial stage of the study, three left the K-group permanently to join the M-group, one left the K-group temporarily, and one ceased to be observed.

Tausi, a senior daughter of the most dominant female, Wantangwa, left the K-group and transferred to the M-group at the age of 11, in 1968. Since then, she has not once returned to the K-group, although it is unknown whether she still continues to stay in the M-group. Sada, the daughter of Wangala, was observed copulating for the first time at the estimated age of 10 in 1970; she left the K-group to transfer to the M-group at the age of 12. Ndilo, a junior daughter of Wantangwa, began to show slight swelling of the ano-genital region at the age of 7.5 years in 1971. She temporarily joined the M-group only to return to the K-group after a 1-month stay at the age of 8 in 1971. She moved again to the M-group at the age of 9 around September 1972. She seemed neither attractive to, nor copulated with, any adult males of the M-group until September 1973, when she was first observed copulating with a young male of the M-group. She has remained in the M-group for more than 4 years. It has been confirmed that both Ndilo and Sada have moved together with M-group chimpanzees to the known southernmost part of the home range of the M-group.

Chausiku, the senior daughter of Wabunengwa, was not observed to show the maximal ano-genital swelling or to copulate until 1967. She began to show the maximal swelling at the age of 10 in 1968, attaining sexual maturity, when she moved to the M-group, but returned to the K-group after a 1-year stay. She has now moved with the core-subgroup of the K-group for more than 7 years since her temporary transfer and gave birth to an infant in December 1974. Wahanse, the daughter of Wangulu, became sexually mature at the age of 10 in 1973 and left the K-group in September 1975. Thus, only one female among six young females who attained the age of 11 in the K-group still remains—Tatu, the daughter of Wamikambi.

Moreover, one subadult female named Wamkime who was observed with K-group chimpanzees at the feeding station for a few days of the study period of 1967 transferred permanently to the M-group, although now it seems very probable that she was originally a member of the M-group. She has been moving with the M-group for 9 years since then. She gave birth to a male infant in 1971.

The above data seem to indicate that young females tend to leave their natal group at least once when they reach the later adolescent stage of development; moreover, most of them transfer to the adjoining group permanently.

Transient Mix of Strange Females—An Ecological Phenomenon?

Full adult (including mother-infant) or strange adolescent females have sometimes been observed moving temporarily in the core-subgroup of the K-group (Nishida 1968; Nishida and Kawanaka 1972). Although some of the strange adolescent females have been estrous, mothers with infants did not show any sign of receptivity. They moved and fed with members of the K-group, but disappeared after spending several days and were not seen again (therefore, they were not individually identified).

Such transient transfers were almost always observed when both the K-group and the M-group were situated close to each other in the over-lapping area, during the migration season of the two groups, which occurs twice a year. Consequently, this transient transfer may be a phenomenon occurring only near the feeding station (a spot rich in food supply) when food is locally short. It may represent an aspect of the social adaptation of chimpanzees against local heterogeneity and im-balance of the food supply (Nishida 1974). It is important to remember that this phenomenon is confined to females. It is certain, however, that this type of transfer cannot occur unless it is based on a specific social relationship between the strange female and the adjoining unit-group.

Unit-Group As Outbreeding Population

All immigrant females that have remained in the K-group for more than a year, had periods of maximal ano-genital swelling during the year, as well as at the time of immigration. More than half of the adult or subadult females who have been members of the K-group since the initial stage of the study had either long periods of nonreceptivity with dependent infants, or periods in which they were estranged from the core-subgroups.

This situation may characterize the gene composition of a unit-group. If each year we compare the number of receptive resident fe-males with that of immigrant females who joined the K-group after 1969, we find they have been equal since 1972. However, if we compare the frequency of copulation of newcomers with that of resident adult females, we find that the newcomers have exceeded the residents since 1972 (Table 6). The newcomers have been selected as mates by K-group males with a probability of more than 50%. This comparative frequency of copulation may give us an idea of the extent of outbreeding in the unit-group, even if it does not give the actual outbreeding ratio. This was calculated on the assumption that all adult males and all resident females were born in the K-group. In reality, however, it is likely that many resident females originally came from outside the group when they were subadults. If this is the case, the outbreeding ratio would be much higher. A true estimate could be made if the origin of all reproductive members of a unit-group were known.

Although five of seven newcomers had not delivered infants by 1974, Wakasila who joined in 1969 had an infant in February 1972. Considering that the gestation period of chimpanzees ranges from 196 to 260 days (Gavan 1953), it is probable that Wakasila's infant is the offspring of a K-group male, because Wakasila moved with the core-subgroup of the K-group from June to August 1971 and was observed copulating with K-group males between June 4-7. Wantendele delivered a male infant in June 1974, who was also an offspring of a K-group male (based on the same process of reasoning).

On the other hand, among three young females that transferred permanently from the K-group to the M-group, Sada and Ndilo gave birth to infants in November 1973 and May 1975 respectively. It is also certain that their infants are offspring of M-group males.

The unit-group has thus proved to be very open genetically. We can also state that the breeding unit among wild chimpanzees is far greater than the size of the social unit.

BASIS FOR THE FLUIDITY OF THE UNIT-GROUP

Nature of the Social Bond Among Adult Females

The aggregative ability among adult females is much lower than among adult males, as stated earlier. This may explain why most social interactions among members of a unit-group occur either among adult males or between adult males and adult females, in remarkable contrast to the infrequency of interaction among adult females (Nishida 1970). If we take the frequency of grooming relationships as an index of the strength of the social bond among individuals, we find that 46% of the total occurs among adult males, 39% between adult males and adult females, in sharp contrast with 10% among adult females (Table 7). We must admit, however, the presence of observational bias due to more frequent encounters with core-subgroups than with mothers' groups. It is important to note, also, that both mutual greeting and sharing

TABLE 7
Grooming relationships among chimpanzees of the K-group[a]

| GROOMER | GROOMEE | | | |
	ADULT MALE	ADULT OR SUBADULT FEMALE	JUVENILE OR INFANT	TOTAL
Adult male	218	96	6	320
Adult or subadult female	91	49	13	153
Juvenile or infant	1	3	0	4
Total	310	148	19	477

[a] The tabulation represents the frequency of grooming interactions among chimpanzees of the K-group, recorded mainly at the temporary feeding area during the period from November 1973 to February 1974, in which 5 adult males, 10 adult or subadult females, 2 juveniles, and 1 infant were observed to interact with one another.

behavior, which characterize the high sociability of chimpanzees, are very rare among adult females (Nishida 1967; 1970).

Such data may imply that females are egocentric and individualistic, and males are highly sociable (Nishida 1967). Generally, social bonds among females may, if anything, be *passive* homogeneous aggregations based on physiological homogeneity in activity, occasioned by similar age, maternity, and sexual receptivity. We can also state that females gather in the process of trying to follow adult males. A female gathering is generally an accidental aggregation rather than a unit based upon clear principles of grouping.

Only two types of rather long-lasting female bonds have been recognized, one of which can be termed as *babysitter relationships*. Young nulliparous females often follow a particular mother-infant dyad, as long as the infant is 0.5 to 2 years old. Although the nulliparous female persistently grooms the mother of the infant, she seeks the opportunity to carry the infant away to groom or play with him. The babysitter relationship sometimes continues for more than 3 months, although the young female returns to the core-subgroup when she becomes highly estrous. Such a relationship was recognized between Chausiku and Wakasila-Milembe, Wakiluhya and Chausiku-Katabi, Wantendele and Chausiku-Katabi, and Gwekulo and Wantangwa-Milongo. It must be remembered that this relationship is only secondary in terms of female relationships, because nulliparous females are not always attracted by the mother, but by the infant.

Another type of female association has been observed only once but it continued much longer. This is friendship between two adolescent females. Fatuma and Ndilo of the M-group almost always moved together for a period of 1 year in 1974.

Dominance relationships among adult females cannot be defined as clearly as among adult males. Interesting changes in dominance rank have been observed among adult females. There is a remarkable rise in a female's dominance rank when she becomes estrous. Young adult females in estrus often threaten high-ranking females and may even snatch food from some nonconsorting adult males. In the first place, estrus seems to eliminate an adult female's shyness and makes her bold and aggressive. In other words, estrus *masculinizes* females, however paradoxical it may seem. Some young adult females tend to show threat displays (branching, slapping, branch-dragging) toward other females, when they become estrous. One particular female (Gwekulo) occasionally engaged in charging displays, generally peculiar to adult males, although in a more clumsy way. In the second place, a rise in dominance rank is related to the fact that estrous females in the K-group were often monopolized by the highest ranking male (Kasonta). The high rank of young estrous females is only a temporary state based on physiology and their intimate sexual relationships with the highest-ranking male.

Another change in female dominance rank was once recognized between two females who had a babysitter relationship. A young adult female (Chausiku) had been dominant to Wakasila since the latter transferred to the K-group in 1969. But, in December 1973-January 1974, when Chausiku became a babysitter of Wakasila, the dominance rank was reversed, possibly because Chausiku had to groom Wakasila continuously in order to take her infant. This reversion still continued after

TABLE 8
Dominance relationships among females of the K-group[a]

YEAR OF MOVING IN	DOMINANT	WT	WB	WA	WS	WI	CH	WN	WO	WL	WD	GW	WP
	Wantangwa		o	o		o	o	o	o	o	o	o	
	Wabunengwa	x					o	o				o	
	Wakasunga	x			o	o	o	o					
	Wasalamba		x			o	o					o	
1969	Wakasila	x		x	x		o			o	o	o	o
	Chausiku	x	x	x	x	x		o		o		o	o
	Wangala	x	x	x		x				o			
1971	Wakampompo	x								o	o		
1971	Wakiluhya	x				x			x		o	o	
1972	Wantendele	x			x	x		x	x	x		o	o
1972	Gwekulo	x	x		x	x	x			x	x		
1972	Wakapala					x	x				x		

[a] The tabulation is based on the agonistic interactions observed during the period from January 1973 to January 1974.

TABLE 9
Change in familiarity index of 12 mother-offspring dyads

DYAD	BIRTH YEAR[a]	OFF-SPRING'S NAME[b]	MO-THER'S NAME	AGE OF OFFSPRING								
				0-1	1-2	2-3	3-4	4-5	5-6	6-7	7-8	8-9
MOTHER-DAUGHTER	1957[a]	TA	WT									
	1958[a]	CH	WB									73.1
	1960[a]	SA	WN							88.2	64.7	0
	1963[a]	WH	WG					100	100	95.0	85.5	55.6
	1963[a]	ND	WT				100	100	100	93.5	87.8	89.3
	1964[a]	TT	WM			100	100	100	98.7	95.0	14.3	83.3*
	1965[a]	SI	WN		100	100	100	78.6*	80.0	83.3	75.0	33.3
	1966	MD	WS	100	100	100	100	100	100	60.0*†	78.6	76.1
	1967	MW	WB	100	100	100	100	93.8	100	90.9	88.0*	94.9
	1970	NK	WN	100	100	100	100	–	100			
MOTHER -SON	1958[a]	SB	WW									
	1966	MS	WW	–	100	100	100	100	78.9*	78.9†	100	81.2

* The birth of infant sibling
† The death of infant sibling
↗ Moving-out of the unit-group
↘ Returning to the unit group
– Not observed through a year

[a] Presumed

[b] See Figure 3 for the abbreviation of the individual name.

Chausiku's delivery. On the other hand, mothers with newborn infants sometimes became very shy and timid and avoided otherwise subdominant females without infants rather than fight with them. Wakasunga and Wasalamba, for example, when carrying young infants, ran away, grimacing, when young subdominant females approached them while seeking sugar cane in their area.

Physiological conditions of females make female relationships very complicated and unstable. We lack sufficient evidence to determine dominance relationships because of the infrequency with which some females are observed. It is also recognized that aggressive encounters do not often occur among certain adult females and that some past-prime females only pant-grunt to each other when they meet.

We cannot assume, however, that linear dominance hierarchy is never present among adult females. It has become obvious that a past-prime female (Wantangwa) has completely dominated all the other females for the 7 years of the study. Dominance relationships between two adult females should be determined when they are not estrous and when adult males are not in their vicinity. If we follow this procedure strictly, it is possible that a linear dominance hierarchy emerges at least among the majority of adult females of a unit-group (Table 8). It is a general rule that older females are superior in rank to younger females.

TABLE 9, CONTINUED

DYAD	BIRTH YEAR[a]	OFF-SPRING'S NAME[b]	MO-THER'S NAME	AGE OF OFFSPRING								
				9-10	10-11	11-12	12-13	13-14	14-15	15-16	16-17	17-18
MOTHER-DAUGHTER	1957[a]	TA	WT	61.5	58.8	0↗	0	0	0	0	0	0
	1958[a]	CH	WB	75.0*	25.8	16.3↘	30.3	15.3	20.3	32.7	12.9	11.7
	1960[a]	SA	WN	16.7*	2.2	7.1	16.7↗	0	0	0		
	1963[a]	WH	WG	12.0	62.5	47.1	42.9↗					
	1963[a]	ND	WT	37.5↗	5.6	0	0					
	1964[a]	TT	WM	33.3	–	36.4						
	1965[a]	SI	WN	–	70.0							
	1966	MD	WS	92.5								
	1967	MW	WB									
	1970	NK	WN									
MOTHER-SON	1958[a]	SB	WW	55.6	94.6	82.2	47.5	9.3*	13.6†	29.3	24.0	37.7
	1966	MS	WW	98.7								

Mother-Daughter Relationship

Changes in ten dyads of mother-daughter relationships in the K-group have been traced for longer than 6 years (Table 9). If we show the degree of association of a daughter with her mother by a familiarity index so as to categorize developmental stages of female chimpanzees, we can roughly define individuals between 0 to 4 years of age as in the stage of 100% dependence (infant), from 4 to 8 years of age as 85% dependent (juvenile), from 8 to 10 years of age as 50% dependent (adolescence) and from 10 to 12 years of age as 20% dependent (subadult). Thus, the mother-and-daughter relationship maintains a strong social bond until the latter becomes at least 7 years of age, but generally diminishes sharply when the daughter reaches the later adolescence to subadult stage, although much individual difference is recognized. Moreover, the birth of junior siblings generally decreases the degree of association, but the death of infant siblings tends to increase it.

As already mentioned, among five young females whose growth was traced intermittently until they were at least 10 years old, Chausiku, Tausi, and Ndilo suddenly severed association with their mothers by moving into the M-group. On the other hand, when both Sada and Wahanse became 8 to 9 years old, their mothers ceased to move with the core-subgroup of the K-group. Their association with their mothers seemed to be weakened, so to speak, passively. But, their moving-out occurred at the age of 12 years. Thus, all five females who grew into adulthood in the K-group, moved out of the K-group between the ages of 9 and 12, except that Chausiku returned again.

Adult males are either completely tolerant or rather indifferent to infants. Past-prime males even look after 0.5 to 2-year-old infants during the resting period. They hug, tickle, and bite them gently or lie supinely while supporting infants on their feet and hands. Juveniles also are tolerated by adult males, although the adults sometimes threaten when they come too close. Several high-ranking males of the M-group, however, especially tolerated two orphaned juvenile males who habitually followed, fed in close proximity to (or, in physical contact with), and were often groomed by one of them.

Adolescence changes the situation of immature individuals. Since adolescent chimpanzees, both male and female, are often threatened and attacked if they come within close sight of the adult males, they must situate themselves more or less at the periphery of the subgroup. They are not completely expelled, however. If they stay at some distance from the adult males, they will be allowed to remain there.

No case has been observed where mothers become intolerant of their daughters after the latter reach adolescence, although agonistic interactions concerning food are occasionally seen among them at the feeding ground. Wangulu, for example, incessantly robbed her daughter Wahanse of sugar cane, even after Wahanse attained the age of 12. It follows that adolescent or subadult females were not driven out by their mothers nor by adult males but that they moved out to the adjoining group spontaneously.

Goodall (1973) described a case of a family group (named Flo group), in which the old mother, Flo, very often formed a cluster with her adult daughter, juvenile son, and grandson, which two adult sons sometimes joined. Such a big family group, however, has not been observed in the K-group. Moreover, it has become clear that at least two dyads of mother and subadult daughter did not show any remarkably affectionate gesture toward each other even when they met after a separation of several days or weeks. Since Chausiku returned to the K-group after a 1-year stay in the M-group, she has shown only 15-30% familiarity index with her mother Wabunengwa during these 4 years. This index is either average or lower than the usual female-female relationship. Moreover, grooming behavior between Chausiku and Wabunengwa has been very rarely observed, when it was observed it was always Chausiku who groomed.

In Case 2 (section 3), we described what happened when a subadult daughter, who had transferred to the neighboring unit-group, and her mother, who had remained in the home group, met at the overlapping area of the ranges after a long time of separation. Wantangwa (mother) and Ndilo (daughter) did not try to approach or glance at each other. In spite of a 5-month separation, no grooming interaction or even bodily contact was observed between this mother-daughter dyad during the subsequent few days of their encounter.

Male-Female Relationships

The factors connecting the cluster of adult males with each adult female (or biological family) are still unclear. The majority of adult females have sometimes kept away from the male cluster, often moving separately, even though they are moving in the terrain within the home range of the unit-group. Why, then, have they maintained relationships with males for a long time? Why do they share the same terrain with adult males? Some hypothetical factors are female dependence on males, kinship, and sexuality.

1. Female dependence on males.

Reynolds (1966) defined the role of male chimpanzees as *food finders*. It is apparent that the seasonal and daily ranges of males and estrous females are larger than that of mother-infant groups. In seasons when staple fruits localize in a given area, each individual chimpanzee of a unit-group shares the same narrow terrain, but the males and infantless females range in a wider area on a *daily* basis. In seasons of transition between staple foods, males and females with no infants move in a wider area on a *daily and seasonal* basis. Each individual mother accompanied by her offspring continues to stay in her favorite locality, utilizing mainly the remaining resources (previous staple food).

The ranging patterns differ because patterns of resource utilization are different in male-estrous female groups and in mother-offspring groups. Males and estrous females are likely to visit many different food trees, but mothers stay longer in a narrower area to exploit a few localized resources.

The core-subgroup moving in a wide area finally discovers the fruit-abundant location and vocalizes and displays, attracting mothers accompanied by their young; thus, the area of intensive resource utilization gradually shifts from one place to another. It is very probable that mothers and some nonestrous females avail themselves of information obtained by males, since these females are more immobile.

Female dependence upon males is recognized in that females beg and/or snatch food from males, not vice versa, and that males share food with females and not vice versa (Nishida 1967; 1970).

In food transition periods, a core-subgroup moves widely and often to the edge of the home range, where it sometimes has vocal encounters with chimpanzees of the neighboring unit-groups. The core-subgroup then returns toward the center of the home range. Mothers with infants, thus, may also obtain information about the whereabouts of the neighboring unit-group, while they themselves are staying safely near the center of the home range.

2. Mother-son relationship.

Since it is not clear whether there are any pairs of adult males and females with kinship relationships (mother-son or sibling relationships) in the K-group (except one case) it is impossible to decide how kinship affects the social bond between adult males and adult females. The only case known may constitute rather negative evidence against strong social associations between an adult female and her fully adult son.

Sobongo, estimated at 9 years of age in 1967, was almost always following his mother, Wabwema, who had a 1-year-old male infant, Masisa, on her belly. Although Sobongo's familiarity index with his mother was high until the age of 11, the score abruptly declined at the age of 12 (Table 9). The separation from his mother occurred at the age of 13 when her estrous periods resumed after Masisa attained 5 years of age, and this coincided with his participation in the cluster of adult males of the K-group. She may have been moving out into one of the neighboring unit-groups at that time. We assume that the mother deserted the K-group range and her son did not follow her. He began his adulthood by continuously following the last-ranking male, Kamemanfu. Sobongo had a strong bond with him, and finally dominated him.

Sobongo was still dominated by his mother at the age of 11, but he overcame a top-ranking female, Wantangwa, at the age of 13, thus dominating all adult females. His full participation in sexual episodes in the midst of adult males came at the age of 14.

This one-case observation thus indicates that the crucial process from adolescence to adulthood in males lies, on one hand, in the rupture of the close association with his mother and the participation in the cluster of adult males, and, on the other hand, with domination over all adult females and beginning of full-fledged sexual participation. We were able to recognize this dyad of mother-son relationship because when Sobongo and Wantangwa met, they groomed each other very intensely and for a long time. Also, when one of the pair was attacked by a third animal, each helped the other actually or vocally.

It must be pointed out that, since chimpanzee males join an adult-male cluster of their unit-group and most females move out of their own unit-group at their maturity, very close association will be improbable between a mother and her full adult son and between adult brothers and adult sisters.

3. Sexual attraction.

While females are estrous, there is no doubt that sexual factors play the greatest part in male-female associations. Positive correlation between the presence of estrous females in the subgroup and the size of subgroups is admitted. Among 987 subgroups of the K-group recorded at the feeding ground during the study period from 1966 to 1971, 218 contained at least one estrous female and in 204 copulatory behavior was observed. The average size of subgroups containing estrous females (13.5 individuals) is much larger than the average size of total subgroups (8.6 individuals). At almost all times of the year the average size

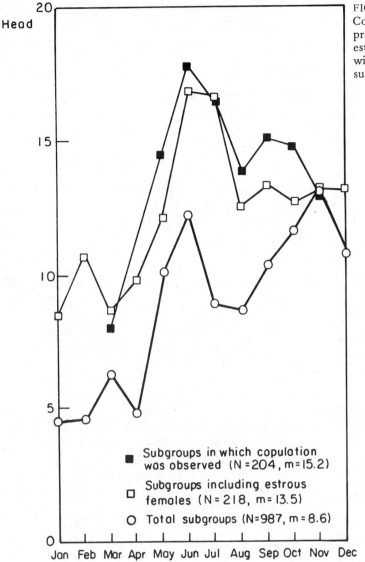

Head

FIGURE 5
Correlation of
presence of
estrous females
with size of
subgroups

■ Subgroups in which copulation
  was observed (N = 204, m = 15.2)

□ Subgroups including estrous
  females (N = 218, m = 13.5)

O Total subgroups (N = 987, m = 8.6)

Jan Feb Mar Apr May Jun Jul Aug Sep Oct Nov Dec

of subgroups in which copulatory behavior was observed (15.2 individuals) exceeded that of subgroups containing estrous females (Figure 5).
The presence of estrous females has a positive effect upon the size of
the subgroups (Nishida 1973).

It is not impossible, however, that a large aggregation caused by any
factor (e.g., localization of abundant food) brings about sexual excitement and estrus in females. A feedback relationship between the group

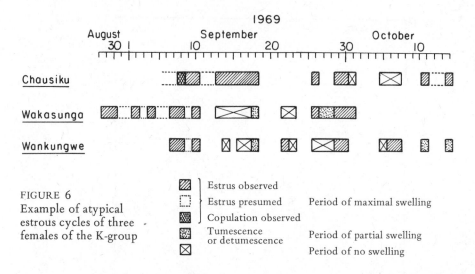

FIGURE 6
Example of atypical
estrous cycles of three
females of the K-group

⬚ Estrus observed
▢ Estrus presumed
▨ Copulation observed
▣ Tumescence or detumescence
⊠

Period of maximal swelling

Period of partial swelling

Period of no swelling

size and estrus in females may be plausible, since the sexual-physiological condition of females seems to be influenced by a social factor (probably the presence of another estrous female). There should be a period of partial or no swelling lasting more than 20 days between the end of the first maximal swelling and the beginning of the next maximal swelling. In reality, however, I often observed that the next maximal tumescence began much earlier than expected; the period of partial or no swelling sometimes lasted only one week or two weeks (Figure 6). In such a case, another estrous female was almost always present in the same subgroup. Moreover, the length of the period of maximal anogenital swelling was sometimes difficult to define and varied remarkably according to circumstances. It ranged from 4 days to 2 weeks. Such phenomena may imply the presence of a mechanism in which the sight of the sexual condition of the females acts as visual stimuli (e.g., posture and gesture of estrous females, or the white tumescent sexual skin) and/or the odor of genitalia of estrous females acts as an olfactory stimulus, thus causing the cerebral cortex to secrete pituitary hormone, which brings about estrus (Nishida 1973).

Goodall (1971) described a case where an old female, Flo, with a 2-year-old infant, suddenly showed a genital swelling on the day after she witnessed a strange young estrous female followed by many adult males at the feeding site: "Her swelling disappeared the next day—nor did she show any signs of swelling again for the next four years."

Mutual stimuli among females may sometimes cause the synchronization of estrus; i.e., when many females gather in one subgroup, they can become estrous. It sometimes happens that as many as five females show the maximal genital swelling at the same time. Some of these estrous females obviously show extremely irregular patterns of estrous cycles (Figure 7), which might conceivably be affected by the presence of the other estrous females. McGinnis (personal communication) recorded a very clear-cut example of synchronization of estrus in female chimpanzees: a particular dyad of adolescent females of Gombe National Park always moved together and synchronized their estrous cycles.

Goodall (1968b), McGinnis (this volume), and Tutin (1974) have reported examples of a rather fixed sexual relationship between a particular male and a female, which was termed *escort* or *consort* relationship. According to Tutin, many males of Gombe National Park often choose particular females as their mates *(partner preference)*, maintain constant proximity to these females, and interrupt mating between their females and other males *(possessive behavior)*. The particular dyads, moreover, sometimes go on safari for as long as 28 days *(safari behavior)*. McGinnis recorded an example of a relationship that lasted for 3 months.

This kind of relationship, safari behavior especially, may constitute an example of the disruptive effect of sexual drive upon the male bond, hence the size of core-subgroups. Partner preference and possessive behavior have been occasionally observed among K-group chimpanzees. High-ranking males have been observed to interrupt the mating between their females and the subdominant males on 10 of the total 383 copula-

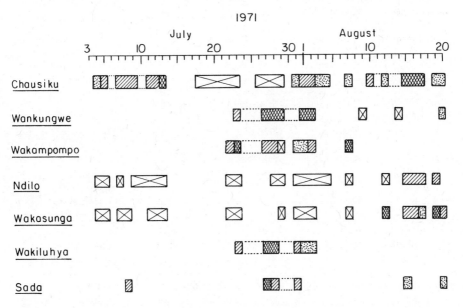

FIGURE 7
Example of simultaneous estrous state among females of K-group (see Figure 6 for the key to schematic representations)

ations (Table 10). On the other hand, estrous females have been observed to force themselves into the mating pair and to push aside the mating subdominant female by either or both hands or to press their own genital area towards the penis of the male and thus to interrupt on 11 occasions of the total 383 copulations. Wantangwa, thus, pulled apart the mating pair on six occasions, on three of which Kasonta stopped mating with the younger females and instead copulated with Wantangwa (Table 10).

TABLE 10
Cases of interrupted mating among chimpanzees of the K-group[a]

| INTERRUPTER | MATING PAIR | | BEHAVIOR OF | RESULT |
	MALE	FEMALE	INTERRUPTER[b]	
Kasonta	Kaguba	Wakasunga	B	Original mating interrupted.
	Kaguba	Wankungwe	B	
	Kamemanfu	Wakiluhya	B	
	Kamemanfu	Wantangwa	B	
	Sobongo	Gwekulo	B	
	Sobongo	Chausiku	B	
	Kasanga	Chausiku	B	
Kajabala	Kaguba	Tausi	B	
	Kamemanfu	Wakiluhya	B	
Kasanga	Sobongo	Wamikambi	B	
Wantangwa	Kasonta	Chausiku	A,D,E,G	Wantangwa mated with Kasonta.
	Kasonta	Chausiku	A,D,E,G	Original mating interrupted.
	Kasonta	Chausiku	A,D	
	Kasonta	Chausiku	A,D	
	Kasonta	Chausiku	C	
	Kasonta	Chausiku	A,E,F,G	Wantangwa mated with Kasonta.
	Kasonta	Wakiluhya	A,D,E,G	
Wakiluhya	Kasonta	Wakapala	C	Original mating not interrupted.
	Kasonta	Chausiku	A,E	Original mating interrupted.
Wabwema	Kasonta	Wantendele	A,F,G	
Gwekulo	Kasonta	Wakapala	A,G	Original mating not interrupted.

MALE (first group), FEMALE (second group)

[a] The data was obtained during the period from July 1966 to February 1974.

[b] A: rushing to the mating pair. B: charging to hit or stamp on the male. C: glaring at the pair and/or holding the male by one hand. D: pushing the female by one or both hands.
E: forcing oneself between the pair. F: pulling aside the pair by both hands. G: pressing the genital area to the penis of the male.

TABLE 11
Sexual relationships among chimpanzees of the K-group[a]

	MALE							
FEMALE	KASA-GULA	KASON-TA	KAJA-BALA	KASAN-GA	KAME-MANFU	KAGU-BA	SOBON-GO	TOTAL
PAST-PRIME								
Wamikambi		5	4	6	1	3	2	21
Wankungwe		13	7	12	4	6		42
Wabunengwa		1	3			1	*	5
Wangala		2						2
Wabwema	*	1	2	2	1	*		6
Wantangwa		9		6	3	1	1	20
YOUNG PRIME								
Wasalamba	*	12	2	1		*		15
Wakasunga		8	3	3	3	6		23
Wamkime			3	1		1	*	5
YOUNG								
Chausiku	*	55	24	3	6	6	4	98
Sada	*	4			2	3		9
Wahanse	*				2	*		2
IMMIGRANT								
Wakasila	*	6	8	3	1	2	1	21
Wakampompo	*	2	2	2		*	1	7
Wakiluhya	*	22	11	4	5	*	5	47
Wantendele	*	11	3	1	1	*		16
Wakapala	*	17				*	1	18
Gwekulo	*	12	2	3		*	4	21
Unidentified Females	*	1	2	2			*	5
TOTAL	0	177	80	49	29	29	19	383

[a] The tabulation represents the total frequency of copulation in each dyad, recorded throughout the study period from July 1966 to February 1974, during which only Kasonta, Kajabala, Kasanga, and Kamemanfu shared almost the same observational rates relevant to the direct comparison.

* Asterisks imply that these dyads had no opportunity to mate, since one partner of each dyad was absent, nonreceptive or immature. Age-grade of females is as of 1973.

The continuous escort behavior by most dominant males has been observed. For example, Kasonta monopolized Wantendele for 5 days in December 1972 and Wakasila for 6 days in January 1973. He was then observed to copulate with each female 6 times and 17 times respectively. No other male was observed to copulate with these females during this period. It is usual for all adult males to gather and move in a single sub-group even when four or five estrous females are present, as well as when only one estrous female is present. Only recently was safari behavior of longer duration also observed in the K-group chimpanzees.[4] Kamemanfu and Wakapala moved on their own independently of other members three times during the period from March 1975 to April 1976.

Twice, the relationship lasted for more than 3 months, and once for about one month. Sobongo and Wakiluhya moved on their own at least twice during the period of June 1975 to May 1976. The relationships lasted for 2 days and about 1 week respectively. The consortship continued even when the female ceased to be estrous as McGinnis pointed out (this volume). While the consort relationship continues, the male incessantly showed courtship display toward the female. Thus, the continuation of the consort relationship is responsible for an adult male of the dyad who courts his female by the behavioral mechanism analogous to neck-bite behavior found in hamadryas baboons (Kummer & Kurt 1963).

The accumulative data throughout the study period, however, seems to indicate that any male of the K-group can mate with any female of the group with the possible exception of the mother-son dyad, although particular females may be more often chosen by particular males than by the other males (Table 11). It must also be pointed out that chimpanzee females can be estrous even after conception. Most females show maximal swelling 5 months before their delivery. A young adult female, Chausiku, was observed to show maximal genital swelling and to copulate only 2 months before her first delivery. In chimpanzees, as in human beings, sex itself deviates from the original sole function of reproduction.

Attitude of the Unit-Group Towards Strange Females

When chimpanzees of a unit-group recognize the approach or presence of an unknown chimpanzee, they panic and become extremely tense as mentioned previously. But, if the stranger proves to be a female, K-group chimpanzees soon become calm. Adult males are very tolerant and females are rather indifferent to a strange female, even when they seem to see her for the first time.

Among seven newcomer females, Wakampompo and Wakiluhya were observed to be highly attractive to Kasonta, and copulated with him during the few days of peak estrus after they joined the K-group for the first time. Kasonta, moreover, prevented the other males from approaching these females by attack and threat. On the other hand, Gwamwami, an adolescent newcomer, followed Kasonta continuously for the first 4 months after she transferred to the K-group, though she was not courted by him. Their familiarity index remained extremely high (71.0 average, 42.5 to 88.5 in range), although she showed only adolescent ano-genital swelling and did not copulate with any male.

A newcomer female holds the lowest, although stable, status among adult females for a long period after joining a new unit-group. Five females who joined the K-group from 1971 onward occupied the cluster lowest in rank order among adult females. The dominance relationship among six newcomer females seems to have been determined directly by the relative length of the period that elapsed since their transfer to the K-group; that is, the earlier the date of transfer, the higher the position a female occupies. Although agonistic encounters were not seen in all dyads among these newcomers, it seems that a linear dominance hierarchy is established among them (Table 8).

DISCUSSION

The Concept of Unit-Group

There has been much controversy concerning social structure among wild chimpanzees. Reynolds (1966) and Goodall (1965; 1968b) denied the existence of any *permanent* group except those of mother-infant (or, juvenile), although some have insisted upon the existence of the social unit (Itani & Suzuki 1967; Izawa 1970; Nishida 1967, 1968; Sugiyama 1968).

Permanent does not always mean *cohesiveness* of the group. No one will deny that the human family is a permanent group, although members of a family may not always stay together. The chimpanzee social unit is a permanent group, at least in the sense that the unit-group is connected closely with a given territory and lasts permanently and independently of individuals, even if members of the group change constantly.

Recent advances in the field study, however, have revealed much more similarity of social structure between chimpanzees of the Mahale Mountains and those of Gombe National Park (and possibly of the other areas). Some controversial points are examined here.

1. *Social Unit.* Goodall (1965; 1968b) previously mentioned: ". . . they (chimpanzee groups) cannot be divided into separate communities. It seems likely that only a geographical barrier would constitute a limiting factor on the size of a community." But she recently stated that the chimpanzee population of Gombe Park is *divided into communities* of individuals who recognize and may interact with each other (Goodall-Lawick 1971; 1973a) and that her term "community" corresponds to our term "unit-group" (Goodall-Lawick 1973b). Teleki (personal communication) mentions that there are definitely three distinct communities within the park. Moreover, Bygott (this volume) observed that their study community experienced fission into two (sub)communities. No evidence more unequivocally shows the existence of social units. His description of each subcommunity agrees with what we defined as a unit-group (Nishida 1968; Nishida & Kawanaka 1972). Albrecht and Dunnet (1971) have stated that their study population in Guinea corresponded to our unit-group.

2. *Territorialism and Antagonism Between Unit-Groups.* It has been pointed out that the two unit-groups of the Mahale Mountains are territorial and antagonistic to each other (Nishida 1968). On the basis of two definite encounters between members of different communities, Bygott (this volume) showed that a similar principle of avoidance is operative among unit-groups of Gombe Park. Although chimpanzees have often been referred to as an example of nonterritorial animals, especially to refute inborn territorial drive in mankind (e.g., Crook 1968; Montagu 1968), such a line of reasoning is not appropriate.

3. *Dominance-Subordination Relationship Between Unit-Groups.* I have mentioned that one unit-group (the M-group) as a whole is dominant over another group (the K-group) as a whole (Nishida 1968). None of the researchers mentioned whether this kind of relationship exists among unit-groups of Gombe Park; however, this does not imply that such relationships do not exist. Perhaps no event has been observed that would provide evidence of the existence of this relationship. The term *displacement relationship* among unit-groups, which Teleki (personal communication) proposes may thus be applicable to situations observed both at Gombe and Mahale.

4. *Mass Migration of the Unit Group.* The mass migration which has been well documented in the Mahale Mountains has not been recorded at Gombe or in Budongo Forest. This behavioral difference must be ascribed to the difference in the extent of the heterogeneity of the environment and probably in the population density of each habitat. The habitat of the M-group and the K-group is obviously much more heterogeneous than either of the Budongo Forest and Gombe Park, in that Mahale contains both Miombo woodland-forest mosaic and the wide belt of the semi-deciduous forest (Kasoge Forest). Local heterogeneity of seasonal abundance in food supply may cause the northward migration of the M-group, which in turn greatly influences the movement of the K-group. Great difference in the size of both unit-groups may facilitate freer movement of the M-group. It is plausible that high population density, and hence small size of home ranges, may relate to the lack of large-scale migration of unit-groups in the Budongo Forest (Sugiyama 1968).

5. *Transfer of Females.* Transfer of females from one unit-group to another has been recorded both at Mahale and Gombe. At Gombe, four adolescent females *temporarily* left their home communities during periods of estrus and mixed and mated with males of neighboring communities. On the other hand, four young strange females *permanently* joined the habituated community, but no habituated adolescent female has *permanently* left her range (Goodall, personal communication).

The rare occurrence of the permanent transfer (only four cases) at Gombe contrasts remarkably with many cases (seven permanent moving-in and four permanent moving-out females) recorded in the K-group. The contrast will be clearer if we consider the larger size of the Gombe study population. This difference may suggest the variability of relationships among unit-groups. In reality, the greatest attention must be paid to the fact that three of four subadult females of the K-group who transferred permanently, and two K-group females who temporarily transferred, selected the M-group. This fact may imply not only that female transfer is obviously confined to neighboring groups, but also that a female can shift only to a specific group among neighboring groups. The B-group is a unit-group ranging to the north of the K-group. What little surveys we made of the B-group have not produced evidence that any members of the K-group ever moved out to that group. Moreover, scarcely any overlapping area of home ranges was found between the B-group and the K-group, although it occurs to a great degree between the M-group and the K-group. The extent of the member flux between unit-groups may be determined by the extent to which the two unit-groups share the overlapping area. Further study will settle the problem of whether the *mating sphere* in chimpanzee females is confined to one or several particular neighboring groups.

Teleki (personal communication) stated: "Provisioning caused an influx of chimpanzees from the southern portion of the Gombe Park, adding perhaps as many as 10 chimpanzees from another community to the study population.[5] . . . These immigrants, who were there for 6 or more years during provisioning, went south again soon after bananas were no longer provided freely."

Although I admit that ecological factors must be considered in some cases of transfer, *provisioning determinism* cannot possibly be accepted as an explanation. It cannot explain why transfer is confined to females (at least in my data); furthermore, it conflicts with the fact that strange females joined the K-group in 1972 during the period when the K-group was moving free of provisioning in the Miombo forest-mosaic area, and with the fact that some females without infants left the K-group for many months while most of the K-group chimpanzees were regularly visiting the permanent feeding station. Even if we accept the explanation of member influx caused by provisioning, it simply *reflects the nature* of social characteristics of chimpanzees, because this rarely occurs among the provisioned troops of Japanese monkeys who share the same feeding ground.

Chimpanzee Social Structure

The social unit of chimpanzees is a group with fluid membership, which has as a nucleus a cluster of strongly bonded adult males connected with a given annual range and some prime and past-prime females closely associated with them. Included in the unit are other females who are much more variable in the extent of their association with adult males, but who, in principle, share the same annual range.

Inter-unit-group antagonism is based mainly on the antagonism among adult males who belong to different unit-groups. On the other hand, the fluidity of a unit-group is based on the dispersive, egocentric tendency of adult females (especially mothers), the habit of moving-out among females, and the tolerance of the unit-group toward strange females.

Two types of member transfer among unit-groups have been recognized: permanent and temporary transfer (Nishida 1973). The former is confined to subadult females and the latter is seen mainly among fully adult females, although it is seen also in adolescent and subadult females. The temporary transfer of adolescent females may be an exploratory visiting, after which, if *satisfied* with the new unit-group, she will sooner or later transfer permanently, but if she is not, she may return to her natal group. On the other hand, temporary transfer among some fully adult females roughly coincides with the period when their dependent young have reached the juvenile stage and the females recover sexual receptivity. Estrus is a passport that females must carry with them in both temporary and permanent transfer. It should be noted that sexual drive itself does not always attach the females to domestic males. I have presumed that temporary disappearance of fully adult females might have been due to a kind of temporary visiting on a basis of the nexus of kinship relationship covering plural adjoining unit-groups (Nishida 1973). The mother-daughter relationship observed so far, however, would seem to negate my assumption, although the problem has not yet been settled.

The data collected to date clearly show that chimpanzee society never has a *matrilocal* mating system, which prevails among Cercopithecine society (e.g., Nishida 1966; Rowell 1966; Sugiyama 1967). A matrilocal mating system is here defined as the system in which females are in principle connected with the home range of their natal unit-group from generation to generation, males generally move out of the natal group, and strange males come from outside to join the group. The *patrilocal* mating system, defined as the reverse of the above, has not been known among feral subhuman primates studied so far. Whether chimpanzees have a patrilocal mating system has not yet been settled. Four males disappeared from a unit-group at Mahale and it is certainly unlikely that all would have died, because two left the unit-group when they were in the prime of life. There is no evidence that they transferred to any other neighboring unit-group. In addition, no adult male outsider has ever been reported to have *joined* the habituated unit-group of either Gombe or Mahale. This seems to be a very significant fact. It is plausible that these males are living singly or with one or several females somewhere in the high altitude of the Mahale Mountains or far from the study area, independent of any unit-groups. Also, there is no evidence yet that any male ever transfers from his *natal* unit-group to another. It is conceivable that a juvenile male, accompanied by his mother who has no other younger offspring, may transfer to a neighboring unit-group and spend the latter part of his life in that group. At any rate, the home range of a unit-group can remain unchanged, although new recruits are continually coming in from outside.

I would point out one more outstanding characteristic of chimpanzee sociality—clusters of adult males. Recent field studies have revealed much diversity in the social structure among great apes. The mountain gorillas live in a one-male group, of which blackbacked or young silverbacked males, sons of the great silverback, must finally move out. Solitary silverbacked males never join a one-male unit; instead they kidnap one or two young females from the one-male unit, thus establishing a new one-male unit (Fossey 1974; personal communication). Orangutans, both male and female, live solitary lives in principle. Although adult females sometimes move together temporarily, and males and females establish consort relationship during estrus of the female, field workers have not recorded two males making contact with each other peacefully (e.g., MacKinnon 1974; Rodman 1971). Exclusivism among adult males characterizes the social structure of two genera of great apes. On the other hand, what little evidence is available seems to suggest that the social structure of pygmy chimpanzees may be a compromise between that of the gorillas and chimpanzees (Kano, personal communication).

Chimpanzee male clusters function chiefly in food-finding and territorial defense. As such, a male cluster must be stable for a long time. It remains for future study to determine whether the male bond is only based on ambivalence of subdominant males towards dominant males or on some other factors as well, such as kinship or co-socialization. We should follow the life history of individual chimpanzees from early infancy to seniority on a more extensive scale. More 20-year studies are needed.

Acknowledgment

I should like to express gratitude to the Tanzanian government for permission to do research. I am grateful to Professor Msangi, who has encouraged our work, and the University of Dar-es-Salaam. The research would have been totally impossible without many facilities given by local government officials and village people of Kasoge and Katumbi who have welcomed us as Wenyeji (residents).

I could not have carried out the work without the help of Tanzanian field guides and assistants who have helped us collect information and who also accumulated the field data themselves during our absence from the field station. In this revised version, I added some data obtained during my very recent research, which I carried out as Game Research Officer of the Ministry of Natural Resources and Tourism, Tanzania in 1975-1976. I sincerely thank the officials of the Game Division for their cooperation.

I should like to express my thanks to Professor J. Itani for his continuous advice and encouragement; to Dr. K. Kawanaka, Mr. and Mrs. M. Kakeya, Mr. S. Uehara, and my wife, who stayed with me in the field station during each of the later periods of my study and who not only provided unpublished data, but helped me in the various aspects of the study as well as in the maintenance of the field station. The staffs of the Department of Anthropology, the University of Tokyo generously permitted me to undertake this long study in Africa. I am indebted to Dr. G. Teleki for his invaluable comments and suggestions. The research has been financed mainly by the Overseas Scientific Research Fund of the Ministry of Education, Japan and partly by the Wenner-Gren Foundation for Anthropological Research, the Japan Society for the Promotion of Science, and Japan International Cooperation Agency. I am thankful to the above institutions.

Notes

1. This paper was first written in October 1974 and revised again in February 1977, adding some relevant data obtained during a full-year study carried out from May 1975-May 1976 (17 months, October 1965-March 1967; 4 months, July-November 1967; 3 months, August-November 1968; 3 months, August-November 1969; 4.5 months, June-August, October-December 1972, and one week in April 1972; 3.5 months, October 1973-February 1974).

2. (J. Itani, frequent short-term studies from 1966-1974; K. Kawanaka, 6 months from August 1969-February 1970, and 3 months from September 1972-January 1973; A. Mori, 4 months from September 1972-January 1973; and S. Uehara, 4.5 months from October 1973-March 1974).

3. No agonistic interaction was observed between Kasagula and Kasonta (Nishida 1970), but the detailed examination of the data of the banana test shows the dominance of the former.

4. Kamemanfu was once recorded to move about always with a mother-juvenile group (Wangulu and Wahanse) for 46 days: In January 1969, the K-group was still moving in Myako Valley and the M-group in its southernmost area of the home range. On January 10, these three chimpanzees appeared at the permanent feeding station for the first time since the northward mass migration of the K-group in September 1968. Either or both of them visited the station 35 days; on 25 of which, Kamemanfu and the family group came to and left

the station together. But, Wangulu was not observed to show any sign of receptivity throughout the above period. This association ended on February 25, when the other members of the K-group joined them.

5. According to the recent paper by Teleki (1976), as many as 6 *males* and 11 females *joined* the study community in 1964-1965. But, Goodall (1973a) mentions that not a single male has transferred to the study community. Moreover, Teleki (1976) questions the occurrence of fission described by Bygott (this volume) and interprets it only in terms of emigration. This very important discrepancy remains to be explained by Gombe scientists, although it seems that Teleki's interpretation is inconsistent with our data obtained at Mahale.

Takayoshi Kano

A Pilot Study on the Ecology of Pygmy Chimpanzees, Pan paniscus

The genus *Pan* can be divided into two species: *Pan troglodytes* and *Pan paniscus*. *P. troglodytes* inhabits a stretch of land across a mid-section of Africa. Its distribution area extends from the western coast, across central Africa, to the middle of eastern Africa. The southern boundary of the area is marked by the Zaire (Congo) River. *P. paniscus* occupies an area to the north, and is distributed across the tip of an imaginary arc connecting the left banks of the Zaire and Lualaba Rivers. The southern boundary has not been clearly defined, but is presumed to include the territory around the Lukenie and Kasai rivers (Figure 1). *P. paniscus* recently appeared as a new species in a taxonomic classification

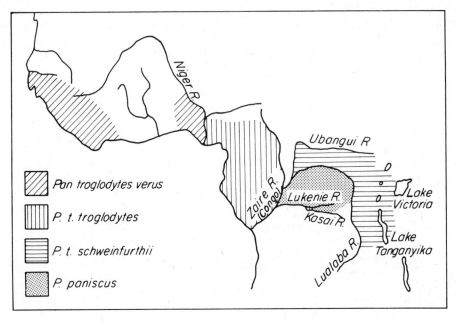

FIGURE 1
Distribution of genus *Pan*

of the anthropoids. Previously, the species had been regarded as a sub-species of *P. troglodytes,* (also known as *P. satyrus paniscus);* but H. J. Coolidge has assigned them special status as an independent species in his taxonomy.

Numerous morphological studies of *P. troglodytes* have been under-taken; but of the few that were carried out concerning *P. paniscus,* the monograph by Coolidge in 1933 is still regarded as one of the best. According to that study, members of *P. paniscus* are smaller than *P. troglodytes;* consequently, they are referred to as pygmy chimpanzees. In addition to their small stature, they portray many other character-istics more infantile than *P. troglodytes.* Because of their primitive characteristics, Coolidge thought that *P. paniscus* was closer than all the other existing anthropoids to the common ancestor of the genus *Pan.* This supposition may be correct, and a study of the environment and social groupings of *P. paniscus* may still provide many clues to help solve the mystery of Pongidae evolution and reconstruct the ecology and the society of the early hominids.

Only a few ecological studies that refer to *P. paniscus* have been available. For example, the brief reports by H. Schouteden and E. Schwarz, published as far back as 1930, partly described the habits of *P. paniscus* but were aimed primarily at naming the localities where they could be found. Almost all the information in the reports was obtained from the natives rather than through scientific observation.

The field studies of *P. paniscus* were hampered because of political disturbances in the area, which still prevail today. At the same time the primatologists of the world began to seriously study the African anthropoids at the beginning of 1960, there was a disturbance in the Congo. The continued civil war and the political instability of the Republic du Zaire made a survey of the area impossible for a long time.

In the spring of 1972, Dr. T. Nishida flew alone to Zaire and made a brief survey along the Lake Tumba shore at the west end of the distribution area. During that survey, he observed three beds, collected some bones of *P. paniscus* that had been preyed upon by the native villagers, and brought home much information.

In July 1973, Nishida and I left for Salonga National Park, which he had concluded would be the most promising place to find *P. paniscus.* Contrary to our expectations, however, our survey ended in disappointment—we did not find a single pygmy chimpanzee bed. In the middle of September, after a two-week stay in Mbandaka along the Congo River, Nishida started for Tanzania while I returned to the Congo Basin. I found *P. paniscus* initially on November 1 in the Ikela zone near the Ikele village. When I left, at the end of December, I had had a total of 14 direct encounters with these chimpanzees, in the Ikele, Bokungu, Befale, and Djolo zones; however, I recorded beds only in the Bolomba zone.

HABITAT

The vegetation of Congo Basin forests can be classified into two main types, which I will refer to simply as A and B. A-type forests are the swampy regions submerged during the flooded season. B-type forests are low-to-medium altitude regions supported by relatively hard soil. Another type of secondary vegetation has developed on the swidden field, composed of *Musanga smithii* trees, which are peculiar to these forests. Swidden fields are interspersed within the B-type forests; so apart from the *Musanga smithii* trees, the secondary forest and the most arid section of the B-type forest have common species components (Figure 2).

P. paniscus is scarce or even nonexistent in the A-type forest. These large swampy regions have developed in the Mbandaka, Ingende, and Bikolo zones of the western Congo Basin. I could not find any proof that pygmy chimpanzees inhabit the Ingende zone. Some of them were found in the Bikolo zone that Nishida visited, but their population density is extremely how. In addition to these, the zones around the main rivers in the Congo Basin are surrounded by belts of swampy forest several kilometers wide; however, no chimpanzees are living there.

FIGURE 2
Forest of type B near Yasongo village at western Befale district

Groups of pygmy chimpanzees were encountered 13 times in the B-type forest and secondary forest. Their beds were situated mostly in the B-type forest, whereas a considerable amount of their food remnants were found in the secondary forest. It can be concluded that the chimpanzees prefer to inhabit the B-type forest but also make use of the secondary forest.

The three types of forest mentioned—A and B type and swidden with secondary forest growth—are found in the vicinity of the Ikele village. In the western part of the village, among the swampy type-A forest vegetation, no pygmy chimpanzee remnants were discovered. In the southeastern part, where the darkest and thickest B-type forest flourishes, very few remnants were observed. It was in the northern, southern, and northeastern part of the village, in a B-type forest, that large gatherings of them were directly observed and most of their beds and food remnants were found. The forest was comparatively light and sparse under a canopy of leaves. *P. paniscus,* then, predominantly uses the relatively light and arid section of the B-type forest and the secondary forest.

FOOD HABITS

The foods of *P. paniscus* can be divided into three groups: (a) juicy fruits or fruits with rich pulp, (b) fibrous foods such as shoots, stalks, and leaves, and (c) hard nuts such as beans.

Of the fruit group, I found only three species: *Anthoclitandra robustior, Landolphia owariensis,* and *Dialium* sp. The *Dialium* sp. is a giant Caesalpiniaceae. The chimpanzees do not eat the hard beans themselves —just the soft, juicy starch and fiber on the surface. The other two species are fruits of Apocynaceae and both are very juicy. The many discarded fruit remnants indicated that pygmy chimpanzees prefer the fruits of *Tricoscypha acuminata, Anonidium manni,* and *Aframomum* spp. In addition, some fruit remnants of *Musanga smithii* and *Myrianthus arboreus* were also recorded.

Many leavings from the fibrous foods group were also found in the area. *P. paniscus* especially likes the shoots of the giant herbs, *Haumania liebrechtsiana* and *Sarcophrynium macrostachyum.* Stalk remnants of *Aframomum* spp., *Costus* spp., and *Palisota aubigwa* were also scattered around.

Various kinds of beans belong to the third group, categorized as hard nuts. According to native information, *P. paniscus* definitely eats most of the Caesalpiniaceae beans and most likely eats the very big beans of *Pentaclethra macrophylla.*

During my survey (October to December) I had the impression that, of all three groups of foods eaten by these chimpanzees, the first and the second groups were the most important ones, since remnants of both groups were continuously encountered. (It might be noted that the pygmy chimpanzees eat both the frugivorous foods of *P. troglodytes* and fibrous foods of gorillas.)

SPECIFIC FEEDING BEHAVIORS

My survey revealed some interesting information about the feeding behaviors of *P. paniscus:* Unlike *P. troglodytes,* they dig in the ground and fish.

Digging in the Ground

On November 1, I found some holes dug in the forest to the south of the Ikele village. It was my first such sighting. I was surprised to hear from trackers that the holes were made by pygmy chimpanzees. I found it difficult to believe that they could dig holes as big as 30-40 cm in depth and 50 cm in diameter. Many times after that, similar holes were found, always with pygmy chimpanzee footprints around them. On one occasion, my tracker pointed them out commenting that they were the *farms* of pygmy chimpanzees.

On November 16, I finally came upon a group of pygmy chimpanzees working on one of these farms. At 2210 that evening, pygmy chimpanzees were heard around the Bottendo River basin to the north of the Ikele village. Despite my stealthy approach, I was instantly detected by two chimpanzees who then disappeared among the branches, sending keen alarm calls. In my crouched position I started toward them once again, but was startled by a chorus of violent alarm calls from 30-40 pygmy chimpanzees in the underbrush. I heard a number of them rapidly scurrying away through the underbrush.

Nearby, there was a newly dug farm of quite sizable dimensions where they had been staying. On the turned-up clay, numerous footprints, which were still fresh and clear, were found. I was not fortunate enough to observe their digging directly, but it became obvious that when the pygmy chimpanzees dig in the ground they are searching for something, which, according to native information, is a mushroom *(?Langermannia fenzlii)* called Shimoukiro.

Fishing Behavior

Many villagers remarked that fish is one of the foods of *P. paniscus.* If verified, this information about feeding behavior would be valuable; however, it seemed unlikely since pygmy chimpanzees avoid water. The information was very ambiguous and had not been directly observed; however I could not dismiss it without some consideration.

Twice at different streams, near Ikele and east of Limongo, I observed the mud in which pygmy chimpanzees had supposedly been digging. The object of their search could have been fish, since some kinds of fish of the Congo basin inhabit the mud of small streams.

Meat- and Insect-Eating Behavior

It is often reported that *P. troglodytes* eat meat and insects, but the same information has never been reported about *P. paniscus.* According to the Isunguna villagers, several pygmy chimpanzees once cornered a small duiker and felled it with broken twigs. They played with it but did not kill or eat it.

I once noticed a nest of termites *(Cubitermes intercalatus)* that was broken and scattered near the Ikele village. My tracker claimed that the pygmy chimpanzees had tampered with it. In many villages it is believed that the chimpanzees break down the termite nests, which are constructed by the termites at the roots of trees. The villagers report that pygmy chimpanzees sit around the crushed nests afterward, as if warming themselves at a campfire. The villagers do not know if the pygmy chimpanzees eat the termites. However, if it is true that they do crush the nests, it is possible that actual termite-eating might follow.

Bed-Making Behavior

I recorded 1292 *P. paniscus* beds throughout the survey. Their beds averaged 80-120 cm in diameter and were almost spherical in shape.

Only once (on November 14) was the entire bed-making procedure observed, at the upper stream of the Bottendo River, to the north of Ikele. At 1728, a male chimpanzee climbed a *Polyalthia suaveolens* tree 8 cm in diameter and about 10 m in height and sat there for 2 minutes. Then he bent the tree with his weight in the direction of a neighboring *Anonidum manni* tree 3 m away, pulled the branches of that tree toward him, and began making a bed by intertwining the branches of the two trees. After working diligently for only 5 minutes, he finished the bed. Twenty minutes before this, I had heard sporadic snapping sounds of branches here and there among the trees and had surmised that the branches were being collected for building. After seeing the bed-making activity, I was convinced that the earlier sounds indicated a similar activity. The next morning 28 new beds were recorded around that area, indicating that had indeed been the source of the sounds.

The pygmy chimpanzees make beds a short time before and after sunset, apparently because they can be finished quickly, just before they are needed. Most of the time, their beds are not too far from the ground. They rarely use the tall trees of the first layer of the forest. Most of the 910 beds recorded in Ikele and Yaloshidi were concentrated in trees no more than 15 m in height and 20 cm in diameter.

I cut the trees down several times and examined the beds in them to determine what kind of branches were used as bed materials. I found only the branches from the bed trees themselves; no extraneous branches from other trees were used.

Seventy-one beds were found directly on the ground. They were made from shrubs growing in an area about 2-4 m². The trunks were snapped downward 20-50 cm from the base toward the center of a circle. Usually four to seven shrubs were used and never more than 13. In the case of larger shrubs, one or two were enough for making a bed, since the long branches were coiled into a spiral after they had been snapped downward.

We found only one bed that was made without trees—it was spotted in a place where fallen leaves had been piled up. Fallen leaves were often piled up on top of deserted ground beds, and were also found in the new beds. Perhaps leaves had been collected within the reach of the chimpanzees and used as their cushions.

A few ground beds were also observed around the *P. troglodytes* habitat. In contrast to *P. paniscus,* their beds were usually made in the daytime for the purpose of taking a rest.

VOCAL SOUNDS

P. troglodytes are said to be the noisiest mammals in Africa. *P. paniscus* is probably the next most clamorous mammal living in the same area. Their screaming immediately reveals their location, making it unnecessary to search for them. Other wild animals do not usually shriek so incessantly; for instance, the black mangabeys and guenons, often found in the same area, scream only when they sense danger.

Both species of the genus *Pan* frequently utter sounds even when the group is peaceful, but there is a great difference in volume. The voices of *P. troglodytes* are so loud that they sometimes carry a distance of up to 2 km. Their loud, shrill cries create a great wailing uproar. The pygmy chimpanzees, however, barely murmur. As far as I was able to determine during my survey, the pygmy chimpanzees seldom screech. They sometimes whined together, but their individual voices were so feeble and soft, described as "hou hou —— ," that they could not be heard 500 m away, sometimes not even 200 m away. On October 28, I heard them near the Isunguna village for the first time. My tracker said that pygmy chimpanzees were chattering in the forest just ahead of us, but to me it sounded like hornbills twittering in the distance.

Another difference between the two species besides volume is that pygmy chimpanzees have a special alarm call. The word equivalent can be written as "hii——," symbolizing a very high-pitched and delicate metallic sound. At first, I thought the strange sounds were being emitted only by infant chimpanzees, since I had once observed them sending the warning cries. Later I noted an adult male chimpanzee making the same shrill call and realized that almost every *P. paniscus* is likely to make these alarm calls, regardless of age or sex. The differences in the vocal sounds of *P. paniscus* and *P. troglodytes* will be clarified, after we succeed in determining exact discriminations of vocal sounds among individuals through observation in the future.

GROUP SIZE AND GROUPING

Larger groupings were usually observed among *P. paniscus* than among *P. troglodytes.* The information obtained through inquiries further reinforced the data. It was impossible to count the exact group numbers in the forest. Judging from the number of individuals estimated by the volume of noise they made when encountering men, and by the degree of tree-shaking, however, it was estimated that the pygmy chimpanzees are usually in groups of from 15 to 30 (Table 1).

TABLE 1
Group size

ENCOUN-TER NUMBER	DATE	TIME	LOCATION	NUMBER WITHOUT AN OVER-LAP IN COUNTING	ESTI-MATED SIZE OF GROUP
1	Nov. 1	1035-1105	Ikele	7	15-27
2	Nov. 11	1415-1435	Ikele	3	c. 10
3	Nov. 13	1545-1640	Ikele	4	20-30
4	Nov. 14	1140	Ikele	10	c. 20
5	Nov. 14	1709-1735	Ikele	2	c. 30
6	Nov. 15	0546	Ikele	?	c. 30
7	Nov. 16	1029	Ikele	2	30-40
8	Dec. 4	0850-0920	Djefela	2	15-20
9	Dec. 8	1345-1410	Yama	5	15-20
10	Dec. 11	1323-1517	Esangani	22	60-90
11	Dec. 14	1600-1612	Wamba	2	10-15
12	Dec. 15	1010-1045	Wamba	12	c. 30
13	Dec. 20	0836-0950	Bongoli	4	15-20
14	Dec. 20	1555-1615	Yasongo	17	c. 30

Further evidence that pygmy chimpanzees form a sizable group is the large number of beds recorded within close proximity. Usually after one was found, other beds that had probably been constructed about the same time were also discovered. Although the beds were usually grouped collectively, we must acknowledge the occasional existence of small populations and solitaries separate from the large groups. A few new beds were sometimes observed in complete isolation, and in the Ikele village we discovered a small quantity of new food remnants far from the nomadic range of the large group.

For survey purposes, a group of pygmy chimpanzees always includes both male and female adults, as well as immature members. The groups are organized for reproducing and nursing; sometimes, in their intra-group associations, very special separation by sex can be found.

In southern Esangani, I watched a parade of 18 pygmy chimpanzees (composed of 11 females, five infants, a male, and a juvenile) traveling through the branches of tall trees of *Pitadenia africana* from 1412 to 1415 on December 11. From 1420 to 1440 of the same day, I spotted four more pygmy chimpanzees on three trees a little way off. At 1530, I noticed four more—all males—on three other trees.

Near Wamba village, I counted seven distinctly individual pygmy chimpanzees at 1010 on December 15. There were four adult males and two adolescent individuals (probably males). At first I thought there were no females in this group; but at about 1040, I saw three of them jump from a *Brachystegia laurentii* to a *Dialium* sp.—first, a female,

131

FIGURE 3
Juvenile pygmy chimpanzee caught near Euri, Bolomba district

followed by a male and another female. At almost the same time, I saw four females move one after the other through the canopy of trees which was 30 m away.

Near the Yasongo village in the Befale zone, I observed two males and later another three males coming down separately from the tree *(Scrodophloeus zeukeri)*, around 1610 on December 20. Seven females were seen moving from a tall tree *(Cynometra sessiliflora)*, 40 m away from there, to a lower tree. At 1612, a juvenile and a male, followed at 1615 by a female with an infant, hopped down to end the parade.

ARBOREAL OR TERRESTRIAL?

Coolidge introduced the observational data made by Derforge, who had kept one young *P. paniscus* and one young *P. troglodytes* at the same time, and who found that *P. paniscus* was better at climbing trees and was usually more cautious. He noted that the pygmy chimpanzee never put his weight on dead branches, but the *troglodytes* chimpanzee often found himself falling down unexpectedly when the branches broke. This seems to indicate that, of the two species, *P. paniscus* is the more arboreal.

According to my observation, too, pygmy chimpanzees seemed much quicker in their movements in the trees than the *P. troglodytes* I observed in western Tanzania. During our observations, when the pygmy chimpanzees sensed us, they did not leap down from their trees to escape like the others, but instinctively hid themselves among the thick-leaved branches of very big, tall trees. *P. troglodytes,* on the other hand, apparently regarded the ground as a place of safety.

The natives, however, regard pygmy chimpanzees as terrestrial animals. They believe that pygmy chimpanzees climb trees only to sleep at night. There are indications of the terrestrial tendency of *P. paniscus,*

FIGURE 4
Chimpanzees in progression, in the forest near Essangani. Young adult male (right), adult female carrying a baby on her belly (left below), adult female (left above).

including digging in the ground, fishing, and construction of ground beds. The villagers' observations were not based on rare or exceptional cases, but on frequently seen behavior of the pygmy chimpanzees.

The evidence shows, first of all, that the ground beds recorded in the Ikele area account for a little less than 8% of all *P. paniscus* beds, compared to practically no ground beds for *P. troglodytes.* Secondly, pygmy chimpanzees obtain much of their major foods on the ground. All the fibrous foods mentioned earlier grow near the ground. Many of the juicy fruits that the pygmy chimpanzees eat—such as *Dialium, Antho-clitandra,* and *Landolphia*—are found only on high branches; however, the fruits of *Aframomum* and *Sacrophrynium,* of course, are found on the ground. *Tricoscypha acuminata* is a big tree, but its fruits hang down like clusters of grapes on short branches attached to the tree trunk near the root. The fruit of *Anonidium manni,* which falls on the ground when it ripens, is also an important food for pygmy chimpanzees. In comparison, members of *P. troglodytes* do not eat as much of the food they find on the ground, apparently only because they prefer food in the trees, not because it is unavailable on the ground.

According to the report of Goodall (1963), most beds of *P. troglodytes* in western Tanzania are made in trees at the height of about 9-24 m. According to my observation in both the Ikele and Yaloshidi villages, most of the *P. paniscus* beds are made at the height of 6-10 m. The number of beds decreases remarkably at levels higher than 16 m. Pygmy chimpanzees inhabiting the Congo forest, which is more densely grown with tall trees than the forests in western Tanzania, make beds closer to the ground.

In conclusion, even though pygmy chimpanzees seem to be arboreal in their efficient movement among trees, we cannot say that they make less use of the ground than *P. troglodytes,* and so we can safely assume that they are primarily terrestrial.

CONCLUSION

Detailed study of *P. paniscus* has just started. The purpose of the preliminary survey was to find several fields suitable for more intensive survey in the future. Consequently, I did not stay in one place long enough to collect substantial data; but instead collected random information in various areas. I was unable to survey over as extensive an area as I had hoped, because of the road conditions in the Congo basin.

My strongest impression was that this species of anthropoids differs from *P. troglodytes* in behaviors, food habits, and vocalizations. I cannot guess how many more contrasts will be found in the future. Some differences may result from variations in environment, but others might be understood from the phylogenetic point of view.

Acknowledgments

This study was financed by the Overseas Scientific Research Fund of the Ministry of Education, Japan. Facilities were provided by the Government of Zaire. I am indebted to Professor J. Itani, the director of the Kyoto University Expedition to Central and East Africa (1973-1974) for his guidance and encouragement and to Dr. T. Nishida, the University of Tokyo, for his guidance and cooperation in the initial period of the study.

I should like to express my gratitude to Dr. Ntika Nkumu, the director of I.R.S.A.C., who generously permitted our research, granted us research associateship in the Institute, and encouraged us in our work; to Dr. J. Verschuren, the director of Parcs Nationaux Zairois, for his permission to do research in Salonga National Park; and to Professor P. van Leynseele (Kinshasa), Dr. H. Hamaguchi (Bukavu), St. Paulo and Mr. Leo Verheyen (Mission Catholique, Boende) who welcomed us as guests and gave us much information. I am much indebted to Professor C. Evrard for his identification of my plant collections.

This is the revised version of the paper presented at the Burg Wartenstein Symposium (read by Professor Itani and Dr. Nishida).

PART TWO

Gorillas

Some years ago George Schaller (1963) studied the mountain gorilla. In 460 hours of observations, he outlined some of the main features of gorilla social organization, and, far more important, demonstrated that this reputedly dangerous animal could be studied at close range. In his paper, Harcourt apologizes that his observations on the social roles of silverback, blackback, female, and immature gorillas is based on only 900 hours of observations. How times have changed! The stability of the silverback, adult female, and young as one group contrasts with the mobility of young adult males (who may become solitary) and young adult females who shift to other social groups. The moving out of maturing female gorillas and chimpanzees is in marked contrast to the behavior of monkeys—in which case the males leave the natal group.

Dian Fossey's continuing study of gorilla behavior now offers the reader data on the maturation of 32 young gorillas from seven different groups. More than 10 years have gone into this remarkable study, and the reader can only hope that Fossey's work will continue despite difficulties, political and otherwise.

Dian Fossey

Development of the Mountain Gorilla (<u>Gorilla</u> <u>gorilla</u> <u>beringei</u>):
The First Thirty-Six Months

Although fairly detailed studies of the development of behavior in a number of monkey species are now available, there are relatively few quantitative data on the free-living apes. Goodall (1968b) has provided an invaluable descriptive account of mother-infant relations in chimpanzees *(Pan troglodytes schweinfurthi)* with some quantitative material, and Schaller (1963) has described some aspects of development in gorillas *(Gorilla gorilla beringei)*. In neither species, however, are quantitative data available on the extent of individual differences or on many of the factors affecting mother-infant relations. This paper represents only a first step toward answering some of these questions about the mountain gorilla.

Because of the shyness of unhabituated gorilla groups and the difficulties encountered in observations among the dense vegetation of their habitat, an economical study of the mountain gorilla is difficult. Indeed, in a 4-month study conducted by Osborn (1957) no infants were even observed among 4 groups totaling 15 animals.

The development of the gorilla infant is of special interest because of the organization of the unit in which it matures. Gorilla populations are basically divided into small family groups, the composition of which remains essentially constant for prolonged periods. The infant is therefore able to form relatively stable, consistent, and even long-term associations with his peers, siblings, parents, and other group members. It is possible that such long-term cohesion—in particular, the permanency of the attachment of the dominant male to several older, high-ranking females—offers a unique social environment among all great apes.

Although there is obviously an intergradation in all stages of development, I have defined infanthood as that period of time from birth through 36 months of age. The interactions and relationships of the various age-sex classes within the group to the infant, the physical and behavioral stages of development occurring throughout the first 36 months, and some parturient data will be discussed.

SUBJECTS, STUDY AREAS, AND METHODS

A total of 32 infant gorillas, distributed among seven gorilla groups composed of 120 animals, were observed during a 7-year period of an ongoing field study. Of the 32 infants, 24 were born during the study period (19 within the current study area), 8 died, and 22 have, so far, reached 37(+) months of age. Data collected during the early days of the field work from individuals assumed to be older infants, whose ages were not known precisely, have been omitted. In addition, data collected from 2 infants born nearing the completion of this paper have been omitted, with the exception of being recorded as to their birth month (Figure 1), which fell within the 7-year study period.

The first 6½ months of the study took place within the Parc des Virungas of Zaire (formerly known as the Parc Albert of the Congo) from January to July 1967. At that time, eight infants were found within three groups totaling 50 animals. In a former paper (Fossey 1972), ten infants from this area were recorded, but two of these have since been reclassified as juveniles. When Schaller conducted his notable research work within the same area, he found 19 infants among 69 animals belonging to three groups (Nos. 6, 7, and 8). These groups closely matched the descriptions and range areas of my three groups, which I numbered 1, 2, and 3. In 1972, census work carried out in the same region found only four infants (an additional two were newly deceased) among a total of 36 animals considered to have been the same three groups previously studied.

The second part of the study was undertaken in September 1967 within the Parc des Volcans of Rwanda where 24 infants among four groups were observed for 6½ years. The membership within the four groups varied from 69 to 42 individuals throughout the study period.

There were no major differences in habitat or basic group structure between the two study areas. The second region, however, had been subjected to far more harassment by encroachers, which substantially reduced the possibilities for observing details of mother-infant behavior until habituation had been thoroughly achieved. During the last 3 years, two of the study groups became totally accustomed to the observer, permitting observations at a range of a few feet or less. Other groups also became much more tolerant of an observer's presence.

Field methods consisted of daily attempts to establish contact (visual observation) with a group or groups whose approximate whereabouts were usually known from the previous day's contact. Methods of field procedure have been described previously (Fossey 1972). Provisioning techniques were not used. Field observations were concerned with all types of behavior, with no emphasis on any one category.

In addition to direct observations, regular counts and examinations of night-nests contributed a significant amount of information concerning the time and date of parturition.

The preliminary analysis of field data involved categorizing field notes under specific subject headings, one of which was concerned with infants. The known and/or presumed birth month of each infant was first abstracted. The infant data were then combined into 2-month units from months 0-1 (referred to as 1) through months 36-37 (referred to as month 37). This was necessary since the time that each infant was observed per month varied widely (range 0-10.08 hours; mean = 1.47). Of a total of 3568 hours spent in contact with the gorillas, 413 hours contributed direct observations of infants.

PHYSICAL AND BEHAVIORAL DEVELOPMENT

The stages of development throughout the first 36 months of the gorilla infant's life have been defined by two separate criteria: (a) physical characteristics; and (b) behavioral patterns. As previously stated, all stages obviously intergrade and, of equal importance, there were individual variations, a factor apparent even in carefully documented zoo records (Gijzen & Tijskens 1971). The physical characteristics described for each stage were those most outstanding during any one period. Since there were considerable changes during the first 6 months, that particular age period has been divided bimonthly. Subsequent physical

141

changes were of a longer transitional period, approximately a year. All body weights were estimated based on the actual weights of two newly deceased infants, aged 9 and 11 months respectively.

The period of infancy has not been subdivided into stages 1, 2, and 3 as this would not aid cross-species comparisons currently available (Carpenter 1965; Gartlan & Brain 1968; Hall & DeVore 1965; Jay 1963; Reynolds & Reynolds 1965; Goodall 1968b). Definitions of physical criteria should prove beneficial to future gorilla field workers but of small consequence to those working with other species of non-human primates. Ideally, it would seem that both physical and behavioral stages of development should be stated separately in accordance with the absolute ages in which they were observed. In this way comprehensive, interspecific comparisons could be made. A concrete example of successive stages of behavioral development of both field and laboratory animals has been summarized by Hinde (1971).

Physical Development

Newborn: The body skin is pinkish-gray in tone, with pink concentration on the skin of the face, ears, palms, soles, and tongue. Body hair is medium brown to black and is sparsely distributed except for the dorsal surfaces. Head hair is jet-black and appears slick. The white tail tuft is undefined. The face is wizened with a pronounced protrusion of the nasal region giving a pig-snout appearance. The ears are prominent and the eyes are either squinted or closed. The limbs are thin and spidery with digits usually tightly flexed. A portion of the umbilical cord, up to 6 inches, may remain intact throughout the first 24 hours. The body weight is approximately 3½ pounds.

Schultz (1969) has estimated the birth weight of gorillas as 2.6% of the nongravid maternal weight. Groves (1970) gave an average birth weight as between 3½-5¼ pounds (1.6-2.4 kilograms), based primarily upon captive records.

1-2 Months: The body skin is pinkish-gray, with pink to pinkish-gray skin on face, ears, palms, and soles; the tongue is pink. The body hair is black, short, and shiny, very sparsely distributed on the ventral surface and insides of legs and arms. The head hair is black, short, and wavy. There is a faint definition of a white tail tuft. The face retains a wizened appearance with the nose protrusion gradually receding; ears are prominent and the eyes small. The limbs and digits remain thin and spidery. The body weight is approximately 5 pounds.

Schaller (1963) stated that the skin was dark brown to black between the ages of 1½-2 months. He also noted pink spots on the soles of the feet of several infants, a characteristic seen in several infants of this study and felt to be an individual variation. He said that there was a noticeable increase in hair by the age of 2 months; however, I would be inclined to attribute the hair increase to a later stage.

2-4 Months: The body skin is gray with pinkish-gray skin on face, ears, palms, and soles; the tongue is pink. Body hair is beginning to fluff and lighten in color with thicker distribution on the dorsal surface, but it is still sparsely distributed on the ventral surface and insides of legs and arms; the head hair is long, wavy, and reddish brown. The white tail tuft is defined. The ears are now partially obscured by hair growth. Although gradually flattening, the nasal section is wide and remains the predominant feature of the face with very faint beginnings of indentations above the nostrils. The eyes are small and rounded. The limbs and digits still remain thin and spidery. The body weight is approximately 7 pounds.

Schaller also mentioned the brown crown hair at the age of 2½ months, but stated that the abdomen and chest were covered with hair by 3 months, a condition I did not find noticeable until the following stage.

4-6 Months: The body skin is gray; the facial skin and ears are gray-black; the palms and soles are pinkish-gray; the tongue is mottled pink. The body-hair is long and fluffy and reddish-brown to black in color, sparsely distributed on the ventral surface and insides of legs and arms. The head hair is nearing maximum length with a cap of reddish colored hair growing out in tufts. The white tail tuft is clearly defined in color but not yet in size. The ears are obscured by hair growth; the nasal region is flattened and rounded. The eyes are larger, rounded, and becoming the predominant feature of the face. The limbs and digits are somewhat spindly in appearance. The body weight is approximately 11 pounds.

Schaller also noted a "wild fringe of brown hair" between the ages of 17 and 30 weeks.

6-12 Months: The body and facial skin is gray to gray-black; the palms and soles are grayish; the tongue is mottled pink. The body-hair is fluffy and reddish brown to black in color but still sparsely distributed on the ventral surface and insides of legs and arms; reddish-brown head hair is at maximum length and growing in straggly tufts. The white tail tuft is clearly defined in color and nearing maximum size. The head appears exceptionally large in proportion to the rest of the body. The large rounded eyes are the predominant feature of the face with the nasal wings thickening and beginning to take a definite shape. The limbs show evidence of roundness and musculature. The body weight is approximately 18 pounds. By the eighth month, the visible teeth (incisors), gums, and back surface of tongue show traces of tartar deposits accumulated from the foliage eaten.

Groves (1970) has listed the average weight at one year as being 15 pounds and Schaller (1963) lists it as between 15-20 pounds, all referring to zoo infants. Schaller has stated that infants and some juveniles have white teeth, but although the teeth certainly lack the extent of black tartar deposit of the adults, brown stains are beginning to form during this early stage of development.

12-24 Months: The body and facial skin appears gray-black to black; the palms and soles are gray-black; the tongue is mottled pink. Reddish-brown to black body hair is becoming thicker, shorter, and straighter and is still sparsely distributed on the ventral surface and insides of legs and arms. Reddish-brown to black head hair is still fluffy, but becoming thicker and shorter. The white tail tuft is clearly defined in color and of maximum size (diameter up to 4 in.). The head-body proportions are beginning to approach those of an adult. The eyes have obtained their maximum roundness and are the predominant feature of the face with eyelashes long and abundant. The nasal region is well defined. The musculature development of the limbs is becoming much more pronounced. The body weight is approximately 30 pounds.

From zoo records Groves has estimated the body weight at 2 years as 30 pounds; Schaller estimates it as 35 pounds. In addition to their large, rounded eyes, their most conspicuous feature is the fluffy, white tail tuft, which attracts immediate attention to their specific age group. Goodall (1968b) notes the tail or rump tuft as being conspicuous in the chimpanzee between the age of 6 months-2 years.

24-36 Months: The body skin is gray-black to black; the facial skin is black; the ears, soles, and palms are gray-black; the tongue is mottled flesh-gray. The reddish-brown to black body hair is much thicker and becoming noticeable on ventral surface and insides of legs and arms. The reddish-brown to black head hair is thick and losing its fluffy appearance. The white tail tuft is diminishing in brightness of color and size. Rounded eyes remain the predominant facial feature. The nasal region becomes less outstanding in size, with strong linear wrinkles forming along the nasal bridge. The musculature is well developed, and the protrusion of the abdomen is noticeably apparent for the first time. The body weight is approximately 45 pounds.

Based upon zoo records, Groves lists the average 3-year-old weight as between 55-60 pounds, and Schaller lists it as 60 pounds. Schaller noted that the tail tuft was present on the rump of infants at birth and persisted at least until the age of 4 years. By that age, however, this study has noted the tuft appearing only as a faint, nearly imperceptible lightening of the hair around the rump region, although there were slight individual variations.

Behavioral Development

The stages of behavioral development indicate the maintenance activities and capabilities of the infant; the general nature of the interactions with its mother, and some aspects of its social awareness of others within the group. All these subjects are discussed in more detail in the next sections. Again, it should be stated that there were intergradations in all stages of behavioral development as well as individual variations. These were more pronounced in the behavior category than in the physical stages.

Newborn (first 24 hours): The infant is capable of clinging to the ventral surface of the mother, unsupported, for at least 3 minutes, and is able to hold its head upright when the body is vertically supported. The grasping reflex is strong. The limbs sometimes exhibit a spastic type of involuntary thrusting movement especially when searching for the nipple. These movements help to bring the infant's head into line with the nipples; rooting and nuzzling responses are evident. Most of the time it appears asleep. The infant is always carried in a ventral position when the mother travels; when the mother is seated, the infant is always cradled against her chest or held in her lap. Vocalizations consist of weak, puppy-type whines.

Schaller (1963) felt that newborn infants appeared to lack the strength to clasp the mother's hair securely so that the mothers had to support them continuously with at least one arm. Lang (1960a, cited by Schaller) noted that the grasping reflex in one newborn infant was weaker than that of a chimpanzee or an orangutan. In a zoo-born female, Hess (1973) noted considerable interest in genital inspection of a newly born male infant, which has not been observed in this study.

1-2 Months: The infant is capable of clinging unsupported to the ventral surface of its mother for longer periods. Under arduous circumstances, i.e., when the mother is tree or rock climbing, it tends to slip into a low abdominal or inner thigh position. The infant's limbs may exhibit random, flailing motions when being groomed or searching for the nipple, still accompanied by rooting and nuzzling motions of the head. The combination of thrusting and pushing motions of the legs and body appear more coordinated and stronger than the pulling actions of the arms when seeking the nipple or trying to reposition the body. Finger and precise arm movements remain uncoordinated. Eye-focusing ability appears limited only to stationary objects within 10 feet of the mother. The infant flinches in response to loud noises, usually those made by other animals within the group. Solid food intake is limited to mouthing or prolonged chewing of vegetation debris deposited on mother's body. Approximately 75% of the time, it appears asleep. The infant always travels in a ventral position but, when in its mother's lap, the infant is only supported about 50% of the time, especially when she is feeding. Vocalizations consist of whines and loud, high-pitched wails.

Observations of newly-born zoo gorillas made by Hess (1973) stated that the infant, during its first few weeks, explored its own body; however, this type of exploratory behavior was not observed among the wild animals at this age. Other observations of captive animals cited by Groves noted that they could roll from their backs to their stomachs and begin to creep, pushing with the legs, by the fourth week. This study has not yet observed such behavior at the same age, although possibly it does occur during the night-nesting periods. Groves said that by the age of 8 weeks the head could not be held up for any length of time. Schaller observed infants at the age of 1 month clinging to the ventral surface of their mothers, without assistance, by grasping with both hands and feet in the vicinity of the armpits and abdomen respectively for about 10 seconds. He also saw an infant of 1½ months on the back of a female, but seemingly insecure in this dorsal position. This same infant disappeared shortly thereafter. He noted that infants between the age of 6-7 weeks began to raise their heads and look around while in a prone position.

2-4 Months: The infant clings unsupported in a ventral position for prolonged periods of time with increased strength of grasp, which possibly bothers the mother. It may kick, whack, or push at its mother's body when being groomed for prolonged periods. It is able to obtain the nipple without a preliminary period of seeking and will strongly pursue efforts to obtain the nipple if thwarted by its mother. The infant begins exploratory play activities with surrounding vegetation and on the mother's body. Arm movements, directed toward nearby foliage, are gross, jerky extensions with the fingers widely spread before the object is actually contacted, and then it is usually only grasped. Play on the mother's body consists mainly of unbalanced crawling, sliding, patting, and hair pulling. The mother may reciprocate play with gentle, repetitive shoves and may also discipline by mock-biting or pushing back at the infant. The infant can focus at least as far as 40 feet, and is able to follow moving objects. It flinches visibly at loud noises. Solid food intake consists primarily of bits of vegetation and wood remnants picked off the mother's body. It rides dorsally for very brief periods of time in a sideways, sprawled position if the mother pushes it onto her back. When the mother is sitting, the infant is either on her lap, by her side on the ground, or clinging to her legs. Distress and play facial expressions are clearly defined. Vocalizations consist of whines, wails, screeches, and panting play chuckles.

Groves, citing observations of captive gorillas, states that the infant can crawl steadily by the ninth week and that, by the fifteenth week, it can push itself into a sitting position when lying on its stomach; it still cannot sit up unaided for any length of time. Among the free-living gorillas, Schaller observed an infant between the age of 10-12 weeks pulling at a vine and eating it and also biting into an herb plant. He also observed one infant crawling slowly behind its mother at this age, a behavior which has not been seen in this study. He noted infants, by the age of 16 weeks, riding frequently on their mothers' backs, although they were transferred to a ventral position when the mother sat or moved rapidly.

4-6 Months: The infant now travels either ventrally or dorsally on the mother, with the ventral position used about 60% of the time. When in the dorsal position the infant rides high on the mother's neck or shoulders, tightly clutching her hair with its fingers; and with its body usually stretched out prone along the top of her back. The infant objects less to grooming, but if a session is prolonged, it will strenuously break away from the mother's grasp. The infant continues to seek vigorously to suckle, although the bouts are of shorter duration. The range of play behavior is expanded from exploration to manipulation of vegetation. The infant's solo play also involves patting, clapping, and the whacking of its own body. Play continues on the mother's body but is more strenuous and includes mock-wrestling with her extremities. Social interactions with other young animals consist of mild whacking or patting; most of this is done either while on the mother's body or near her side. When body contact with the mother is broken for the first time, the infant always remains within arm's reach; and restrictive measures taken by the mother are more frequently observed. Focusing ability of the infant extends beyond 40 feet and there is a noticeable increase in the coordination of head movement when following another animal's movements. Plucking at leaves and vines and gnawing on stalks are common. The first quadrupedal walking is attempted within arm's reach of the mother but is clumsy and very uncoordinated. Vocalizations consist of whines, wails, panting chuckles, screeches, and screams.

Groves has listed the following behavioral capabilities at the ages of 17, 19, and 20 weeks respectively: (a) sits up steadily; (b) stands and walks on knuckles; (c) stands erect bipedally, may have temper tantrums and begins to climb. Schaller observed among the free-living animals that by 4 and 5 months infants sought social contact with others and crawled away from their mothers; but they were not usually permitted to venture more than 10 feet from her before being pulled back. He stated that they exhibited good control of their limbs and once observed an infant of 16-18 weeks stand bipedally (shakily) and beat

FIGURE 1
A 6-month-old infant, Cleo, enjoying a play session using her mother's body hair as a play object. Photograph by Dian Fossey. © National Geographic Society.

its chest. The latter has not been observed in this study at such an early age.

6-12 Months: The infant still continues to travel, both ventrally and dorsally, using the dorsal position about 80% of the time and with greater security. Reluctance to follow and travel independently now result in a third form of travel called *rump-clinging,* in which the infant follows the mother with one or both hands clasping the hair on her rump. Suckling efforts continue but are more frequently checked forcibly by the mother. Solo play now includes tree climbing. Although locomotor patterns advance more rapidly than in any other period, agility in trees or running attempts is still poor. Social play with other infants and juveniles increases greatly but is still not as frequent as solo play. There is still no preparation of food items, which are either plucked or gnawed on. Vocalizations consist of whines, wails, panting chuckles, screeches, hoot cries, and shrieks.

At approximately 6 months of age, Hess (1973) observed in captive lowland gorillas that a mother sometimes used genital touching as a training method by which the infant was urged to crawl, climb, walk, and run. He stated that it was also used to get rid of the infant for short periods of time. Similar behavior has not been observed among free-living gorilla mothers. Groves has stated that the infant, at 34 weeks, is able to walk well in an upright position. Schaller noted that some infants of 8 months were 20 feet or more away from their mothers, which was never seen in this study. He also observed mild rebuffs from the mother; once when an infant of 8 months sought the nipple, and another time when an infant of about a year attempted rump-clinging.

12-24 Months: The infant nearly always travels in a dorsal position when long distances are covered, but otherwise usually follows its mother independently or clings to her rump hair. It may still travel in a ventral position in a stressful situation. Suckling efforts continue, but the infant is becoming more responsive to the restrictive measures taken by the mother when she wishes to terminate a bout. Play interactions with other infants and juveniles are much more vigorous and begin to

FIGURE 2
A 7-month-old infant, Cleo, suckling while mother, Flossie, casually grooms its arm. Photograph by Dian Fossey. © National Geographic Society.

FIGURE 3
The typical high dorsal travel position of the young gorilla infant at 7 months.
Photograph by Dian Fossey. © National Geographic Society.

outnumber solo play sessions. Swinging, twirling, and semi-brachiating activities are practiced in trees although agility and dexterity are still not fully developed. Obtaining food items now involves awkward attempts at preparation such as stripping leaves from central stalks or wadding of vines. Social responses toward adults now include grooming the mother (first seen at 12 months) and attraction toward the silverback. Response to the observer, which up to now has mainly consisted of staring, now includes chest-patting, foliage-whacking, clumsy attempts at strut-walking, and a compressed-lip facial expression. Vocalizations consist of whines, wails, panting chuckles, screeches, hoot cries, shrieks, temper tantrum screams, howls, and pig-grunts.

Five out of 11 observations made by Schaller of suckling behavior occurred with infants between the ages of 1 and 1½. Because it was easier to observe older infants suckling, such a low count suggested to him that partial weaning had taken place by that age. However, I have observed suckling frequently during the second year, and it seems likely that intensive weaning occurs after that age.

24-36 Months: The infant may continue to ride dorsally during a prolonged travel route but is more often seen following independently. It may travel briefly in a ventral position during stressful situations. Suckling efforts continue, but the mother may rebuff the infant severely if the attempts continue for too long. The infant seeks play opportunities much more frequently, both with a wider range of individuals as well as in a larger group. Tree play becomes more inventive and daring, although the infant becomes cautious when juveniles approach him in a tree. Locomotor abilities are not fully developed, but a type of canter gait permits more rapid ground travel. Plucking of food items decreases as nearly all leaves and vines are eaten following a brief period of preparation (wadding, stripping, peeling). Selectivity as well as a small degree of competition for favored food items is seen for the first time. Awareness of other animals within the group is indicated, not only by grooming of the mother, but also by grooming of siblings, peers, and silverbacks; much more time is sought near the silverback; curiosity is expressed in copulations of adults. Vocalizations consist of whines, wails, panting chuckles, screeches, hoot cries, shrieks, temper tantrum screams, howls, pig-grunts, and basic disyllabic variants of the belch vocalization. Throughout all these periods, the infant continues to night-nest with its mother and may continue to do so until the age of 5 years.

Hess (1973) felt that none of the instances of genital touching he observed among captive lowland infants could be called masturbation during the first 2 years of life. One female infant, aged 26 months, was seen for the first time assuming a copulatory position against the back of her mother. Schaller's wife noted a 2-year-old captive male assuming a copulatory position behind a 7-month-old female. Similar behavior at this age was not observed in this study. Schaller noted one instance of an infant, ages 2½ years, grooming another. He noted only two observations of younger infants grooming mothers. He also observed that infants over 2½ years sometimes ran behind their mothers even when the travel was rapid.

PREGNANCY

Napier and Napier (1967) have included the fetal phase as one of the life periods of primates and a process that is continuous with postnatal growth. The period is ended "when the size of the head is consonant with a safe delivery."

Eleven records of gestation periods of captive gorillas vary between 238 and 295 days (mean = 263; median = 258). Results from this study have observed copulations from a minimum period of 212 days prior to parturition to a maximum period of 284 days. Observations were not consistent enough during the assumed estrous cycles, however, to draw any positive conclusions about gestation periods except that they appeared grouped around the 260th day.

Physical Description of Mother During Pregnancy

In only two of nine cases where the mother was observed before parturition was pregnancy suspected by the observer: the abdomen is ordinarily vastly extended due to the tremendous amount of coarse vegetation matter consumed, so the additional swelling of pregnancy is inconspicuous. In neither of the two cases were the breasts noticeably swollen. In two other instances, the females were considered pregnant because of abdominal distension, but no births occurred.

Among captive females there may be edema of the ankles (Thomas, cited by Schaller 1963), but this was not seen among the free-living gorillas of this study. Also, with captive females, distension of the abdomen is usually noticed in addition to swelling of the breasts. Lang (cited by Schaller 1963) recorded two instances where the female squeezed milk from her breast 3 months before giving birth.

Behavioral Description of Mother Before Parturition

In four of six cases (where the mother was *consistently* observed before parturition) signs of unusual behavior were noticed. Only one of these four females was a primiparous mother, and her behavior differed in an interesting manner from the other three females, each known to have been multiparous animals. At least 10 days before parturition she formed a close relationship with an older female who had the youngest infant of the group (7 months of age). Her proximity was well tolerated by the older female for prolonged periods, especially during the day-nesting sessions. On these occasions the younger female was allowed to groom the infant either while it was being held in the arms of its mother or when it was nestled between the bodies of both. When not grooming or gently cuddling the infant, the younger female either groomed the infant's mother, occasionally herself, or stared intently at all movements of the infant. Throughout the 10-day period preceding parturition, the young female traveled consistently with the older animal, often walking beside her with one arm extended across the other's shoulders.

The three multiparous females were not seen to behave in this way. On the contrary, they seemed to prefer to maintain a distance from other group members both when traveling or during the day-nesting periods. One of the females, accompanied by her juvenile (41 months of age), maintained distances of over half-a-kilometer from her group during periods of travel and nesting. The distant, and usually rapid, travel routes of the three multiparous females appeared to influence both the speed and the direction of movement of their groups, who traveled in a parallel line with them. The three females also seemed nervous and excitable and fed little. A recent observation was made by Stewart (personal communication) of a multiparous female of this study, approximately a week before her parturition the female assumed a copulatory position once with another female, in addition to copulating with the group's silverback three times.

Schaller did not notice any irritability among the free-living females during his study. Captive gorillas have exhibited some variances in temperament before parturition: 3 weeks before giving birth, a female from the San Francisco Zoo spent more time in the artificial trees of her enclosure as though wishing to avoid social encounters with her four cagemates (Bingham and Hahn 1973); a female observed by Crandall (1964) became quieter and more contented; Reed and Gallagher (1962) reported the female as becoming more irritable and fond of sweets; Thomas (1958, cited by Schaller) also observed an increase in irritability.

BIRTHS

Degree of Seasonality

Figure 4 presents the total number of births per month in the two study areas and includes three records from two adjacent mountains. The figure shows 24 births where the month was known precisely and 13 where it was estimated from later observations. The estimates refer to infants under 18 months of age, when the study began, and were based on the physical and behavioral criteria given in a previous section. The figure indicates that births occurred throughout the year with a peak between July and August and a trough between September and December.

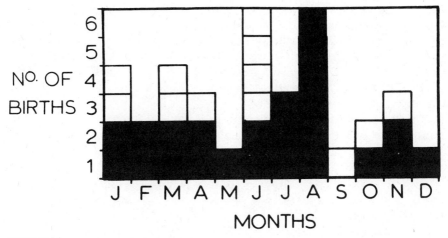

FIGURE 4
Frequency distribution of months in which births occurred over a 7-year period (1967-1974). Known births are represented by black squares; estimated births are represented by white squares. (No births were estimated for infants over about 17 months.)

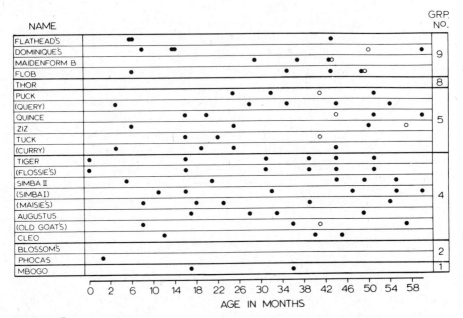

FIGURE 5
Relative ages of other infants and juveniles within each group when an infant was born. Black circles show ages of all individuals under 5 years in each group. White circles indicate known siblings. Names of deceased infants are bracketed.

Impressions gained from the field indicated that food resources did not have a direct relation to a possible birth peak, as food was abundant throughout the year. Weather was a more likely correlate. June through August constituted the driest season, while the wettest period usually occurred from February through May. During the rainy months, many of the gorillas, within different groups, were observed sneezing, coughing, and wheezing, which was not the case in the dry months. It is possible that, during the wetter seasons, a higher mortality rate occurs among newly born infants.

Figure 5 shows the number of peers and siblings present in each group when an infant was born. Of the 24 infants born within the study groups, only ten had one sibling under 5 years of age still within the group. Even if an additional 22 months (from 60 to 82) was included in Figure 5, only one of the 24 infants (Tuck) would have been definitely known to have had two siblings when born, although older siblings were strongly suspected in two other cases. We have concluded, based on marked facial resemblances, night-nesting associations, and prolonged daily periods of proximity with their mother, Effie, and younger siblings, Puck and Tuck, that the young female Piper (born approximately 41 months before Puck) was the older sister of Puck and Tuck (born 41 months after Puck).

The evidence presented in Figure 5 would seem to indicate that there has been no noticeable aggregation of births within any 4-month period in the past 5 years within the main study groups.

The shortest period of time between the death of an infant and parturition has been 10 months: Female Flossie lost a 6-month-old infant in May 1968 and gave birth in March 1969. The shortest period of time between viable births has been 3 years: Female Flossie gave birth to Cleo in August 1971 and to Titus in August 1974. The longest known period between viable births was more than 6 years and 11 months: Female Old Goat gave birth to Tiger in November 1967, had a possible nonviable birth in April 1971, and was unimpregnated when deceased in October 1974. The mean period of time between births of infants surviving beyond their third year was 45 months; the median, 41 months.

An estimation of 54 birth dates was made by Schaller from the results of his study on an adjacent mountain, Mt. Mikeno. Thirteen births occurred during his study period and 14 others were considered accurately dated to within the span of a month. The remaining half were thought to vary in reliability up to 3 or 4 months. His table showed a slight peak in December, a stronger peak between July and September, and that births occurred throughout the year. From this evidence Schaller felt there was no preferential breeding season.

In another field study in the eastern section of the Virunga Mountains, Kawai and Mizurhara (1959b) found six infants of similar ages among four groups totaling 39 animals. On this evidence they questioned whether a breeding season might possibly exist within that area. Within the Kayonza forest of Uganda, however, they found no similarities in the dung sizes of infants.

Infant Mortality

Of the 24 known animals born during the study period, and within the study area, two died in their second year; three in their first year, and three within the first 24 hours (two of these possibly nonviable births). The above figures, in addition to the evidence presented in Figure 5, suggest a heavy mortality rate in the first 2 years of life. Also, four infants were known to have died in regions outside the study area during the same time period (three of these were included in Figure 4). Three of the latter died within their first 3 months and the fourth was stillborn.

Within the study areas, and excluding the two possible nonviable births, the causes of death of three infants are known to have resulted from interactions of their groups with either lone silverbacks or a subgroup: (a) Infant (Phocas), within the first day of life, suffered fatal injuries when his mother was attacked by a lone silverback who had entered the infant's group with violent displays after following them for

155

2 days. (b) Infant (Curry), in his ninth month, received ten bite wounds, one of which was instantly fatal, during the interaction of his group with a lone silverback. The interaction was not observed but the subsequent trail evidence indicated it had been violent in nature. (c) Infant (Thor), in her 11th month, received two bite wounds, either of which could have been instantly fatal, following the interaction of her subgroup with the silverback and other members of a larger group. Visibility was hampered by foliage during the 1¼-minute violent outbreak of screams and chasing in which Thor was killed (Harcourt, personal communication).

The bodies of two deceased infants were known to have been carried or dragged approximately 300 feet before being left by their mothers on the same day they were killed; bodies of two others were transported for 3 days before their mothers left them; one was deposited in a day nest.

Schaller estimated a mortality rate of 23% among gorilla infants in their first year and between 40 to 50% in their first 6 years. The results of this study found an infant mortality rate of 27% in the first year and 25% in the first 6 years within the study area. Precision in the evaluation of mortality rates is complicated by the fact that the nesting sites of some parturients were undoubtedly missed, so that nonviable births, and even infants living a short time may have been overlooked.

Neither the writer nor Schaller found evidence of leopards preying on gorillas within the Kabara study area. In 1963 Alan Root (personal communication) did retrieve an infant gorilla toe from the dung of a leopard near Kabara, but that alone is not conclusive evidence that the leopard was responsible for the infant's death.

Of interest is that none of the three deceased infants in Schaller's study were mentioned as having been observed dead following group interactions, all of which he stated were peaceful. He also noted only two instances of subgrouping which occurred in one group (Group IV), and an infant was born on both occasions.

Schaller also noted a female carrying her deceased infant for about 4 days before leaving it.

Time of Delivery

Actual parturitions have not been observed. Six observations of either the nesting sites or newly born infants indicated that all births occurred at night. For the free-living gorilla female, the advantages of nocturnal births may be similar to those suggested by Bowden (Bowden, cited by Jolly 1972); that is, being a diurnal animal, the female does not have to travel at night and the other members of her group are also usually inactive at this time. On one occasion a lone silverback night-nested only some 300 feet from the nesting site of a parturient female's group after taking a female from her group. Running displays and chest-beats continued throughout the night between the lone silverback and the silverback of the female's group, exposing the newly born infant to possible physical trauma from the displays of the group's silverback. Indeed, this same infant was killed the following day by the lone silverback. Infrequently a group may travel at night (usually rapidly and for distances up to approximately 3 kilometers) if disturbed by an interacting lone silverback or another group. Although births have never been known to have occurred on nights when this type of travel was recorded, field impressions indicated that a parturient female would be left behind by her group on such occasions. (A dying female was left behind by her group under similar circumstances.)

Schaller encountered two females several hours after they had given birth and mentioned one as having delivered during the night. Only three of 11 records for captive females were night births; seven occurred in the morning, and one around midday.

Birth Sites

Four birth sites were examined approximately 2-30 hours following birth, and two newly born infants were observed as the groups departed from their night-nest areas. In two cases, the delivery process occurred within one night-nest and the majority of blood was found deposited within the nest, thoroughly saturating the leaves. The other two cases were believed to have been nonviable births. In one case, both day- and night-nests were saturated with blood over a period of 3 days; in the second, the female, a primiparous mother, built a succession of four nests during one night. Each was separated from the previous one by only a few feet and the placenta was deposited in the fourth and last nest. (This was the only placenta ever found in a nest.) In both cases of multiple nests, an abnormal amount of blood was found within the nests and on the ground between them.

Out of seven births in one group over a 6-year period, three occurred in one location about 50 yards in diameter. The site, a nearly flat hill top with steep sides, was situated far back in the middle of a large and deep ravine. The area appeared to offer maximum visibility of surrounding terrain and was seldom frequented by other groups. There were no obvious distinctions among the other three known sites.

Captive gorillas usually do not eat the placenta (Brandt & Mitchell 1972), although they may nibble or suck on it.

Responses of Other Group Members to the Newly Born

In seven instances when it was possible to observe mothers and infants within the first week following birth, the mothers usually occupied a peripheral position to the group during day-nesting periods, feeding, or traveling, so that the chances for interactions with other individuals were lessened. In three of these cases, the silverback of the group showed a high degree of nervousness, which was directed toward the observer, and the contacts had to be terminated. Before the groups were thoroughly habituated, it is possible that the observer's presence was partially responsible for the distance maintained by the mother. On those occasions when it was possible to watch a group without their knowledge, it was seen that the group members showing the most interest in the newly born infant were juveniles, young female adults, and the silverback, in that order. With the exception of siblings, other animals were seldom observed approaching closer than 10 feet to the mother, and other females appeared to ignore the mother. Curiosity was expressed by direct staring, especially when any sound was heard from the infant, and by attempts to gain good views of the infant. New mothers most frequently traveled and night-nested nearest to that female with the next youngest infant in the group; however, an exception to this was observed with two females who habitually occupied a peripheral position to their group.

One interesting observation of a female with an infant less than 48 hours old occurred when her group was interacting with an all-male group of five animals. She climbed about 30 feet into a fruiting tree, which apparently subjected her ventrally clinging infant to distress, as whines were heard for several minutes during and following her ascent. Upon hearing the whines, all the males of the second group, who had been spread out feeding about 50 feet away, immediately gathered together to sit and stare intently at the female for 8 minutes.

Schaller cited one observation of a female with a newly born infant who approached and then rested against the back of the group's silverback; he later fondled her or the infant intently.

MOTHER-INFANT INTERACTIONS

Quantitative Changes in Proximity to Mother with Age of Infant

Among the free-living gorillas, it was noticed that the mother was mainly responsible for the increase of an infant's independence during the first 3 years. The increase in independence depended more upon environmental factors than individual personalities. To take only one example, an infant with a close sibling or peers was far more inclined to leave its mother at an earlier age, and for a greater distance, than an infant without similar social incentives. This section considers only quantitative data concerning the three distance categories pertaining to the infants' spatial relationships to their mothers:

1. Infant in body contact with mother.
2. Infant in reach of mother's arm (3 feet).
3. Infant beyond arm's reach, but within 15 feet of mother.

The time in body contact was expressed as a percentage of the total observation time for each infant. The time within arm's reach and the time less than 15 feet away were expressed as percentages of the time the infant was observed out of body contact with its mother. The data are divided to answer two questions: What percentage of observed time was the infant in contact with its mother? When the infant was not in contact with its mother, what percentage of the time was spent in distance categories 2 or 3? The median values for all infants in their respective spatial categories are shown in Figures 6, 8, and 10.

Body Contact: With only two exceptions from Group 5 (Tuck and Curry), all infants from all groups maintained body contact continuously with their mothers until 5 months (4-6 months old). The median values for all infants did not fall below 50% of time observed until 13 months, and by the age of 36 months, all infants spent more than 70% of their time off their mothers with the exception of the Kabara groups.

Tuck, of Group 5, had two siblings aged 41 and 82 months older than he. Both siblings spent a considerable amount of time near their mother, an extremely tolerant female, so that by the age of 5 months, Tuck was observed in multiple social interactions involving grooming, play, and even transport with his two siblings.

Curry, also of Group 5, was 3 months younger than Tuck but had no siblings. His primiparous mother had been a transfer from another group; perhaps for this reason she usually maintained a wider spatial

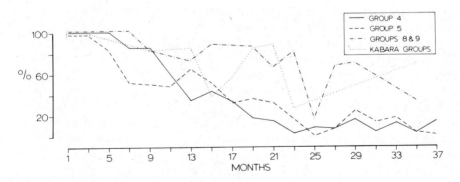

FIGURE 6
All Groups: Median percentages of time in body contact with mother. Each group's line represents the median values for all infants within that group. Interruptions of successive monthly observations are not represented.

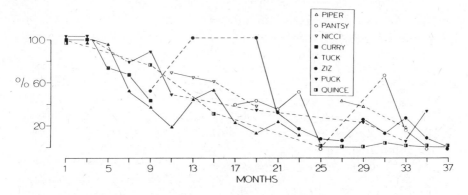

FIGURE 7
Group 5: Percentage of time in body contact with mother. Each infant's symbol represents percentage of observation time seen in body contact with mother for each 2-month period observed. Interruptions of successive observations are represented by a dashed line. White symbols represent females; black symbols represent males. Half-filled symbols represent unknown sexes.

distance between herself and other group members. Observations suggested that Curry went farther away from his mother at a younger age in order to establish social relationships with other infants.

The relative ages at which the Kabara infants broke body contact with their mothers seemed, according to Schaller's (Schaller 1963) descriptions of mother-infant behavior, to have occurred at younger ages than I observed.

Within Arm's Reach: Infants showed more variation with age in this classification than in any other. Environmental factors as well as varying degrees of tolerance among the mothers were considered the factors responsible for the pronounced variations (Figure 8). Up to the age of 17 months, all individuals of all groups (except for Curry, Tuck, and Puck) remained within arm's reach more than 30% of the time that they were off their mothers. Again, the increase in distance from their mothers by Curry and Tuck most likely was attributed to the reasons given previously. Puck, with one older sibling, only spent one period (point 7) almost entirely outside of arm's reach of his mother (90%); subsequent data for Puck, after point 11, were very sparse (Figure 9).

As can be seen in Figure 8, the decrease of time spent within arm's reach of mother by Group 4 infants was consistently more gradual than that of Group 5. The maximum variability between Group 4 individuals for the 16 time periods that produced data was 54% (point 19) with a median range of variability of 36% and a mean of 35%. It is felt that the median values shown by Group 4 present the most typical picture of time spent within arm's reach of mother.

Within 15 Feet: The median values of time spent within 15 feet of the mother showed more consistency with age than in any other positional category (Figure 10). In only two instances was more than 50% of infants' observed time spent beyond 15 feet of the mother before point 25 (Figure 11). Simba I (deceased) and Cleo were both infants of female Flossie; an earlier infant of Flossie's was missing and assumed dead at the age of 6 months. (Observations of Flossie's last born, Titus, aged 5 months, indicate that he also will maintain increased distances from his mother at an early age.) With all her infants, Flossie exhibited a relatively high degree of "disinterest" which was indicated by infrequency of grooming, inattention to their whereabouts or activities, in addition to unusually early restrictive measures (mock-biting and rejection by hitting or pushing) used either to check or terminate suckling and playing efforts of the infant.

Individual data from Group 5 (Figure 12) indicates that Quince regularly occupied more time 15 feet from his mother than did most infants of the group after point 25. The reasons suggested for this were similar to those seen in the case of Curry (Figure 7). The mother of Quince, a low-ranking female, spent most of her travel, feeding, and resting time apart from the bulk of the group. In order for Quince to associate with his peers, it was necessary for him to leave her, usually at distances over 15 feet, during his second year.

Within each of the three median figures (6, 8, and 10) representing the values for all mother-infant distance categories, a large amount of variation was found in the results obtained from Groups 4 and 5 and those obtained from Groups 8, 9, and from Kabara. The principal reason for this was that the data from the infants of Groups 4 and 5 amounted to roughly 10 times the data obtained from Groups 8, 9, and Kabara

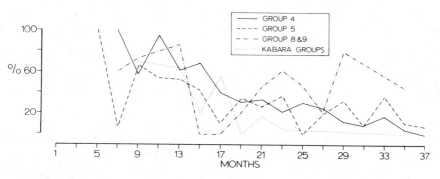

FIGURE 8
All Groups: Median percentages of time spent within arm's reach of mother.
Each group's line represents the median values for all infants within that group.
Interruptions of successive monthly observations are not represented.

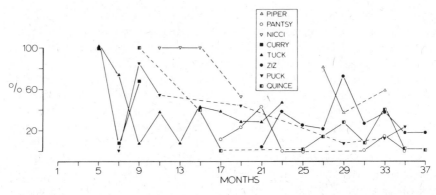

FIGURE 9

Group 5: Percentage of time spent within arm's reach of mother. Each infant's symbol represents time spent within arm's reach of mother as a percentage of time infant was observed to be out of contact with its mother for each 2-month period observed. Interruptions of successive monthly observations are represented by a dashed line. White symbols represent females; black symbols represent males. Half-filled symbols represent unknown sexes.

(374 observational hours as opposed to 39). Although the number of individuals was nearly similar (17 and 15 infants respectively), the first two groups were far more consistently studied. This, in turn, resulted in a higher degree of habituation, which lessened the chances of the infant seeking its mother during the observer's presence (especially true for body contact) and also allowed for far better visibility of all group members. Between individuals of Groups 4 and 5 there was seen a fairly close correlation in spatial relationships among those for whom the greatest amount of data had been obtained. Some outstanding differences between individual infants are explained by circumstances: (a) cases where there were not a vast amount of data (Ziz, Figure 7; Papoose, Figure 11); (b) cases of close sibling availability (Tuck, Figure 7); (c) cases where the mother occupied a peripheral position from the group (Curry, Figure 7; Quince, Figure 12); (d) cases of maternal disinterest (Cleo, Simba I, Figure 11).

Sex Differences

Discounting infants whose sex could not be determined, an equal number of male and female infants were observed over the entirety of the study period (N = 7); however, the bimonthly periods for which female data were recorded totaled only 49 for all seven individuals as compared to 74 for the seven male infants. The female records were also more fragmented because four females were over 17 months of age when the study began (compared to one male). For these reasons quantitative data concerning sex differences were not considered adequate.

The impression gained from field observations suggested that male infants spent considerably more time with their mothers during the first 2 years, after which they more rapidly increased both time and distance spent away than did female infants. One strong factor affecting this impression was the presence or absence of siblings. Depending on the tolerance level of the mother, an older sibling, regardless of sex, was always observed to attempt increased proximity with the mother, and infant siblings (male records only) always severed body contact at an earlier age than infants without siblings. In both of the latter categories, there were more correlations in the spatial distance maintained with the mother following the second year.

Qualitative Changes in Proximity with Age of Infant

Newly Born: During both feeding and day-resting periods, both primiparous and multiparous mothers spent much time looking intently at their infants, often followed by either nuzzling the infant to their face or drawing it closer to their body. Extensive grooming sessions were not observed, even when the umbilical cord was still attached to the infant. It was suggested by observations that grooming most likely occurred during the night-nesting periods, thus was not seen, and that the frequent nuzzling was accompanied by licking.

Short durations of suckling were also observed during the first month and seldom continued over 50 seconds during the first 2 weeks of life. Suckling was usually accompanied by rooting motions of the head; arm flailing movements were more often seen when the mother shifted positions during a suckling bout. A preference for one breast over the other

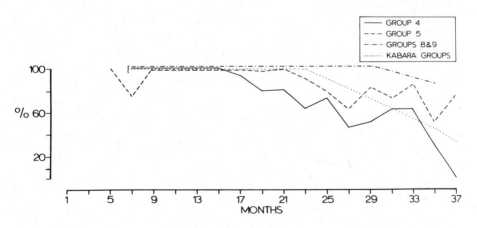

FIGURE 10
All Groups: Median percentage of time spent within 15 feet of mother (includes < arm's reach). Each group's line represents the median values for all infants within that group. Interruptions of successive monthly observations are not represented.

163

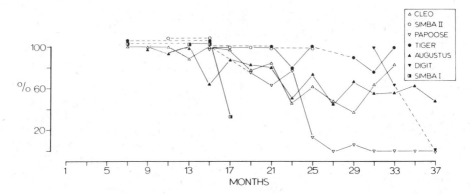

FIGURE 11

Group 4: Percentage of time spent within 15 feet of mother (including < arm's reach). Each infant's symbol represents time spent within 15 feet of mother as a percentage of time infant was observed to be out of contact with its mother for each 2-month period observed. Interruptions of successive observations are represented by a dashed line. White symbols represent females; black symbols represent males. Half-filled symbols represent unknown sexes.

was observed among nearly all infants, and the preference bore a strong relation to the carrying arm favored by the mother. No infants were observed using the nipple as a means of support.

When a group was feeding intensively, mothers with infants under 1 month of age fed less than did other individuals, and, also less immediately before the infant's birth. Several observations of mothers with newly born infants indicated that the mothers appeared to want to feed but were handicapped by the need to use one arm to support the infant. At this age the infants were never deposited in the laps of their mothers without support: rather the mothers cradled them with one arm (usually the left) or their inner thighs, and fed on ground foliage with the other hand without the usual degree of preparation techniques or selectivity. Rare exceptions occurred, such as a case previously mentioned when a female climbed a fruiting tree without attempting to support her infant who was less than 48 hours old.

Nearly all infants were carried or supported themselves in a ventral position until the third month. Primiparous mothers, however, did not carry their newly born infants as successfully as did multiparous mothers. Experienced mothers often traveled with somewhat of a mincing gait, which assured the infant of partial support from all four extremities if it was not being supported manually. This gait was especially noticeable when difficult terrain was encountered. Primiparous mothers initially traveled as they had prior to their infant's birth. As a result, the infant sometimes slid toward or even between the thighs of the mother, which hampered her movement until the infant was repositioned. Impressions received from field observations, particularly those concerned with transportation of the infant, suggested that the primiparous mother was fairly quick to learn the more successful techniques, which would enable her to deal more capably with the needs of her newly born infant as well as her own.

As an infant matured, there were far more individual variations in the manner in which it was handled by the mother. The variations appeared to be a reflection of the mother's temperament rather than her experience. Some mothers appeared very aware of how their infant was positioned before they moved; others moved without seeming to heed the state of readiness of their infant; thus some infants were seen simply clinging with their arms around their mothers' necks with the lower portions of their bodies hanging free, or riding backwards or sideways when in a dorsal position. Such aberrations in travel positions were never observed to have any serious consequences.

The safety of infants, as well as young juveniles, was far more seriously affected by the position of the mother in relation to other group members during stressful situations. Field data suggested that the more dominant female readily retreated to a peripheral position during periods of tension such as group interactions or intragroup outbreaks. Some females separated themselves by as much as 300 feet when an interacting silverback came into the group; at such times they might carry an older infant ventrally or a juvenile dorsally. Most infants, over 2 years of age, usually responded appropriately during these situations and maintained close proximity with their mothers until the disturbance was over; infants and young juveniles without mothers also stayed near such females when the group was disturbed. On those occasions a dorsally riding infant or young juvenile would flatten its entire body against the mother's back and wore a distressed facial expression. In-

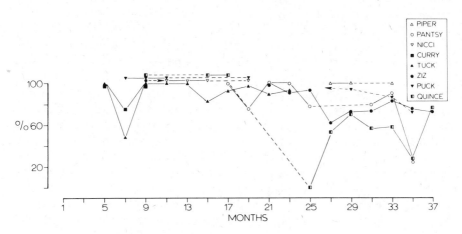

FIGURE 12
Group 5: For explanation see Figure 11.

fants younger than 2 years old seldom gave any indication that they were aware of unusual excitement and had to be collected by their mothers and somewhat forcibly held.

There were marked individual differences in the degree in which various mothers controlled their infants. Both passive and strict mothers might show an increased lack of tolerance toward an infant, if they themselves had been involved in a conflict or altercation with another animal. This was particularly true when the group was traveling relatively slowly but feeding intensively; at such times right-of-way disputes were most frequent. If a low-ranking mother's position was challenged by a more dominant animal (by whacks, mock-bites, or vocalizations), the subordinate mother could readdress her irritability toward her infant; whacking or pushing a dorsal infant from her back, mock-biting the infant if it sought to be close to her, or pig-grunting at it, if it were playing nearby. Many infants gave distress vocalizations such as whimpers or whines; if continued, most mothers then reassured them by an embrace.

The grooming of a young infant was nearly always initiated by the mother and usually terminated by the infant. With younger infants the mother concentrated on genital areas, legs, arms, and back in that order, often turning the infant upside down or positioning it in what appeared to be uncomfortable placements. The infant expressed its objections by wiggling or, at a later age, by kicking and whacking the mother. During the first six months, the mother would then desist from further grooming; during later months she showed less tolerance of the infant's objections and continued grooming after rebuffing the infant. The youngest incidence of self-grooming in addition to grooming of mother was observed at the age of 1 year; peer grooming at 14 months. During the second year the infants more readily accepted grooming from their mothers, and there was a noticeable increase in both the frequency and length of grooming sessions. Because of the body areas involved as well as the concentration of the mother, grooming of the infant seemed to be primarily a functional rather than a social activity. This impression was reinforced when a 3-year-old infant lost her mother and was adopted by the silverback. Although he spent a considerable amount of time grooming the infant, she appeared conspicuously ill-kempt in comparison with other infants of the same age, indicating his lack of maternal experience.

Unlike grooming, nursing was nearly always initiated by the infant and terminated by the mother. It was not always possible to determine precisely how much time was spent suckling when the infant was very young because the mother's body often obscured observations. However, suckling bouts seldom lasted more than a minute and, as the infant aged, involved considerably more switching between nipples with the left nipple usually favored. All suckling sessions observed were terminated by the mothers, and were increasingly observed after the age of 6-12 months. Two juveniles without younger siblings were observed regularly suckling at the ages of 4 and 5 years respectively.

Infants would pull at the ends of vines and leaves that were being eaten by their mothers, but no female was ever seen giving her infant solid food; they did take *nonfood* items away from them. Such items consisted mainly of brightly colored flowers that the infants plucked and then tried to put into their mouths. One 7-month-old infant reached for a lobe of dung that had been deposited by a passing female. Her mother quickly reached ahead of her, picked up the lobe, and threw it out of reach. Schaller noted a female remove a nonfood item from her infant; and, he also once observed a female assist her 5-month-old infant obtain food by breaking a stalk and then laying it on the ground.

During the day-resting periods, especially when prolonged, infants played on their mothers' bodies for lengthy periods of time—sliding down their backs and abdomens, hair-pulling, or wrestling with the mothers' arms and legs. The mothers sometimes reciprocated by gently nudging the infant or by rocking one of their own arms or legs back and forth. The infant tackled the mother's extremities in much the same

FIGURE 13
Female Flossie holding 7-month-old infant, Samba I., during the day-resting period. The mother is nuzzling the infant while the silverback, Uncle Bert, watches them both with a contented expression. Photograph by Dian Fossey. © National Geographic Society.

manner it would later tackle and wrestle with other infants. If the play appeared to become uncomfortable for the mother, she would often simply rest her arm or leg on top of the infant thus hampering its movements. Play with mother made up 4% of all play recorded for all infants (N = 72) and amounted to 9% of all play time (855 minutes). Expressed in percentages of play time per successive year, it was observed 24, 57, and 18% respectively. Among all types of play during the first 3 years, it constituted 8% the first year, 3% the second year, and 1% the third year. These figures are exclusive of social play with peers or other group members. Due to the infrequency of mothers' responses, such play was considered under maintenance activities. Schaller also noted tolerance on the part of adults toward playing infants, but observed only one instance where a female reciprocated play with an infant who was not her own.

As body contact between infant and mother decreased, some mothers maintained a continual interest in the whereabouts and activities of their young. During the first year the infant was nearly always in sight of its mother, so that if play sessions appeared rough, the mother could either give a disyllabic disciplinary type of vocalization to recall her infant to her, or she would physically retrieve it. The mother did not normally retrieve infants of 2-3 years when they gave distress cries or temper shrieks during play sessions with their peers. They did, however, come forward to collect their infants and juveniles whenever the youngsters advanced toward the observer during the early days of the study. On one occasion, after a prolonged observation of a group, the members slowly moved off feeding in a typical fashion. When it was thought the animals were out of hearing distance, an observer imitated the handclapping sound of one of the group's infants who habitually handclapped. The sound immediately brought the infant's mother, a very shy and low-ranking animal, running back alone and screaming to the observers. This incident illustrated the degree to which a mother would protect her young. Of further interest, in this case, was the indication that the mother did not know where her 2½-year-old infant was at the time. Mothers were always observed to go readily to the aid of their infants when they were stranded in a treed position; this was also noted in Schaller's study. One very unusual observation, which almost seemed to contradict the impression of maternal protectiveness, occurred when a mother, carrying her 10-month-old infant under one arm, began to climb a slick, rock-faced ravine wall. She reached for one spindly tree to pull her over the rock; the tree broke, and she immediately threw her infant slightly to the side and above her where there was a slight lichenous cover. She clung momentarily, then appeared to claw her way over the rock and toward the foliage growth. She made no effort to retrieve her infant who slid down between the rock and her side and clung independently until both reached safety.

When infants became separated from their mothers during travel, their distress was often indicated by whimpers and whines that could lead to temper shrieks if the separation was prolonged; mothers usually responded although not always immediately. One infant was especially inclined to give loud distress vocalizations throughout his third year, and his mother became noticeably less responsive to them. Perhaps for this reason, he then passed through two additional stages not observed in other infants: following a temper tantrum he whacked at foliage or beat on the ground wearing a distressed, pucker-lip type of facial expression before going off to search for his mother. During the next stage, he used a new technique, often successful; he climbed into small trees, which gave him the advantage of seeing over the dense herbaceous foliage. Upon spotting his mother, he would either rapidly descend the tree or leap from it and run in her direction. Schaller also noted infants of 1 and 1½ years of age screaming upon losing contact with their mothers and a variation in reactions among individual mothers.

Field observations over a 7-year period have shown that mother-offspring relationships extended well beyond the period of infancy and, at least, into young adulthood. The presence of an older sibling could result in an infant becoming independent of its mother more rapidly, and perhaps was also responsible for increased tolerance by a mother toward her newest infant. The rank of a mother within her group appeared to have some bearing upon the manner in which she handled her infant, although infants, before the age of 2 years, did not appear aware of their social status within the group.

INFANT MAINTENANCE BEHAVIOR

Once they began to sever body contact with their mothers, the infants' immediate responses were directed toward colorful, usually stationary objects, which evoked either a feeding, exploratory, or play response; vegetation elicited all of these. Far more time was spent playing than feeding during the first year when increasing the distance from the mother. Solo play with vegetation involved exaggerated manipulation of leaves and flowers, such as whacking, tossing, patting, bouncing upon, and also included tree acrobatics, such as repetitive climbing, jumping, swinging, branch patting, and breaking and tossing of moss. Among older infants, idiosyncratic and inventive techniques were observed—younger infants were more limited in strength and coordination in such play. For all infants, solo play with vegetation made up 30% of all play records (N = 489), and amounted to 24% of all play time observed (2283 minutes). Expressed in percentages of play time per successive year, it was observed 16, 48, and 36% respectively; among all types of play during the first 3 years, it constituted 42% the first year, 22% the second year, and 22% the third year. The intensity and duration of the play was usually heightened when older youngsters were playing nearby; thus a higher percentage of solo vegetation play occurred during the first year when the infant was less inclined toward social play. Very few efforts were made to climb trees until at least the ninth month, and then only when no other infants or juveniles were nearby. The approach of an older infant nearly always resulted in the younger animal descending rapidly, sometimes half-falling, and returning to its mother. Attempts to climb small stalks or saplings were seen at an early age but were often unsuccessful as the infant appeared to have no comprehension of what could support its weight.

In a unique observation, Humphrey (personal communication) once saw a 28-month-old infant use a gourd type of vine fruit as a play object. The fruit was held in the teeth by its stem and then beaten by the hands, which made a sharp, clapping sound somewhat like a chestbeat. The fruit was also tossed, rolled, and retrieved. Although the fruits were plentiful and there were six other juveniles and infants within the group, no other animal paid any attention to it.

Attempts at nest-building were often interspersed with solo vegetation play, but when they began to show some degree of concentration and effort, nest-building practice was no longer classified as play. Twenty-seven practice nests were observed being built. The youngest individual seen (an 8-month-old male) took 4 minutes trying to bend foliage stalks around him, but when they continued to spring up despite his comical efforts to hold them down with hands, feet, and body, he gave up. During the first 3 years, practice nests took an average of 6, 7, and 6 minutes, respectively, to build. The youngest animal known consistently to build and sleep within his own night-nest near his mother was 34 months old.

Solo vegetation play and solo self-play (whacking and patting of body, somersaulting, exaggerated tumbling, hand-clapping, leg kicking, arm waving) served the common function of allowing the infant, particularly in the first 18 months, to share the proximity of older playing animals without becoming involved themselves, unless prompted. Both forms of solo play decreased in the third year when participation in social play became more frequent. Solo self-play made up 11% of all records for all infants (N = 186) and amounted to 9% of all play time (855 minutes). Expressed in percentages of play time per successive year, it was observed 23, 59, and 18% respectively; among all types of play during the first 3 years, it constituted 22% the first year; 10% the second year, and 4% the third year.

Infants spent very little time grooming themselves; only 30 self-grooming records (totaling 74 minutes) were observed during the first 3 years (2, 13, and 15 respectively). Observations of idle picking or scratching were not included under grooming. Occasionally infants came into contact with the sticky, resinous fluid from the lobelia plant and showed much concern, as evidenced by their facial expressions and frantic efforts to remove it. The infant would rub the dirtied part on the ground surface; this was especially awkward when it involved the dorsal surface of the arm or hand. They were never seen trying to pick off the fluid.

As previously stated, feeding activities occupied only a small percentage of the young infant gorilla's time and consisted of chewing and simple plucking of leaves mainly during the first 18 months; preparation of food items became far more frequent during the second and third years.

Chewing of twigs and stems made up 18% of all food technique records (N = 60) and amounted to 19% of timed techniques (198 minutes). Expressed in percentages of techniques timed per successive year, it was observed 30, 42, and 27% respectively; among all types of feeding techniques during the first 3 years, it constituted 32% during the first year; 17% the second year, and 16% the third year.

Plucking of leaves or vines made up 24% of all food technique records (N = 83) and amounted to 28% of timed feeding techniques (284 minutes). Expressed in percentages of techniques timed per successive year, it was observed 41, 53, and 6% respectively; among all types of feeding techniques during the first 3 years, it constituted 61% during the first year; 31% the second year, and 5% the third year.

Preparation methods involved wadding, stripping, and collection; they made up 52% of all food technique records (N = 454) and amounted to 45% of all timed techniques (454 minutes). Expressed in percentages of feeding techniques timed per successive year, these were observed 3, 40, and 57% respectively; among all types of feeding techniques during the first 3 years, they constituted 7% during the first year; 38% the second year, and 75% the third year.

Gnawing (prolonged use of the incisors primarily on rotten wood or bark) made up 6% of all food technique records (N = 20) and amounted to 8% of all timed techniques (81 minutes). Expressed in percentages of feeding techniques timed per successive year, it was observed 0, 85, and 15% respectively; among all types of timed feeding techniques during the first 3 years, it constituted 0% the first year; 14% the second year, and 4% the third year.

The rapid decrease in the plucking technique accompanied by the equally remarkable increase in preparation of food possibly gave some indication of the rapidity of the learning, or imitative, processes within the first 3 years. Plucking was certainly not an economical means of obtaining food, although it is perhaps the most practical method for an infant when virulent food items such as nettles and thistles were concerned. It could not be clearly determined precisely how the infant

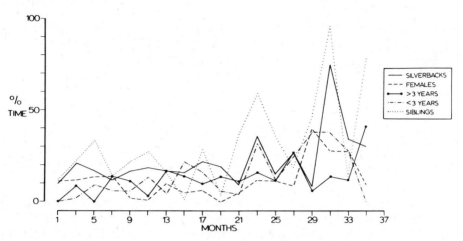

FIGURE 14

All Groups: Median percentages of time spent within 10 feet of infant by animals of various age-sex classes when the infant was within arm's reach of mother. The line for each age-sex class represents the median values for all infants of all groups. Interruptions of successive monthly observations are not represented.

learned what vegetation was acceptable as a food item, or how he successfully learned preparatory methods, other than, of course, practice. Observations gave the impressions that the selection of food items was learned from the mother, but that the methods involved in preparation were learned and imitated from the examples of older peers and/or siblings.

The nonsocial activities in which the infant was involved during its first 3 years appeared to have been concerned primarily with familiarizing itself with its environment while, at the same time, developing coordination and strength to cope successfully with the terrain. Even when alone, however, the infant gained knowledge of socially acceptable behavior by his awareness of the actions of others within the group, his peers in particular.

OTHER SOCIAL INTERACTIONS OF THE INFANT

In order to determine how much time each infant spent near individuals other than its mother, the data on each infant were classified according to whether:

1. The infant was in body contact or within arm's reach of its mother.
2. The infant was beyond arm's reach but within 10 feet of its mother.
3. The infant was more than 10 feet from its mother.

In each case, the amount of time other individuals were within 10 feet of the infant was recorded. These times were expressed as percentages of the times the infant was observed in each degree of proximity to the mother (1, 2, and 3 above). Thus the data are divided to give some indication of the various age-sex classifications that associated with both infant and mother and, as well, to differentiate between infant preferences with age as it increased the spatial distance from the mother. The median values for all associations within 10 feet of the infant in the above spatial categories are shown in Figures 14, 15, and 16.

Infant-Infant Interaction

Overall, medians for infant-infant proximity (within 10 feet) were higher than those for any other age-sex classification except for siblings when the infant was within arm's reach of its mother (Figure 14). It may also be noted that, as the infant's distance from its mother increased, especially in the second year, so did its associations with other infants increase (Figures 15 and 16). At the 7th month point other animals also seemed attracted to the infant; however, maternal restrictions at this age tended to curb interactions. As the infant did not move far from its mother during the early months, play occurred only when others approached. Infant-infant play made up 27% of all social play records (excluding mother) (N = 350) and occupied 30% of observed play time (1797 minutes). Expressed in percentages of play time per successive

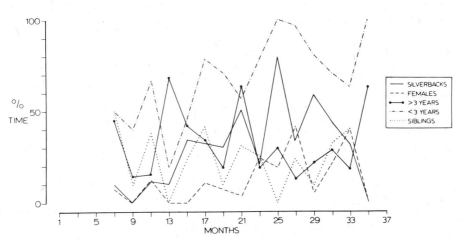

FIGURE 15

All Groups: Median percentages of time spent within 10 feet of infant by animals of various age-sex classes when the infant was beyond arm's reach but within 10 feet of its mother. The line for each age-sex class represents the median values for all infants of all groups. Interruptions of successive monthly observations are not represented.

year, it was observed 9, 44, and 47% respectively. Among all types of social play with other age-sex classifications (excluding the mother), it constituted 56% during the first year, 25% the second year, and 33% the
third year. When an infant in its first year was involved in play with an
individual in its third year (or older), the latter nearly always treated
the infant gently; this did not appear to result from the proximity of
the younger infant's mother. When there was less than a year's difference in age between two infants, similar care was not observed. The tendency to play with only one other individual was seen consistently
throughout the first year and a half unless siblings were present. In
addition to play, another very strong reason for infant-infant proximity
was their common attraction to both juveniles and the dominant silverback of the group; both of these categories served to draw infants
together as they increased the distances from their mothers and was
stronger than a mutual attraction.

Infant-Juvenile-Subadult Interaction (Includes Siblings)

As distance was increased from the mother, infants spent more time
close to animals between 3 and 5 years of age, although the silverback
still continued as a common magnet for all these age groups. By the
second year, the infant appeared to prefer play with juveniles over that
with other infants. The frequency of such play was determined by the
availability of sufficient juvenile play partners who, in turn, preferred
play with those slightly older than themselves. Infant-juvenile play
made up 52% of all play records for all infants (N = 681), and occupied

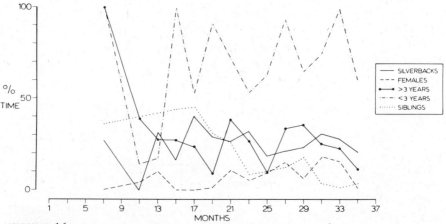

FIGURE 16

All Groups: Median percentages of time spent within 10 feet of the infant by
animals of various age-sex classes when the infant was 10 feet from its mother. The
line for each age-sex class represents the median values for all infants of all groups.
Interruptions of successive monthly observations are not represented.

25% of all play time (3149 minutes). Expressed in percentages of social play time observed per successive year, it was seen 2, 56, and 42% respectively. Among all types of social play (excluding the mother), it constituted 24% during the first year, 56% the second year, and 52% the third year. Siblings were more consistently near infants and their mothers than any other category of animal, and the association was maintained throughout the first 3 years of the infant's life. This maintenance of proximity on the part of siblings was felt to be due basically to two reasons: (a) the mother was more likely to accept the sibling when she was with the infant than she was any other age-sex classification; (b) the sibling's attraction to the mother rather than to the infant itself. (See Figures 14 and 16 where, as the distance between infant and mother increases, the sibling association gradually decreases.) In seven cases, siblings were present (the eighth died the day of birth): in one of these, the time spent with the sibling was greater than with any other individual in eight out of eight 2-month periods ($p < 0.004$); in another, in nine out of ten periods. Fewer data were available for the others but, the combined medians for all infants through periods 1-35 (total of 17 with no data for period 3) show that the time spent near siblings was greater in 12 periods, equal in two, and less in three than the time spent near other juveniles or subadults. Infant-sibling play only made up 10% of all play records for all infants ($N = 133$), and occupied 8% of all play time (492 minutes). Expressed in percentages of play time per successive year, it was observed 7, 81, and 12% respectively. Among all types of social play (excluding mother), it constituted 11% during the first year, 13% the second year, and 2% the third year.

The fairly close associations with subadults (3-6 years), especially after the second year, were probably due primarily to the infant's attachments to juveniles and siblings, which promoted subadult accessibility. Interactions with subadults were more complex than those with juveniles and siblings and, in addition to play, included maternal behavior such as grooming, transport, and even copulatory behavior. Infant-subadult play made up 8% of all play records for all infants ($N = 106$), and occupied 7% of all play time (422 minutes). Expressed in percentages of play time per successive year, it was observed 3, 34, and 63% respectively. Among all types of social play (excluding mother), it constituted 4% the first year, 5% the second year, and 10% the third year. For both the sibling and the young subadult, the infant appeared to provide training in maternal behavior, especially when there was an increase in distance from the mother. On several occasions, when the subadult was not a sibling, there appeared to be some degree of jealousy between two infants if one was not included in the older animal's attention; this was indicated by facial expressions, distress cries, and persistent following, all forms of behavior usually no longer tolerated by the infant's mother.

175

Infant-Adult Female Interaction

Medians for infant-female proximity were lower than for any other category and showed no significant change with age. Infants and adult females were usually only in proximity during the day-resting periods, when both shared the presence of the group's silverback; or, on those occasions when one infant elicited an interaction with another who was near its mother. Only two observations were made of a female, with an infant of its own, seeking another female's infant for grooming and cuddling purposes. There were also only two cases where an infant (2 years of age) suckled a female, not its mother, and several observations of infants under the age of 1 year investigating other females, especially their breasts. Infant-female play (excluding mother) made up 45% of all play records for all infants (N = 6), and occupied 35% of all play time (21 minutes). Expressed in percentages of play time per successive year, it was observed 24, 76, and 0% respectively. Among all types of social play, it constituted 2% during the first year, 51% the second year, and 0% the third year. Generally speaking, it appeared that infants over the age of 2 years went out of their way to avoid the dominant females but tended to ignore the lower-ranking females.

Infant-Blackback Interaction

Data concerning infant-blackback associations were very sparse and were not included in any figures. Infant proximity with blackbacks was usually limited to sibling associations or made possible because of black-back interactions with older female siblings. Another factor responsible for the infrequency of blackback-infant relations was that the adult males usually remained outside the bulk of the group, especially during the day-resting period, thus considerably reducing opportunities for interactions. Two individuals did interact on occasion with infants; one, because of an infant sibling attachment; the second because he had no peers of his own to play with. Infant-blackback play made up 2% of all play records for all infants (N = 32), and occupied 2% of all play time (101 minutes). Expressed in percentages of play time per successive year, it was observed 4, 48, and 48% respectively. Among all types of social play (excluding mothers), it constituted 1% during the first year, 2% the second year and 2% the third year.

Infant-Silverback Interaction

Median values for associations between infants and mature males, or silverbacks, increased with the age of the infant (r_s = 0.44; p = < 0.05), and were significantly higher than the values obtained for infant-female

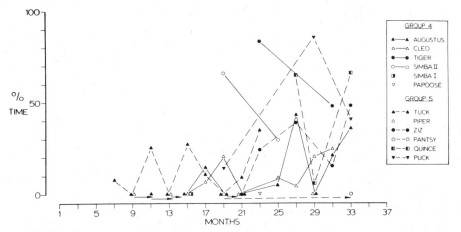

FIGURE 17
Groups 4 and 5: Percentage of time females spent within 10 feet of infant when infant was beyond arm's reach but within 10 feet of mother. Each infant's symbol represents percentage of observation time seen within 10 feet of adult females when infant was beyond arm's reach but within 10 feet of mother. White symbols represent females; black symbols represent males. Half-filled symbols represent unknown sexes.

proximities ($p < 0.001$); in no one period did the difference reach significance.

As previously stated, both infants and females spent a large amount of time near the silverback during the day-resting periods. When these were prolonged, the infants continued to play with their peers, or by themselves, very close to the silverback after the females had begun to feed at a further distance. All silverbacks exhibited a high degree of tolerance toward the play and proximity of younger animals and allowed themselves to be kicked, hit, and tumbled against. If the play between infants became exceedingly rough, the silverback was able to check it with a low, disyllabic belch vocalization. Infant-silverback play made up only 30% of all play records for all infants (N = 4), and occupied 13% of all play time (8 minutes). Expressed in percentages of play time per successive year, it was observed 25, 37, and 37% respectively. Among all types of social play it constituted 68% during the first year, 9% the second year, and 9% the third year. On many occasions when the silverback went off to feed near the day-nesting area, his nest became a favored play site for infants and juveniles; when three or more infants tried to play in the nest at the same time, quarrels often developed.

Field observations conveyed the strong impression that both the tolerance and the consistency of temperament were factors that attracted infants to the silverback. Paternalistic potentialities were apparent in the case of an orphaned 3-year-old, Simba II. Immediately following her mother's death, the infant was observed eating her dung (first time seen), ceased her play activities with others, reduced her feeding, and spent much time simply sitting alone in a dejected manner. For nearly a week she built her own night-nests, very rudimentary structures consisting of only a few folded leaves, before beginning to share the silverback's night-nests. He slowly assumed a maternal role toward the infant, in that he spent considerable time grooming and cuddling her and became somewhat overly protective of her when she eventually began to interact with other infants or when she approached the observer. (It was interesting to note that many younger infants could come within inches of the observer without attracting any attention from the silverback, but the moment Simba II expressed curiosity or interest, the silverback became instantly defensive.)

Figures 17 through 20 clearly illustrate the percentages of time spent in proximity (within 10 feet) to the infant by the various age-sex classes in order of ascending frequency and within one spatial category. It may be seen that adult females associated, by far, the least with infants, particularly in the second year. The rise that occurred after the second year was due to infant-infant association rather than infant-adult female attraction. The juvenile and subadult proximities showed

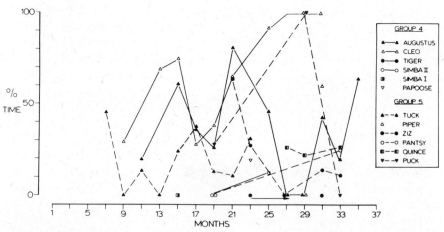

FIGURE 18

Groups 4 and 5: Percentage of time juveniles and subadults (> 3 years) spent within 10 feet of infant when infant was beyond arm's reach but within 10 feet of mother. Each infant's symbol represents percentage of observation time seen within 10 feet of juveniles and subadults when infant was beyond arm's reach but within 10 feet of mother. White symbols represent females; black symbols represent males. Half-filled symbols represent unknown sexes.

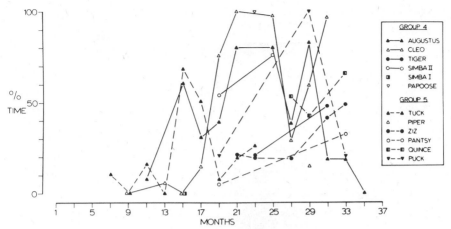

FIGURE 19

Groups 4 and 5: Percentage of time silverbacks spent within 10 feet of infant when infant was beyond arm's reach but within 10 feet of mother. Each infant's symbol represents percentage of observation time seen within 10 feet of silverbacks when infant was beyond arm's reach but within 10 feet of mother. White symbols represent females; black symbols represent males. Half-filled symbols represent unknown sexes.

more individual variation than was seen in the case of females, primarily due to variations in sibling availability as well as in the degree of tolerance of individual mothers. For example, Tiger had no siblings, but a very dominant mother whom juveniles and subadults usually avoided; hence his low record of proximities with those age categories. Within the same group, Cleo, with a very permissive, if not overly casual mother, had a high record of associations with the same individuals who avoided Tiger's mother. There was a slow but fairly consistent rise in silverback associations with the age of infants. This was considered a result of the attraction of the silverback for mother and/or infant. For example, the mother of Ziz and Pantsy was considered a relatively independent female who did not usually seek the proximity of the silverback, but who did maintain a close relationship with her offspring. For this reason, apparently, they were not as frequently observed near the silverback as were Puck and Tuck (within the same group) whose mother was nearly always seen day-resting near the silverback. In all but one case (Puck), infant-infant associations showed the most consistent rise with age than any other age-sex category. It is possible that Puck's extreme drop in associations with other infants at the age of 33 months was due to his mother's conception during this month. To a lesser degree, similar evidence exists in the cases of Tiger at the age of 32 months, and Cleo at the age of 27 months.

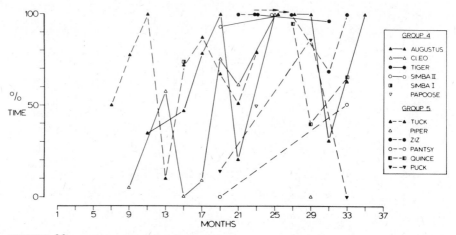

FIGURE 20

Groups 4 and 5: Percentage of time other infants (< 3 years) spent within 10 feet of infant when infant was beyond arm's reach but within 10 feet of mother. Each infant's symbol represents percentage of observation time seen within 10 feet of other infants when infant beyond arm's reach but within 10 feet of mother. White symbols represent females; black symbols represent males. Half-filled symbols represent unknown sexes.

DISCUSSION AND SUMMARY

Gorilla groups offer an exceptionally interesting opportunity for studying the development of relationships that form between the infant and others within the group. The social unit in which the infant matures is sufficiently small for individual relationships to be studied, but also large and complex enough to demonstrate the infant's ability to discriminate and form specific attachments. It is felt that their social structure is unique among all the great apes, in that there is a nexus of relationships involving recognition of familial ties, perhaps even including paternity;[1] a characteristic apparently not present among chimpanzees.

The development of the gorilla infant has been divided into both physical and behavioral categories; even then individual variations were noted among the 32 infants studied. On the whole, both physical and behavioral stages of development observed in this study occurred at a later age (between 6-12 months) than those described by Schaller (1963). Differences in physical descriptions between his study and mine concerned mainly skin tone and hair length and occurred within the first 18 months. Also, observations from this study have noted the tail tuft as diminishing in size and coloration during the third year, while Schaller stated that it persisted until at least 4 years. Differences between the stages of behavioral patterns recognized by Schaller and myself, were concerned with locomotor ability and spatial distances from mother that took place during the first year. Many of the behavioral stages of development, as described by Goodall (1967), were similar to those observed in the infant gorilla; the most significant variation appeared to exist in the mother's reactions to their infants during the first year at least; chimpanzee mothers exhibited a higher frequency of play and grooming with their infants. One of the biggest physical differences noted was the full development of the white tail tuft in the chimpanzee infant between the age of 6 months to 1 year; this did not occur in the gorilla infant until the second year. The most significant behavioral similarity was the breaking of physical contact with the mother, which occurred between 16 to 24 weeks in both the chimpanzee and the gorilla. Neither were observed to use nipple-clinging as a means of support although this is commonly seen in young monkeys.

Pregnancy was difficult to determine in free-living gorillas, but not in the chimpanzees (Goodall 1968b), where pregnancy was pronounced during the final 2-2½ months. Among some individual gorilla and chimpanzee parturients, however, a tendency was observed toward decreased interactions with other animals both before and following birth.

Data suggested that there was a birth peak between July and August although mounting behavior was observed all year. True seasonality, as seen in the langurs, savanna, and hamadryas baboons in particular, was not observed among the gorillas. It is possible that the slight increase in births between July and August was related to their both being dry months.

Birth spacing was similar to that observed among the Gombe chimpanzees, with a minimum period of 3 years usually between births. The chimpanzee data (based upon three older females) suggested that the time interval could lengthen with the age of the mother. Data from this study (based on two older females) gave the opposite impression and provided the shortest birth intervals (36 and 41 months respectively).

The mortality rate of infants within their first year of life was found to be about the same in this study as in Schaller's study (27% and 23% respectively). Three infants died as a direct result of their groups' interactions with other groups or lone individuals. In all three cases, a strong amount of evidence suggested that infanticide resulted from competition over the infant's mother. The mother of infant Phocas was severely wounded the day her infant was killed (visible evidence), disappeared from the group within the week and was never seen again. The mother of infant Curry went off with the lone male thought to have been responsible for the death of her infant (tracking evidence). The mother of infant Thor was taken by the silverback thought to have killed her infant (vegetation partially obscured visible evidence) 5 months following the death of her infant and is known to have remained with him for a period of at least 8 months. No females with infants were observed transferring to a male outside of their own group (eight females have been known to transfer at least once). Killing of infants has also been reported in adult male patas *(Erythrocebus patas)* (DeVore, personal communication), langurs *(Presbytis entellus)* (Hrdy 1974), purple-faced langurs *(Presbytis senex senex)* (Rudran 1973), hamadryas baboons (Kummer et al. 1974), chimpanzees (Bygott 1972), and crab-eating macaques (Washburn and Hamburg 1968). Although it is possible that a high population density, leading to an increase in group interactions, might be a partial cause for infant mortality (Sugiyama 1960), it seems, within this area at least, that the increasingly high proportion of males to females might become a more likely reason for the intensity and frequency of aggressive competition for females.

Infants were never observed to have been used as shields during agonistic interactions.

Curiosity was the most common response of other group members to the newly born infant and was usually manifested by staring, only because of the protectiveness of the mother and her usual avoidance of proximity with other group members. Observations of infant handling by other individuals, such as was seen among the langurs within a few hours after birth (Jay 1965); black and white colobus infants beginning at 3 months (Leskes and Acheson 1971), or vervets (Lancaster 1971), were not observed in this study.

During the course of infant maturation, individual differences were seen in the decrease of proximity with mother; however, it appeared that the degree of independence was primarily regulated by the mothers. Similar findings were observed among captive rhesus monkeys (Hinde 1974). He found that, during the early weeks of life, the proximity between mother and infant was regulated by the mother; later both contact and maintenance of proximity were due to the infant. The rough restraint of infant by mother among the pigtail macaques (Kaufman & Rosenblum 1966) was not observed among gorilla mothers. One of the most important factors responsible for the degree and the duration of mother-infant proximity was the presence or absence of siblings: the presence of an older sibling appeared to result in a beneficial situation for both mother and infant, as the infant had a maternal substitute available, which served as an incentive for increasing distance from mother; and, the mother appeared more permissive and secure when her young infant was interacting with its siblings than with its peers. Goodall (1968b) also observed strong sibling relationships among chimpanzees, which included protection, play grooming, adoption, and kidnapping. Among the gorillas these forms of behavior were not restricted solely to siblings. An animal, strongly suspected to have been her older brother, failed to adopt an orphaned infant.

Aunt-infant interactions such as described by Rowell, Hinde, and

Spencer-Booth (1964) were seldom observed among the gorillas in this study. Prolonged proximity to adult females other than the mother was initially regarded as aunt behavior, until it became clear that the reasons for the associations were infant-infant interactions and not a maternal inclination on the part of the adult female toward another's infant. On the few occasions when behavior akin to aunt behavior was observed, the female was always subordinate to the infant's mother. Totally unlike Spencer-Booth's findings (1968) with captive rhesus macaques, older males interacted far more with gorilla infants than did older females; however, similar to her observations, younger females were more attracted to gorilla infants than adult females.

Paternal behavior was most conspicuous in the case in which a silver-back male adopted an orphaned infant, but it was also consistently seen in the frequencies of grooming and cuddling activities on the part of the silverback toward 3- and 4-year-olds with newly born siblings. Similar observations were made among the Japanese macaques (Itani 1963; Alexander 1970). The factors increasing paternal behavior among gorillas were very similar to those listed by Mitchell (1969) and cited by Hinde (1974): in particular, kinship, sex of infant, dominance of mother, death of infant's mother, and one-male groups. Also in common with Mitchell's findings were the factors decreasing paternal behavior—crowding and increased age of infant. Unlike one of Mitchell's conclusions, consort relations did not appear to increase paternal behavior among the gorillas, and there was no evidence that paternal behavior was influenced by the mother's treatment of the infant.

It was clear that all infants were attracted to and well tolerated by the silverbacks of their group. In several instances, especially when an infant was the offspring of a subordinate female, it appeared that the infant's prolonged proximity with the group's silverback would increase its familiarity to him. Such behavior included not only following and resting near the dominant male, but also frequent grooming of the silverback by the infant.

Infant-peer interactions were more frequent than any other kind in all degrees of proximity with the mother and consisted primarily of play activities. With age, an infant's play activities expanded from play on the mother's body and solo play to parallel play near other young. During the second year, social play interactions were preferred with slightly older partners. Male infants played more roughly than did females and also became slightly more aggressive in nature. Play patterns were similar to those observed among chimpanzees and monkeys; chasing, wrestling, whacking, biting, and use of objects within the natural environment. Tickling, such as was observed among the chimpanzees, was infrequently seen.

The social environment in which the infant gorilla develops during its first 3 years has been seen to grow gradually more complex, in that it involved more interactions with a wider range of individuals existing within a relatively closely constructed unit. The status of the mother within the group might well be the most crucial factor determining an infant's potential to develop relationships with other individuals. Paternal and sibling ties were also felt to promote the welfare of the infant and its eventual security, as well, within the group structure.

Acknowledgment

I would like to express my gratitude to the National Geographic Society for their continued support of the Mountain Gorilla research project; without their invaluable assistance this long-term study could not have been realized. I remain grateful also to the Wenner-Gren Foundation who made it possible for me to work up data accumulated from the field at Cambridge University. I wish to pay very special tribute to Professor Robert Hinde for his constant suggestions, encouragement, and patient advice during the writing of the paper, also to Les Barden of the Sub-Department of Animal Behavior, Madingley, Cambridge who very kindly drew the figures for the text. In addition I remain grateful to the students and my African staff who maintained observational records and camp during my absence.

Notes

1. . . . ."perhaps even including paternity;" I am concerned about the subjectivity involved when stating that recognition of paternity might be possible. This statement is based on two main factors other than simply field impressions: 1) No infant is known to have been killed by its own father; 2) Strong physical resemblances between silverbacks, older infants, and juveniles (when the study began) have coincided with much closer relationships as the animals matured than those between animals who bore no physical resemblance whatsoever to one another. There were several examples of this, the most outstanding one would be Peanuts and Samson as the offspring of Rafiki as opposed to Pug and Geezer of the same group (and nearly the same ages). The latter two bore no resemblances to Rafiki, more resembled each other, did not share his proximity as did Samson and Peanuts, and also left the group far sooner than Peanuts (who never did leave) and Samson.

A. H. Harcourt

The Social Relations and Group Structure of Wild Mountain Gorilla

Schaller (1963) established that most gorillas lived in cohesive groups consisting of a dominant, older male (a silverback), who was clearly the leader and social focus of the group, and a variable number of other adult males, adult females, and immatures. Usually there were more adult females than adult males. He found also that a few adult males traveled alone and only occasionally associated with groups. It has become clear that the degree of cohesiveness of gorilla groups, especially as indicated by the long association periods of numbers of adult males and females, is a facet of social organization not shown by other apes. Orangutan adults associate only temporarily (see authors in this volume); chimpanzee males live in a loose, although stable, community, and the adult females seem to be solitary and to have no community structure (see authors in this volume); gibbons live as family units and only rarely does more than one adult female consort with an adult male (Chivers 1972). This study is an attempt to investigate the causes of the structure and relative cohesiveness of gorilla groups by examining the interactions between individuals. From the content, quality, and patterning of the interactions, an idea of the relations between individuals is obtained (see Hinde, in press). From an idea of the relations between individuals, an extrapolation is made to the relations between age-sex classes. This is only a preliminary report of some aspects of the study, and I mention only a few of the individual differences in behavior that were found.

METHODS

The study was carried out in the Virunga volcanoes of Rwanda and Zaire. (See Fossey 1974; Fossey & Harcourt, in preparation, for details.)

Most of the report is based on about 900 hours of observation of gorillas during 20 months between September 1972 and September 1974.

The size and composition of a group must affect its cohesiveness and its structure. Recent recorded group sizes (Fossey 1972; Harcourt & Groom 1972; Groom 1973) range from 2 to 20. The median group size differed between the west and east halves of the park: in the west half (in which the study area was situated) the median group size was 10 (N = 17), and in the east half it was 5 (N = 11) (Schaller 1963). The adult male to adult female ratio of known groups was 2.7:4.2 (N = 10). I studied a number of groups, but concentrated on two—Group 4 and Group 5. Throughout the study, both groups contained one silverback (adult male of 11 years, or more) and one blackback (adult male of 8-11 years). The number of adult females (8+ years) changed due to transfers and one death (4 to 2 in Group 4; 6 to 3 in Group 5) and the number of immatures changed due to death, maturation, and transfer of females (5 to 4 in Group 4; 8 to 4 in Group 5). There were three other immatures in Group 4, and four others in Group 5, for most of the study period; in addition a new infant was born in each group in the last month of the study.

Field Methods

On most days the animals were found by tracking them from where they had been left the previous day. The two main study groups were well habituated and the animals could be watched from a distance of less than ten meters with little or no discernible disturbance of their behavior. Observations of interactions were recorded on check sheets under behavioral categories, and the participants and the details of the interactions were indicated in a code. Often less than half the group was in sight at any one time, and some animals were seen only for a short time or not at all on any one day. Thus, the only deliberate sampling done was to record interactions between adults in preference to interactions between immatures, if there was not time to record both.

RELATIONS WITHIN GROUPS

The Silverback

The role of silverbacks as leaders of their group was obvious. In groups in which there was more than one silverback, one (the oldest) was clearly the leader, and the other(s) occupied peripheral positions. The leading silverbacks were the protectors of the group; it was they who most frequently stopped intragroup fights; it was they who most frequently controlled the timing and direction of group movement; and it was the leading silverbacks around whom most animals most often congregated, even when the groups were undisturbed. Only silverbacks, not blackbacks, were seen to mate with primiparous and multiparous females, but both mated with adolescent females, and with adult females who had not yet given birth. The amount and frequency of grooming given and received differed between silverbacks, e.g., the silverback of Group 4 had one of the highest records in the group for amount of grooming given, but one of the lowest for amount received, but the silverback of Group 5 had the opposite ratio.

The Blackback

Blackbacks tended to spend much time far from all the other animals, but they were still obviously part of the group. Despite their peripheral position, they sometimes had close and lengthy interactions with other animals, although very rarely with adult females. In Group 4, the silverback and blackback were very rarely close to each other, but in Group 5 the blackback was not infrequently close to the silverback and even groomed him. As blackbacks neared silverback age, they seemed to become dominant over adult females in agonistic interactions. Blackbacks mated with juvenile and adolescent females, but they were never seen to mate with primiparous or multiparous females. Blackbacks, especially older ones, played more than females of similar age, probably because the females were producing, or had produced, an infant.

The Adult Female

Adult females interacted very little with each other and very little with blackbacks. Adult females were attracted to silverbacks, and those with young offspring spent the most time near them. Silverbacks were attracted to some females, e.g., the silverback of Group 4 was clearly attracted to a female who had newly joined the group. Some females spent as much time far from other animals as blackbacks, but were sometimes close to the silverbacks for long periods. There seemed to be a dominance hierarchy in agonistic interactions between females, but instances were rare because of mutual avoidance. Females first showed sexual swelling and overt sexual behavior at 7½-8 years. The amount of sexual activity was markedly different between the two main study groups in the last months of the study: there were more than 20 observations of mating in Group 4, but only one in Group 5 (records from two adult females in each group). Only one female in each group had produced an infant by the end of the study. At least one female mated when pregnant and did so into the last month of pregnancy. Female-female "matings" were seen, but they were rare. For about 10 days after a parturition, the spatial pattern of the group differed from the usual pattern as adults and, especially, immatures were attracted to the mother and her new infant. Mothers with offspring did not play, but primiparous females who lost their infants played again after the death.

By the end of the study, the oldest animal still in the same group as its mother was an 8⅓-year-old female (i.e., an adult by our classification), and at that age she still associated with her mother, sometimes closely, more than she did with other adult females. This female, in fact, stayed in her natal group to produce her first offspring, which was sired by her own father or half-brother. Other animals, aged between 4 years and nearly 7 years, also associated with their mothers more than with other adult females. The relation between mothers and their offspring continued even after a younger offspring had been born, although the mothers then associated more with the younger than with the older offspring. As the offspring became older, the association with their mothers became less close, and the offspring became more responsible than the mothers for maintaining the association. Adult females groomed their offspring more than they groomed any other animal. (For details of the relations between mothers and infants, see Fossey, this volume).

Immatures

Immatures, especially old infants and juveniles, were attracted to the leading silverback of their group, even when their mothers were present. When the group was moving and feeding, they at times followed the silverback, not their mother: a group of immatures near the leading silverback was not an uncommon sight. This relation between immatures and the leading silverback could well be important for survival of the immatures, if their mother died or left the group. However, even with this relation, immatures were adversely affected if they were orphaned (Fossey, D., personal communication; Stewart, K. J., personal communication). Immatures groomed their mothers, younger siblings, and, depending on the immature, silverbacks, more than they groomed any other animals. Immatures tended to play more with those immatures nearer in size to themselves, although idiosyncratic preferences for playmates caused some exceptions to this generalization. Siblings played together more than might be expected, considering their difference in size. (These play generalizations are substantiated by K. J. Stewart [personal communication]).

RELATIONS OUTSIDE GROUPS

In this section I will use the term *unit* to mean either a group, or a lone silverback.

When two units were near each other, the interactions between them varied from almost friendly to aggressive. The nature, quality, and patterning of interactions between any two units varied, but not markedly, i.e., there seemed to be a more or less stable relation between particular units. When units were near each other, females (never males) occasionally transferred from one unit to another: males who left their parent group (not all did so) became solitary, as far as is known. Only once, and then for only a day, was a female known to have been far from other gorillas, and she was then in the process of transferring from one unit to another. Five of the seven females who are known to have transferred between units were nulliparous when they first transferred. The others were primiparous. Subsequently, two of the nulliparous females transferred again as primiparous females. The number of the transfers by the different females varied from at least one to seven, with a median of four: some transferred back to their parent groups for short periods. Movement of females has accounted for more changes in group composition than movement of males. All observed transfers took place voluntarily, and it is difficult to conceive of how an unwilling female could be forced to transfer.

Because blackbacks and nonleader silverbacks had so few interactions with other members of their group, their emigration probably had little effect on the structure of the groups. The movements of females between groups, however, caused changes in the patterning of interactions in the parent groups and in the groups into which they moved, i.e., relations within groups were changed. Such changes must, to a small extent, have affected the structure of the groups. The transfer of a female, or females, to a lone silverback created sets of relations that did not exist before the transfer. If the relations stabilized, then a new, cohesive group would form. The small amount of evidence available suggests that this was how new groups were formed.

SUMMARY

The social structure of most gorilla groups consisted of a stable core of a leading, adult male and one, or more, multiparous females and their dependent offspring. There were close relations between the male and the females, but not between the females themselves. Offspring had close relations with their mother, even into adulthood. Immatures, especially older infants and juveniles, had close relations with the leading adult male also. The unstable elements in the structure were the peripheral, nonleader males, and the primiparous and nulliparous females, particularly the latter. The peripheral males could leave the group (not all did so) to become solitary. The young females transferred to other groups, or to lone silverbacks, and most females transferred more than twice. Movement of females caused more changes in group composition than movement of males. It is suggested that new, cohesive groups are likely to be formed by the transfer of females to lone silverbacks, with subsequent stabilization of relations within the new group.

Acknowledgment

The study was financed by a Royal Society Leverhulme Studentship and a Medical Research Council Studentship, and I am most grateful to both these groups for their support. I will be forever indebted to Miss Fossey and the National Geographic Society for allowing me to work from and use the facilities of the Karisoke Research Centre. I thank the conservateurs of the Parc des Volcans of Rwanda and the Parc des Virungas of Zaire for permission to work within the parks. I thank Miss K. J. Stewart for much interesting discussion, and I am especially grateful to Professor R. A. Hinde for his continual advice and encouragement.

PART THREE

Orangutans

Long-continued field studies of the behavior of orangutans followed the studies of the African apes by about 10 years. Biruté Galdikas, David Horr, John MacKinnon, and Peter Rodman all started their fieldwork with a far greater understanding of the structure of primate behavior than had previously been possible. In this section Galdikas is primarily concerned with gathering more information, whereas MacKinnon and Rodman are more concerned with interpretation.

All the recent, long-term studies confirm the earlier observations that orangutans are, in marked contrast to other apes, primarily solitary animals. The meaning of this behavior is still very much under discussion, as can be seen from these chapters. It has been widely assumed that loss of estrus and continuing receptivity in human females were major causes of human social systems. The behavior of the orangutans suggests that there are other evolutionary possibilities. Female orangutans are more receptive during part of the month; sexually aggressive males will mate at any time. The least social of the apes is the most continuous in sexual activity, and this sexual activity is not at all limited to the period of female receptivity. A question to consider is whether aggressive human males might be the binding force in human society rather than receptive females.

Obviously more fieldwork is needed to interpret the behavioral adaptive mechanisms of the orangutan.

Nick sitting on the ground eating termites. Photograph by Rod Brindamour.
© National Geographic Society.

Biruté M. F. Galdikas

Orangutan Adaptation at Tanjung Puting Reserve: Mating and Ecology

The initial descriptions of free-ranging orangutan as an essentially solitary species (Schaller 1961; Harrisson 1962; Davenport 1967) have been amply confirmed by a number of subsequent, more detailed field studies (MacKinnon 1971; Horr 1972; Rodman 1973a; Rijksen 1974). Whenever orangutan populations have been sampled, the consistent features of their social organization have included a highly dispersed population and a relative absence of social interaction. This absence is not only striking when compared to the social behaviors exhibited by other pongids but also anomalous when viewed in the context of the range of behavior found among higher primates as a whole.

Clearly, orangutan adaptation is unique among the Anthropoidea. Unlike other monkeys and apes, orangutan populations are basically structured in terms of incomplete reproductive units. With the exception of dependent offspring accompanying the adult females, both orangutan males and females remain essentially solitary throughout most of their adult lives.

Furthermore, among all pongid species, sexual dimorphism is most pronounced among the orangutan. Not only is the male twice the size of the adult female, but the anatomical structures most fully developed in mature males include large cheekpads, which enhance the size of the head, and a balloon-like throatpouch, which is sometimes kept inflated. In addition, most mature males have prominent beards (although some females have these as well).

Within recent years there has been a revival of interest among ethologists in the causes of sexual dimorphism and the role of sexual selection in the evolution of vertebrate social systems. The importance of interaction between sexual and natural selection in the evolution of the various forms of social systems extant among the higher primates has been emphasized (e.g., Crook & Gartlan 1966; Crook 1970). It is obvious that even solitary primates such as the orangutan must be organized and behave in ways that facilitate successful reproduction for at least some members of each sex. The mating system will be intimately bound to the ecological adaptation exhibited by the species.

The purpose of this paper is to examine the mating system among a population of wild orangutan in a study area at the Tanjung Puting Reserve, Kalimantan Tengah (Central Borneo), Indonesia in an attempt to assess this system's adaptive significance. The data are based on over 5000 hours of direct observation collected by my husband and myself during the first 41 months (November 1971-April 1975) of a long-term study (still in progress at this writing). The bulk of the observation time consisted of whole days—from when the orangutan first left the nest in the morning to when the animal nested for the following night. We logged 352 of such whole days of observation.

STUDY AREA

The study site is located in the Tanjung Puting Reserve (also known as the Kotawaringin-Sampit Reserve) in Kalimantan Tengah about 20 kilometers from the sea coast. (See Figure 1.) The study area consists of approximately 3500 hectares (or 35 square kilometers), which are bounded along one side by the Sekonyer Kanan River and partially along two sides by smaller tributaries. This area can be loosely characterized as bands of lowland dipterocarp forest incorporating limited areas of tropical heath forest (Anderson 1975: personal communication) alternating with shallow peat-swamps, which gradually merge

FIGURE 1

General location of Tanjung Puting Study Area

along the rivers to form deep riveredge swamp. The study area is relatively flat; elevation ranges from 10 meters to no more than 40 meters above sea level.

The large study area includes three *ladangs,* which are abandoned dry-rice fields now covered with alang-alang grass *(Imperata cylindrica),* ferns *(Gleichenia* and *Pteridium)* and some shrubs. These ladangs account for only a small proportion of the overall study area. The largest of the three fields is very old, purportedly made at least 25 years ago. The others are much more recent. These three fields represent the last ladangs interrupting the forest along this side of the Sekonyer Kanan River as one goes upstream.

The study area forest is largely undisturbed. Even today there is no mechanized logging anywhere along the Sekonyer Kanan River, nor is there any permanent population along its length. During a period near the end of 1970-beginning of 1971, however, a small local logging company employing crews of handloggers operated for several months in this region before going bankrupt. Fortunately, because the company operated so briefly, damage to the forest was slight.

We have covered the entirety of the 35-square-kilometer study area with a network of small crisscrossing trails (including several old jelatung-tappers' trails, which we cleared and reopened). The trail system has been extensively surveyed and staked to enable the accurate mapping of orangutan daily movement and home ranges. The combined length of trails in the overall study area exceeds 125 kilometers. (See Figure 2.)

Field Methods

My husband, Rod M. C. Brindamour, and I, beginning our second year of field work, and aided by our local workman, each patrol the trail system. When one of us finds an orangutan, the animal is followed until it nests for the night. We return to spend the night at one of three camps maintained in the study area. The next morning my husband or I arrive at the nest before dawn. Once an orangutan is located, it is often possible to maintain the contact for long periods of time. We observed one mature male for 23 days; another habituated adult male was followed for 65 days in succession. On another occasion a habituated female and her offspring were observed continuously for 31 days without losing contact.

My husband and I frequently alternate observation days. After 1½ years of training, our Indonesian workman also occasionally alternates observation days with us. In addition, the three of us often split up and simultaneously observe two or more independent orangutan units in different parts of the study area. Thus, in addition to the over 5000

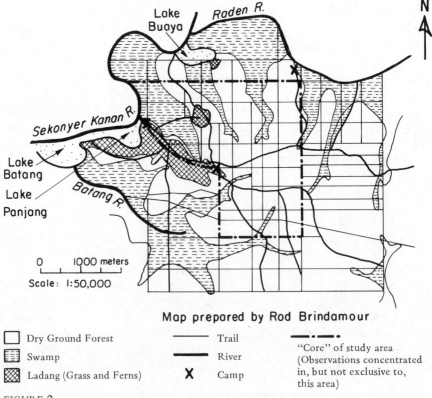

Map prepared by Rod Brindamour

☐ Dry Ground Forest	—— Trail	—··—··• "Core" of study area (Observations concentrated in, but not exclusive to, this area)
▤ Swamp	—— River	
▨ Ladang (Grass and Ferns)	X Camp	

FIGURE 2
Study Area in Tanjung Puting Reserve

hours of direct observation accumulated by my husband and myself, our workman has also logged considerable hours. With the exception of his most recent observations (December 1974-April 1975), however, I will *not* include his observations in the present analysis.

Our observations have been carried out primarily on 54 individuals, all of whom are individually recognized and named. Observations have been concentrated, but not exclusively, in an 11-square-kilometer portion of the study area. Most orangutans encountered in this area are individually recognized and at least somewhat habituated. The only persistent exceptions in terms of individual recognition have been some subadult males.

The degree of habituation varies greatly from one animal to the next. Until partially habituated, the orangutan stayed predominantly in the trees and came down to the ground irregularly and warily. Now, however, the most habituated individuals have ignored our presence less than 3 meters from them while they foraged or walked along the ground.

ORANGUTAN POPULATION UNITS

It has been definitively established (MacKinnon 1971; Horr 1972; Rodman 1973a) that the basic units of orangutan populations consist of (a) solitary adult males, (b) adult females usually accompanied by one or two dependent young, and (c) immature animals in transition between belonging to their mother's unit and establishing their own separate solitary existence. This was also the case in the Tanjung Puting study area.

Adult males were usually solitary. The only time that mature males were in true association with other orangutans is when they accompanied or were joined by an adolescent or adult female with offspring for varying lengths of time.

Although adult females traveled together in small groups, this behavior was relatively infrequent. Multi-unit groupings consisting of as many as nine orangutans were encountered traveling together, but usually not longer than part of one day. Groupings consisting of two adult females and their offspring were sometimes more stable, although generally the two females would separate after about 24 hours of uninterrupted contact. Two adult females (Beth and Cara) and offspring, however, were observed to travel together three times, and at various times they were joined by other individuals. On two of these occasions this particular grouping endured for most of 3 days; however, this was exceptional.

Immature animals were observed alone, in the company of other immature animals, or traveling with their respective mothers. Sometimes immature orangutans were seen in association with adult females who were not their mothers. The social propensities of immature animals seemed considerably greater than those of fully mature individuals. As with the plural adult-female groupings, some of these immature combinations were persistent when viewed over a period of years, in that they tended to recur over a number of sightings. The general pattern was two immature individuals—usually two females or one male and one female. Sometimes three immature orangutans were encountered traveling together as well. Immature animals traveling together did not constitute basic population units, since the associations (although recurring) were invariably temporary and broken off after a number of days or weeks.

The genealogical relationships between individuals occasionally in association (including adult females) need not always be simple ones. The known immature offspring of different mothers were observed traveling together on numerous occasions. Although orangutan populations are extremely fragmented, the recurring nature of small temporary associations among the animals at Tanjung Puting indicates a social organization that is relatively complex and not exclusively based on direct kinship lines. Peer relations seem more important than uterine ties in determining associations among immature animals. Relationships established before maturity continue to affect later life. In one case a recurring association between several initially immature females continued—but with modifications—even after the oldest female attained adult size and gave birth to her first infant.

ORANGUTAN DISTRIBUTION

Female-young home ranges seemed stable, in that during the 41 months of observation reported on here, all females remained in the same general area as when initially observed. The portion of the home range that was intensively utilized and the patterns of this utilization depended to a great degree, however, on the food sources available at any specific time.

Range boundaries seemed to be definite, as it was usually possible to predict the boundaries each individual female would not cross. The home ranges of females overlapped so extensively that there was no area exclusively occupied by any one female-unit. It was possible to find any of perhaps half a dozen females in the same area during the study period. It must be stressed that there was no place in the study area not visited by the orangutans at some time or other. Orangutans, especially males, were even encountered in the ladangs.

The home ranges of four adult females (Alice, Beth, Cara, Ellen, as well as Georgina) became known with some degree of certainty. All occupied ranges that consisted of approximately 5-6 square kilometers in size, which is considerably larger than the female home ranges reported from Kutai (Rodman 1973a) and from Sabah (Horr 1972). Juvenile offspring in the process of assuming independence, whether male or female, traveled essentially within the boundaries of the mother's home range.

The situation concerning male home ranges seemed more complex. It was possible to establish the home ranges of two habituated males over limited periods of time. During the first year of observation one mature male, Throatpouch (TP), was regularly observed. His home range was similar in size to female home ranges and overlapped extensively both with female and other male home ranges. This male suddenly disappeared. Despite extensive searching throughout the study area, he was nowhere to be found. Nor did we hear his long call vocalization, which we had learned to recognize. As time passed, we concluded he must have died (Figure 3).

During the year observed, TP shared his home range with a number of other mature males who periodically entered his range. These males seemed to only stay briefly—sometimes only a few days or weeks—before departing. Over the 41-month period most of these males were repeatedly sighted, although not frequently. One male (HH) was seen twice in early 1972, observed again for a few days in early 1973 (when we watched him leave the study area), briefly glimpsed in 1974 outside the study area, and observed again in 1975. Extrapolation from sightings such as these suggests that the infrequently encountered males occupied home ranges considerably larger than that established for the *resident* male. Attempts to follow these presumably wide-ranging males were invariably frustrated when they entered deep swamp or left the large study area.

After the resident male's disappearance, other mature males continued to visit his home range. Although the frequency of these visits increased, none stayed. Seven months after the resident male's disappearance, another fully mature male, Nick—one that we had never seen before—took over his home range. This take-over was rather specific in that the second male's home range, although somewhat larger, almost totally incorporated the first male's known home range. Although in the past, the first male seemed to be the most obvious presence in the area, despite the other males, the second male now assumed this role. Repeated and intensive observations of the second resident male confirmed that he did not leave his defined home range.

Approximately 20 months after the second resident male had established himself in the area, there was an influx of males into his home range. Most prominent among these was TP—the first resident male! We don't know where he had been for over 2 years, but we can certainly testify with a great deal of assurance that he had not been anywhere in his original home range.

The ranges of subadult or very young adult males presented problems of their own. Subadult males were defined as those males who were intermediate in size between the adult females on one hand and the huge mature males on the other. These subadults were only beginning the development of cheekpads and the large laryngeal pouches that characterized the fully mature males. Some were sexually active. Unfortunately, we experienced some difficulty in identifying a few of the subadult males. Most of them seemed to have home ranges larger than adult female ranges.

PATTERNS OF INTER-UNIT CONTACT

Although variability in orangutan population structure is not great in the sense that orangutans in adult life are always predominantly solitary, there do seem to be differences in the extent of orangutan sociability from one population to the next. The organization of the orangutan population in the Tanjung Puting forest is such that encounters between units are not frequent and do not usually represent everyday occurrences between adult individuals.

Of a total of 5470 hours of observation, 981 hours, or 17.9% of the observation time represented two or more separate orangutan units in contact. Two or more units were defined as being in contact or in the same vicinity when both were simultaneously visible to the observer, and presumably each unit's presence was visually obvious to the other.[1]

FIGURE 3
Throatpouch or TP. First resident male standing on ground looking back. © National Geographic Society.

Two units need not be in the crown of the same tree to qualify as being in contact. Observations of orangutans interacting (e.g., traveling together, chasing, etc.) indicated that they could be a number of trees apart and still maintain interaction. The main criteria for contact was whether the units could identify each other. Two units passing 50 meters apart did not count as a contact, although it was obvious that each unit was cognizant, as we were, that another unit was in the vicinity. Contact was defined only when recognition of the other unit's identity was possible or unavoidable. Two units could be feeding close by, each seemingly totally oblivious to the loud sounds the other unit was making. Then suddenly they would make visual contact and one unit might flee rapidly. Such an incident counted as a contact only for the few minutes that visual contact was maintained and not for the longer period that both units were in the same general vicinity of forest.

Our figure of 17.9% is considerably higher than the 1.65% of all observation time in which aggregations and secondary groupings were observed by Rodman (1973a) in Kutai. This difference can be related to differences in population composition as well as differential ecologies in terms of forest types, food distribution, home range sizes and overlap. Tanjung Puting orangutans, although essentially solitary, were capable of complex social interaction. The picture of Bornean orangutans derived from Sabah (Davenport 1967; MacKinnon 1971; Horr 1972) and Kutai (Rodman 1973a) as exceedingly asocial animals who only very rarely encounter one another does not apply to the Tanjung Puting orangutans. Over 200 unit groupings were observed during the time period discussed in this paper. Although differences exist, the Tanjung Puting data bear some resemblance to the situation described by Rijksen (1974) at a study area in Gunung Leuser Reserve, Sumatra in that adults were considerably more solitary than immatures.

Our figure of 981 hours, however, somewhat exaggerates the extent of orangutan sociability, in that frequently little or no interaction was observed between units in the same vicinity. Two units would sometimes feed close together in the canopy with barely a glance in the other's direction. Some social interactions involved avoidance responses, ranging from very mild reactions, where one unit might withdraw from a tree but remain in the general vicinity when another unit approached, to instances where immediate descent to the ground and flight along the forest floor took place. Occasionally there was aggression. In many instances, however, two or more units would travel together after making contact.

Female-female contact is not of direct concern here; the relationships between any two females varied considerably and females encountered each other in the forest quite frequently. Both adolescent and adult females tended to encounter subadult males more frequently than fully mature males.

In the absence of females, encounters between lone subadult males and mature cheekpadded males were rather rare. These encounters increased in the presence of females. Actual encounters between mature males were very limited. From these patterns of inter-unit contact, the social organization and mating system of Tanjung Puting orangutans can be determined.

Sexual Receptivity and Pregnancy

Orangutan females do not undergo any readily apparent physiological changes correlated with sexual estrus. Sexual receptivity could only be gauged in terms of the female's behavior in the presence of males and, to a lesser degree, the behavior of males towards her. Pregnancy, however, seemed to be accompanied by enlargement and whitening of the perineal region.

The Tanjung Puting evidence suggests that females give birth at a usual interval of about 4½ to 5 years. Thus, none of the females seen with small infants at the end of 1971-beginning of 1972 had yet borne second infants or were visibly pregnant 41 months later.

Male-Female Encounters

The standard adult female response to an encounter with a male was either to avoid or to ignore him. Males likewise frequently ignored females, although on other occasions they would approach and follow any female encountered. Generally, if a mature male were involved, this contact would be brief, with the male not following very far nor for very long while the female vigorously moved away. The vast majority of encounters between males (whether subadult or fully mature) and females did not result in copulation. Subadult males would sometimes check the female's perineal region, either with their fingers or with the mouth. Adult males were never seen to do this.

Both the number of males a female encountered during any one interval of observation and her response to them seemed to vary directly with different stages of her reproductive cycle. Receptive females encountered numerous males (especially subadults) whom they did not actively avoid. Continued observation of the same females after they gave birth indicated that incidence of contact with males dropped considerably. Sometimes they actively avoided any males encountered. For example, one adult female, Cara, was observed in the initial phases of pregnancy. During one 8-day period, she was followed at various times by no fewer than four subadult males who flocked about her either singly or in pairs. She also encountered one large adult male. She did not mate with any of them. After the birth of her infant, her encounters with males became very infrequent. During 31 successive days of observation, she only encountered one subadult male and one mature male. She seemed to shun such contact actively. On one occasion she descended to the ground and ran 300 meters along the forest floor, with her infant wrapped around her neck, to avoid a subadult

TABLE 1
Matings involving a mature male: consortships

| CONSORTSHIPS BETWEEN ADULT MALES AND ADULT FEMALES | | | | INTROMISSION | EJACULATION | LENGTH OF COPULATION BOUT (Thrusting Not Necessarily Continuous) | FEMALE VOCALIZATIONS DURING COPULATIONS | | FEMALE STRUGGLE | MALE VOCALIZATIONS BEFORE/DURING COPULATIONS | |
MALE	FEMALE	DURATION OF CONSORTSHIP	DATE COPULATION OBSERVED				RAPE GRUNTS	KISS-SQUEAK		LONG CALL	GRUMBLING
TP	Priscilla	3/20/72 1052 to 3/22/72[a] 1801	3/22/72	x	x	7'				x	
			3/22/72	x	?	6'		x	x		
		4/19/72[a] 1604 to 4/20/72[a] 1430	4/20/72	x	x	10'				x	
			4/20/72	x	x	14'				x	
TP	Val	4/1/75 1221 to 4/3/75 1845	4/2/75	x	x	7'				x	
			4/3/75	x	x	5'				x	
Fingers	Lolita (also called Hen)	7/14/72[a] 1735 to 7/17/72[a] 1654	7/15/72	x							x
			7/15/72	x						x	
			7/15/72	x						x	
			7/15/72	x						x	
			7/15/72	x	x	3'				x	
			7/15/72	x	x	4'					x
Nick	Noisey	10/23/74 0916 to 10/27/74 1745	10/23/74	x	x	11'					x
			10/23/74	x	x	8'				x	x
			10/24/74	x	x	5'					x
		12/10/74 0835 to 12/14/74 1458	12/13/74	x	x	6'					x
			12/14/74	x	x	15'				x	x
			12/14/74	x	x	9'					x

[a] Indicates that the male and female were already in association when the observation period began or ended. Thus, the true length of contact between male and female is not known.

TABLE 2
Matings involving a mature male: non-consortship

MALE	FEMALE	DURATION OF ASSOCIATION	DATE COPULATION OBSERVED	INTROMISSION	EJACULATION	MINUTES COPULATION (Thrusting Not Necessarily Continuous)	RAPE GRUNT	KISS-SQUEAK	FEMALE VOCALIZATION DURING COPULATION	STRUGGLE BY FEMALE	LONG CALL	MALE VOCALIZATION BEFORE/DURING COPULATION	GRUMBLING
Nick	Val	8/19/73 1547 to 1807	8/19/73	x		9'	x				x	x	

male approaching in the trees. This same pattern was repeated with two other females observed before and after the birth of their infants.

Copulation

Sixteen pairs of orangutans were observed in the process of mating or attempted mating on a number of occasions (See Tables 1 and 2). Copulations generally took place within two contexts. Some males and females established stable consort relationships where they traveled together for several days in succession and mated a number of times. Such consortships sometimes recurred over a period of months. Other contacts were briefer and involved degrees of unwillingness on the part of the female. Such unwillingness ranged all the way from seemingly token resistance with some squealing to fierce battles between the male and female. Usually, in these attempted forced copulations, only one (or no) completed copulation took place per encounter, although the male may have initiated copulations several times or even followed the female for a number of hours.

Copulations were generally face-to-face with one or both individuals hanging from an overhead branch. The male was usually lower than the female. Thrusting lasted from 3 to 17 minutes and was almost invariably preceded by the male's oral contact with the female's genitalia. In the case of mature males, a long call or the grumbling portion of the long call almost invariably accompanied or preceded copulation.

Three mature males (TP, Nick, and Fingers) were seen to copulate with four females (Priscilla, Noisey, Val, and Lolita) over a period of years. Four relationships involved consortships; one was an uncompleted copulation in which the male desisted after the female began struggling.

The first consortship observed involved the first resident male (TP) and an old mature female (Priscilla), who was followed by her large juvenile. Observations on the consortship continued for 3 days during which time the male closely followed the female. Copulations took place on the third day. A month later the pair was encountered again, but observation was possible for only 2 days. Two copulations were seen the second day.

The same adult male and female were later observed traveling together after sexual consortship was terminated but before the female gave birth. These associations, however, lasted only a matter of hours. The female produced an infant within the year.

The second consortship was observed for part of 1 day and then the following 3 days. It involved a mature cheekpadded male (Fingers) and a small eager adolescent (Lolita) mating and traveling together. It is not known how long the consortship lasted after observations were discontinued. Copulations occurred only on 1 day.

The second resident male (Nick), who appeared following the disappearance of the first male, during 1 month was frequently in the general vicinity of a female (Val) with large infant but not in association with her. Although he encountered various other adult females and even followed them briefly, he attempted mating only with Val. On one occasion he located her, followed her for almost an hour and then initiated copulation. She began to struggle and he finally desisted in his efforts. Although intromission did occur, ejaculation did not. The struggle lasted 9 minutes but thrusting was only continuous for 10 seconds at any one time.

Over a year later, the same large resident male (Nick) entered into a consort relationship with a large adolescent female (Noisey) whose home range extensively overlapped with his. The first period of consortship lasted 5 days; copulations occurred only on the first and second days. One-and-a-half months later, Nick and Noisey resumed the consortship for another 5 days. Copulations occurred on the fourth and fifth days. Apart from these two periods of consortship, this male and adolescent frequently encountered one another and sometimes foraged in the same trees but generally did not travel together.

After the first resident male (TP) returned from his 2-year-plus absence, he entered a 3-day consortship with an adult female (Val). Copulations were observed on the second and third days although the adult male and female did not travel together much of the second day.

Subadults were actually more sexually active than the fully mature males and more frequently in contact with females (adult or adolescent) even when copulation did not occur. Mature males generally did not follow females very long outside of sexual consortship, but subadult males followed females for periods extending over 24 hours without mating. This is consistent with MacKinnon's (1971) data; he reported consortships without mating; these involved younger males. Subadults would occasionally investigate the genitals of females encountered without attempting copulation. When copulation did occur, however, it almost invariably occurred without the female's full cooperation.

A number of such forcible copulations or attempted copulations were witnessed. One hapless adult female (Beth) with a large infant was raped each time she met a particular subadult male (Mute). During a period of 5 months, three such encounters were observed. The female's infant participated in the struggle with the male. The struggle was very fierce with the female uttering low, gutteral sounds heard in no other context. In two cases, the subadult male was accompanied by an adolescent female (Noisey) who stayed in the background. After the forcible copulation was consummated, the male did not travel with the female, although in one instance he stayed in her general vicinity for some time. Significantly, subsequent observation disclosed that the female did not become pregnant.

Over 1½ years later, the same subadult male was observed to enter a fruit tree where the same adult female and her offspring were feeding. The male immediately began to attempt copulation with the female. There was a long and intermittent struggle, with the male and female sometimes sitting quietly next to each other. The struggle was accompanied by the female's low, hoarse grunts and groans. Finally the male and female began feeding. Later the female moved away. The male followed and initiated copulation; however, this time the female did not seem to resist. There were no low groans.

A large adolescent female (Georgina) was forced to mate on 2 different days within 1 week by a subadult male and a large not fully mature male. Both males followed her for the remainder of their respective days. Again, although she later became pregnant, it is clear from the timing of the birth that this set of forced copulations could not have been responsible.

Another smaller adolescent female (Fern) was struggling with a newly arrived subadult male who was attempting copulation. The mating was interrupted by the subadult male who had been accompanying the adolescent female all along. The newly arrived subadult moved away. The adolescent and the first subadult continued traveling together as before but did not mate.

A subadult male accompanying two adult females (Dale and Judy), assorted offspring, and extra juveniles in a multi-unit grouping, attempted to copulate, in turn, with each of the adult females. He began

TABLE 3
Matings involving a subadult male: consortship

MALE	FEMALE	DURATION OF CONSORTSHIP	DATE COPULATION OBSERVED	INTROMISSION	EJACULATION	LENGTH OF COPULATIONS (Thrusting Not Necessarily Continuous)	FEMALE VOCALIZATIONS DURING COPULATIONS		STRUGGLE BY FEMALE	MALE VOCALIZATIONS BEFORE/DURING COPULATIONS	
							RAPE GRUNTS	KISS-SQUEAK SQUEALING		LONG CALL	GRUMBLING
Tom	Alice	6/20/74[a] to 6/23/74 1451	6/21/74	x	x	9'	x	x	x		

[a] Consortship began on the morning of 6/20/74 (precise time not available).

positioning them and attempted copulation, but after each female began struggling he quickly desisted. Intromission did not occur in either case.

The sole consortship observed between a subadult male and an adult female consisted of a subadult male (Tom) and a very old lone female (Alice), whose large juvenile offspring was wandering on his own, traveling together for 4 days. Copulation took place on the second day. The female struggled and uttered the low, hoarse sounds heard in no other context. In other instances the female would vocalize the same low sounds whenever the male moved close to her. The pair separated on the afternoon of the fourth day.

On the third day of his association with the old female, the same subadult encountered two females. He left the old female and investigated the genitals and made motions to copulate with one of these new females (Ellen). She resisted but the whole struggle was in a very low-key manner that suggested they were actually play-fighting. The female did not vocalize at any time during this interaction. Later the subadult male fed in the same tree close to her; afterward he went off to rejoin the old female he had been following.

A subadult male (Mute) traveling with his frequent adolescent female companion (Noisey) attempted to copulate with an adult female (Hen carrying infant Hank) on encounter. Despite several attempts, he

TABLE 4
Matings involving a subadult male: non-consortship

| | | SUBADULT MALES AND ADULT FEMALES | | POSITIONING OF FEMALE | INTROMISSION | EJACULATION | LENGTH OF COPULATION (Thrusting Not Necessarily Continuous) | FEMALE VOCALIZATIONS DURING COPULATIONS | | STRUGGLE BY FEMALE | MALE VOCALIZATIONS BEFORE/DURING COPULATIONS | |
MALE	FEMALE	DURATION OF ASSOCIATION	DATE COPULATION OBSERVED					RAPE GRUNT	KISS–SQUEAK SQUEALING		LONG CALL	GRUMBLING
Mute	Beth	12/29/72 0815 to 1213	12/29/72	x	x	x	5'	x		x		
		2/27/73 1330 to 1415	2/27/73	x	x	x	6'	x		x		
		5/17/73 1101	5/17/73	x	x	x	17'	x		x		
		12/6/74 1520 to 1818	12/6/74	x			17'	x		x		
			12/6/74		x	x	14'		x			
Subadult 1/Dale		4/24/73 1252 to 1815	4/24/73	x					x	x		
Subadult 1/Judy		'' '' ''	'' ''	x						x		
Tom	Ellen	6/22/74 1104 to 1148	6/22/74	x						x		
Mute	Hen	9/12/74 0812 to 0856	9/12/74	x				x	x	x		
BWCP	Georgina	5/22/72 0835 to 1800	5/22/72	x					x	x		
			5/22/72	x					x	x		
			5/22/72	x	x	x	13'		x	x		x
Beast	Georgina	5/28/72 1448	5/28/72	x	x	x	16'		x	x		x
Subadult 2/Fern		2/14/73 1611	2/14/73	x	x		6'		x	x		
Subadult 3/Noisey		10/25/74 1006 to 1007	10/25/74	x					x	x		
		10/25/74 1513 to 1550	10/25/74	x	x	x	10'		x	x		
Subadult 4/Noisey		10/25/74 1550 to 10/26/74 0653	10/25/74	x	x	x	7'					

was unable to subdue her and she moved away. Interestingly, the adolescent female twice joined in the struggle, attacking the adult female each time.

Finally, on the third morning of the second resident male's (Nick's) October 1974 consortship period with the adolescent female (Noisey), a subadult male moved towards the adolescent and grabbed her. The female squealed and the mature male raced towards them. The subadult fled. Late that same afternoon while the mature male chased a subadult away, a second subadult appeared as soon as the mature male moved away. He forcibly copulated with the struggling adolescent and then was in turn chased off by a third subadult who also copulated with the adolescent. The return of the adult male the next morning caused this subadult to flee.

Male-Male Competition

Certain aspects of the orangutan breeding system can be inferred merely from the presence of sexual dimorphism in the species. For instance, large degrees of sexual dimorphism are almost invariably associated with polygamous mating systems (Selander 1972), in which there will be strong competition among (usually) the males (Mayr 1972)— whether for territories, places at a common breeding ground, or for females directly. On the basis of evidence gathered from their respective studies, MacKinnon (1971), Horr (1972), and Rodman (1973a) have each emphasized the necessary occurrence of male-male competition in the orangutan populations they investigated.

The Tanjung Puting evidence, both direct and indirect, for the presence of competition between males for females is considerable. First, as noted above, the very existence of extreme sexual dimorphism argues strongly for competition between males who are members of the larger and more aggressive sex. In such competition, size is likely to be important (Crook & Gartlan 1966) in determining dominance and thus access to receptive females.

Second, mature males are spaced in such a way that encounters between them do not usually occur. This contrasts markedly with the adult females and the immature animals who do frequently encounter one another. In the two instances where encounters between two lone, cheekpadded males were observed, there was no physical contact. In the first case, the visiting male (Tusk) sat quietly and very still, low in the trees, while the resident male (TP) called. Then after a fierce initial approach, the resident male slid to the ground and ran off in an opposite direction. A second encounter was observed over 2 years later between another pair of cheekpadded males, in which one male moved in the direction of the other.

Probably the most important mechanism for mediating the spacing of mature males is the *long call,* which serves to indicate the male's location in the jungle (MacKinnon 1971). The call consists of a complex series of grumblings and bellowing vocalizations that may last over 4 minutes and carry, under the right conditions, over a kilometer in the forest. This call is truly the most impressive and intimidating sound to be heard in the Kalimantan forest. Both wide-ranging and resident males were observed making this call. Frequency of calling varied considerably from day to day. Continuous observation of the first resident male (TP) on 23 successive days indicated that he called a maximum of five times per day. There were 2 days when he did not call at all; however, on the average he called twice a day. With the exception of one past prime male, all males were observed to call; some called even more frequently than the first resident male. In addition, males observed in the process of calling sometimes dropped or pushed over large snags as an immediate prelude to their call. The resulting very loud crash certainly increased the effectiveness of the call, in that the direction of the caller could more easily be pinpointed. Occasionally males (and some females) would push over a snag, without calling, to indicate their presence to nearby animals.

The function of this call was gauged primarily from observing the reactions of orangutans to hearing calls. Generally, unless the caller was very close, there was no reaction whatsoever. The individual, male or female, typically did not look up from foraging or look in the direction of the calling male. When the caller was within approximately 100-200 meters, however, mature males reacted either by aggressively approaching the direction of the caller or by rapidly moving away, often on the ground. It is possible that callers were individually recognized. (Certainly, we learned to distinguish the calls of the resident males.) One resident male (TP) was seen to flee from one caller, but later aggressively moved toward another caller in the same general vicinity (even though the callers were far enough away that he could not have discerned who each was by visual identification alone).

Calls sometimes elicited other calls, but not usually. One night was exceptional; we heard three calls emanating from three different directions occurring simultaneously at one point, as the first call elicited a second and then a third.

Males in consortship seemed unusually responsive to distant calls. In one case, the mature male (Fingers) in association with a receptive adolescent (Lolita) violently leaped out of his night-nest shaking branches upon hearing a rather distant call and then called himself. Usually such distant calls went totally ignored.

The long call probably also has a function in attracting females as well as spacing males. During the two mature male-adolescent female consortships witnessed, the adolescent would occasionally wander off or get behind the male as they foraged. When the male called, she would usually hurry toward him. If the female was already nearby, she would often move closer to him and sometimes dangle facing him positioning her genitals towards his face. (This seemed to be a characteristic soliciting position on the part of adolescent females.) The beginning of two of these consortship periods was witnessed. In one case, the adolescent female made her initial appearance seemingly as a direct response to the male's call.

There is some evidence that the movements of calling males, at least the wide-ranging ones, are somehow coordinated. The possibility exists that calling is involved in this as well. It must be stressed that all cheek-padded males were solitary and no such males were ever seen in association. The study area would seem totally vacant of mature males, both in terms of calls heard and males encountered, for long periods of time. Then calling would suddenly begin; at night in our camps we would hear several males calling from different directions. After a period of days or weeks, the calling would cease completely and all the visiting males would vanish. Whatever the reasons for such an exodus, the females and resident male remained unaffected in that they did not leave their home ranges.

We have direct evidence that such an exodus of wide-ranging males did, indeed, occur on several occasions. For instance, during the course of 7 consecutive days in April 1973, we observed three nonresident males (Ivan, Ralph, Tusk) independently move in the same direction (indeed, two used almost the same route for part of the distance) and vanish into the deep river-edge swamp on the edge of our large study area. After their departure, calling ceased and the central part of the study area seemed devoid of all mature males.

When mature males encountered each other in the forest, their behavior seemed to depend on the absence or presence of females. The two encounters between lone mature males involved only minor aggression. The two encounters observed between mature males in the presence of a female, however, resulted in considerable violence.

One such encounter occurred when the resident male, TP, was in consortship with a mature female, Priscilla. A visiting male moved toward the couple. The resident male charged toward him and a fight ensued, which lasted 25 minutes. Fighting consisted mainly of chasing and furious grappling, including biting at each other's face, shoulders, and hands. They fell to the ground numerous times and then climbed back up into the trees to resume the grappling. The bouts of actual physical contact were interspersed with periods when the two combatants, close together, stared intently at each other, their faces motionless and panting. Finally, they separated and sat facing each other 10 meters apart. The resident male threw a snag and called; the other male then disappeared while the victor went off in search of his receptive female (and her offspring) who had slowly moved away foraging while the combat was taking place. It appeared that neither combatant suffered severe injury in the course of this fight.

The second aggressive encounter involved a presumably nonreceptive female who had only recently given birth (Cara). She had been joined by a mature male when there was a nearby call, followed by the violent appearance of the calling male some minutes later. The very large calling male charged in the direction of the first male who came down to the ground and ran off with the second immediately after him. They ran so fast that it was impossible to follow them even though they cut a swath through the undergrowth. About 15 minutes later there was a call from 100-200 meters away. Neither male, however, rejoined the female who had fled in the meantime.

The contacts between lone mature and lone subadult males were more frequent than contacts between solitary adult males. Subadults simply avoided adult males when both were alone in the forest. Five such encounters were observed. In one case, the adult male called moving in the direction of the quickly escaping subadult. In two cases, the mature male gave no indication of even noticing the presence of the subadult male who was very rapidly moving away in the trees. In another case, a subadult male eagerly moved toward the sound of an animal feeding on the ground. He rapidly moved away when an adult male rose up off the forest floor into the trees toward him. In the last case, the subadult male kept well away from the adult (TP) and fled whenever TP moved in his direction, but remained in the general vicinity over half an hour before finally moving away. It is evident from these five encounters that adult male tolerance for lone subadult males was considerably greater than their tolerance for other adult males.

The fact that the focus of male-male competition centered on females was very clearly indicated in other encounters between subadults and adult males. Subadults were frequently in association with females, both adult and nonadult, receptive and nonreceptive. In some cases involving mature females, mature males appeared. One newly pregnant female (Cara) was followed by a number of subadult males; sometimes by two at a time with a third in the vicinity. On two occasions the same adult male (HH) appeared. He dropped snags and called just prior to his appearance. The subadult male in association with the female came down to the ground on each occasion (in one case leaving his night-nest) and disappeared. The mature male did not join the female either time. In most cases of this sort, the subadult male just vanished at the imminent appearance of the mature male by coming down to the ground and running. There seemed to be a marked contrast in the intensity of subadult avoidance reaction. While alone, a subadult moved out of the mature male's way in the trees. When a subadult was accompanying a female, he might not even wait to see the adult male, but would flee as the male approached. The ferocity and violence of the mature male's approach in the trees in these instances was exceptional and left no doubts about the male's aggressive intentions. The aggressive approach of a mature male once separated a subadult and an accompanying large juvenile female.

Among males, a true lack of sociability and the full emergence of male-male competition seems to appear only with adulthood. As many as three subadult males were briefly observed following a receptive female without conflict occurring. Some competition, however, was already evident even among males at this stage of development. The appearance of one subadult male caused another subadult to leave rapidly a multi-unit grouping consisting of two adult females and various immature animals. Another subadult interfered in a forcible copulation between his adolescent female companion and a second subadult. In another case, the approach of two subadult males caused the subadult who had been accompanying a newly pregnant female to retreat very hastily.

An interesting interaction was observed between two subadult males, one of whom, Tom, was traveling with an old adult female, Alice. Tom and Alice had mated once the previous day. While the pair were foraging in a fruit tree, a second, slightly larger subadult male (Balls) entered the vicinity, sitting very still several trees away. He seemed to be unnoticed by the other two. Then he moved closer. Suddenly, the first subadult charged toward him, breaking branches along the way. The fight we expected did not occur however; the two subadult males simply sat about 10 meters apart in the same tree. Later the second subadult moved toward the female in the fruit tree. The first male followed, uttering a low vocalization that sounded somewhat like a long call, but was not as long or as loud. While vocalizing, he moved closer to the second subadult. Some minutes later when he moved onto the same branch, the other subadult moved to another limb and urinated. Later, the first subadult moved closer to the female's fruit tree. Finally, when the old female left, the first subadult followed while the second remained sitting motionless.

The entire incident lasted about an hour and serves to illustrate the beginnings of male-male intolerance and direct competition for females. The fact that at one point the subadult vocalized a minor approximation of the long call is of the utmost significance. It must be stressed, however, that during the afternoon of the same day, a third subadult male entered another fruit tree where the old female and subadult male were foraging without evoking the slightest response.

The indirect evidence probably indicates the scope of mature male-male aggression better than the two instances directly observed. Mature males were distinguished as a class by their disfigurements. Digits seemed especially vulnerable; some males had parts of fingers missing or stiff fingers that were never observed bending under any circumstances. Other males had misshapen or broken fingers and toes. The right eye of one male was missing. Another male had a chunk of flesh missing from his upper lip, with a torn piece hanging down onto the lower lip. One male appeared to have a healed scar on his upper lip. As a final example, one male exhibited an upper canine that projected through his cheek to the outside of his face as well as some fingers and toes that had been gnawed. Rodman (1973a) mentions that one of two mature males resident in his study area had a cut on one cheekpad.

The possibility of congenital defects cannot be ruled out. In the case of the tusked male, this is probably the best explanation. With the exception of one subadult male whose toe was permanently turned up, however, no other subadult male or any female displayed anomalies of this nature.

It might be argued that the defects exhibited by mature males result from specific locomotion techniques not used by other classes of orangutans; however, this does not seem to be the case. Although mature males spent more time moving on the ground than females and immature males, females and immature males also locomote on the forest floor. There is no compelling reason to suppose that ground travel would induce such disfigurements. As for traveling in the trees, all age-sex classes suffered serious falls. Indeed, females and immature animals showed a much higher incidence of small falls (where the animal did not hit the ground but was able to catch itself in the trees) than the mature males, who rarely fell.

The best explanation would be that at least some of the mature male disfigurements represented the results of aggressive encounters with other mature males, presumably in the presence of a female. Only three out of 12 adult males encountered in the study area were without physical anomalies. The prevalence of such anomalies is highly indicative; direct conflict and aggression must have played their part in the lives of most males who survived to maturity.

Selection

Over the course of the observations, it seemed that females as a class tended to prefer fully mature males as sexual partners to the subadult or young adult males who more frequently associated with them. It was also clear that the mature males had preferences related to female age. Nowhere were these preferences more clearly expressed than in the case of adolescent females not yet pregnant for the first time. For example, one large adolescent female (Georgina) in 11 consecutive days of observation encountered four males: two subadults and two fully mature males. Both subadults followed and forcibly copulated with her. Yet the mature males in turn totally ignored her even though she very pointedly approached one of them.

An even more vivid example involved the consortship between a mature male (Fingers) and another smaller adolescent (Lolita). Lolita was first encountered accompanied by a retinue of three subadult males. One subadult soon left, but the other two faithfully followed her at least the rest of the day (5 hours). It was obvious from the adolescent's response to the subadults that she was not interested in copulation. Yet the next month she was observed following and eagerly mating with a large cheekpadded male. This consortship differed radically from the ones involving adult pairs, in that it was clearly the adolescent female who was doing the following and often initiating copulation—sometimes only to be rebuffed. Indeed, at one point she even groomed him (something that rarely occurs among orangutans). Another time she wrapped her arms around his neck in a curiously human gesture—extraordinary behavior for a wild orangutan (Figure 4).

Another instance involved a third adolescent female (Fern), who appeared immediately after a large nonresident male called Ralph. She followed him closely the rest of the day and then nested near him. He ignored her. The next morning she entered his fruit tree and also began eating. When suddenly something crashed in the distance, the male abruptly leaped up from feeding, shook the tree violently, and then grabbed her. He held her for several seconds while she thrashed about squealing loudly. When he released her, she immediately fled. He returned to foraging. Yet this same male had been following an adult female (Cara) only days earlier.

Probably the most convincing evidence for female selection in the mating process was the long-term relationship between the second resident male (Nick) and another adolescent female (Noisey). Although this female regularly traveled with a subadult male (Mute), she initiated consortship with the mature male. During the first consortship period, her eager advances towards this same big male can be contrasted with her total lack of interest in the various subadults who attempted copulation with her in the big male's temporary absence.

About a month and a half later, the adolescent female again initiated consortship. On the first day the adolescent intensively examined the mature male's genitals and groomed him—with no result. Another day the adolescent examined and sucked on the male's genitals without obtaining any response on his part. Only on the fourth day did copulation occur, after the adolescent groomed the male's perineal region. An additional two matings took place on the fifth day. The pair separated

FIGURE 4
Lolita adolescent female (above) and Fingers mature male (below). Fingers is eating termites from a piece of dead wood he brought with him from the ground. Fingers and Lolita formed a consortship pair for several days. © National Geographic Society.

late that afternoon. The male nested while the female continued foraging.

The eager responses of these adolescent females differed considerably from the normal reaction to a mature male's presence, which was to shy away or avoid him. Although adolescent females were observed on occasion to avoid mature males (probably during periods of sexual nonreceptivity), a differential response towards the two age-classes of males was clearly evident. Adolescent females sometimes eagerly approached fully mature males, but never once were seen to approach subadult males in a like manner. As an interesting sidelight, it might be noted that juvenile females were observed to solicit subadult males.

Whether mature females prefer adult males is, admittedly, more problematical, but nonetheless the data suggest that they do. During the course of the study, adult females, as opposed to the eager adolescents, were never seen to initiate contact actively with any male, whether subadult or adult. When the mature male in the two adult-pair consortships initiated copulations, however, the female was usually a willing participant. Data from Rodman (1973a) and Horr (1972) indicate that adult females, whose offspring had become independent of the mother's body, did sometimes respond to calling males by eagerly moving toward them.

The interactions described between adolescent females and mature males illustrate that discrimination in terms of sexual partners was not a female prerogative as it is in some animal species. The preliminary evidence suggests some lack of interest in receptive adolescents and a preference for adult females on the part of adult males. This behavior contrasts with the more indiscriminate sexual behavior of subadult males, who attempted copulation with unwilling females of both age classes.

ECOLOGY

Any explanation of orangutan adaptation and the mating patterns observed must be related to the characteristics of the particular tropical rain forest concerned. Although the ecological aspects of the Tanjung Puting data have not yet been subject to detailed analysis, certain statements can be made even at this point in the study.

A preliminary estimate suggests that at least 600 different tree species are found within the limited confines of the Tanjung Puting study area alone (Anderson 1975). These trees are not gregarious; they occur throughout the forest as complex mixtures of species. Although frequencies vary considerably from one species to the next, most tree species will not be particularly abundant.

In this type of forest, synchronization of fruiting cycles among species does not always occur (with the notable exception of the Dipterocarpacae). Usually only a few species will be in season at any one time and then not all individuals of these species will necessarily be fruiting or flowering. Furthermore, although some species come into season relatively regularly, many tend to have erratic fruiting cycles. Some species may flower and fruit at very irregular intervals that cover a period of many years. On the other hand, lower canopy species might often fruit twice within the space of one year. Even more remarkable, it may be years between fruiting seasons, and then the same tree species may fruit twice within one year, e.g., durian (Holttum 1969). Although there are general overall patterns of fruiting meshing with rainfall (McClure 1966), essentially each tree species appears to have its own fruiting cycle triggered by specific combinations of climatic conditions. At Tanjung Puting seasonal abundance occurred primarily when the fruiting cycles of a number of different trees coincided or when trees of preferred fruit species had particularly good crops.

Orangutans are frugivorous in that fruit seems to be the preferred food. Detailed examination of orangutan diet at Tanjung Puting, however, reveals that over a long period of time the bulk of orangutan diet consisted not just of the seasonally available fruits and flowers of many species, but of the leaves and bark of a relatively small number of species generally available the year round. Depending on the amount and type of fruit available, the contribution of these permanent sources to the diet rose and fell but was never reduced to nil. Some of these permanent sources were among the more common species in terms of their distribution in the forest, but they rarely offered much abundance at one spot (e.g., *Gironniera nervosa; Dischidia rafflesiana,* etc.). Of the species from which leaves were eaten, there always seemed to be some young leaves available throughout the year, although the amounts differed from one period to the next. Termites probably also belong in this category of permanent food resources.

Some of the seasonal food sources—especially the most preferred species—were rather rare but represented concentrations of food in one place, sometimes enough to provide many hours worth of foraging at one time for one or more units. The home range of one female-offspring unit may contain, at the appropriate season, only two or three widely separated, well-fruiting trees of any of these preferred species. When one of these trees came into fruit, it was repeatedly visited by any units exploiting it. Such a unit might be in season 3-4 weeks (e.g., *Durio sp., Arctocarpus rigida,* etc.).

The distribution of other seasonal food sources, when available, approximated the more ubiquitous distribution of permanent sources. These sources, always trees of the middle or lower canopy, were relatively common (at least in terms of the tropical rain forest) and occurred comparatively close together. Sometimes there were enough seasonal trees in fruit to give the forest an appearance of abundance. Although one unit's home range might contain some numbers of each species of this type, frequently one visit exhausted the tree's crop (e.g., *Antidesma cuspidatum, Santiria laevigata,* etc.).

A final category of food sources were species with localized distributions in very specific habitats such as peat-swamp or river-end. Since concentrations of these species might be found in localized areas, the fruiting or flowering of one or more of them generally heralded a period of great abundance with orangutans moving in to harvest the crop (e.g., *Sandoricum emarginatum; Ganua motleyana,* etc.).

The minimal resources available over any long period of time ultimately determine the carrying capacity of the environment. Orangutan adaptation is obviously a function of both seasonal and permanent subsistence resources. The distribution of seasonal abundance is very important, but the population level and even aspects of the social organization are determined also by the presence and spatial arrangement of permanent food resources, as suggested by the importance of leaves in the diet throughout much of the year at Tanjung Puting. Even during periods of fruit abundance, some leaves continued to be eaten. This was further indicated during mid-1973 when no major fruit crops came into maturity. The orangutan were dependent primarily on the permanent food resources in their home ranges for a period of several months. Bark, which is always eaten in small quantities, became the primary item in the diet. Some of the trees whose bark was excessively stripped were species whose fruit provided major sustenance in season. So much bark was eaten that a few trees were actually killed. Had the orangutan been organized into large foraging parties, parts of the forest would have been devastated and future food crops adversely affected.

Briefly, the distribution of resources in the Tanjung Puting forest is such that large concentrations of food are exceptional and scattered. Contrast this with the Budongo forest where stands of one food tree exploited by chimpanzees covered several acres (Reynolds 1965)! Orangutan dispersal at Tanjung Puting is a function of the year-round dispersal of resources in the environment. The scarcity of large figs from the Tanjung Puting study area can be contrasted with the situation in Gunung Leuser, Sumatra, where large concentrations of food frequently attracted large aggregations of orangutans—up to 11 animals at one time (Rijksen 1974).

Permanent orangutan population density throughout the study area forest at Tanjung Puting can be estimated as between two to three individuals per square kilometer. Orangutan biomass-unit area compared well with the biomasses of the other higher primates—gibbons (Hylobates agilis)[2] and red leaf-eating monkeys (Presbytis rubicundus) with whom they shared the inland forest. Pig-tailed macaques have only once been observed in the study area; proboscis (Nasalis larvatus), and long-tailed macaques (Macaca fascicularis) are common only along the rivers. Simultaneous and consecutive observation of two or more orangutan units with overlapping home ranges indicated that the study area forest was very well canvassed, both in terms of seasonally available fruit and permanent resources. Units with overlapping home ranges (including semi-independent offspring traveling alone) would not usually forage in exactly the same trees. The net result was that food sources, particularly seasonal ones, overlooked by one unit would be visited by a second or even third unit foraging in the same area. Any major fruiting trees were invariably visited by at least two independent units—sometimes simultaneously but more frequently consecutively—to harvest the serially ripening fruit. (Indeed, the large fruit of some preferred species, which dropped and ripened, were sometimes taken by orangutans right off the forest floor.) This pattern of orangutan foraging indicated that forest resources were being intensively and continually exploited by subsequently appearing and reappearing units.

Orangutans are exceptionally large for an arboreal primate, and their energy requirements are appropriately high. Orangutan arboreal locomotion, however, is slow and even laborious. Chimpanzees, who mainly locomote along the ground, are able to range widely and have been recorded traveling as much as 8-10 miles (approximately 13-16 kilometers) in one day (Goodall 1963a). Orangutan females commonly traveled in the trees and either carried or were accompanied by dependent offspring. Orangutan female day-travel distance never exceeded about 2¼ kilometers. An average day-travel distance was about 800 meters or less. No doubt, the female's home range is circumscribed to some extent by the distance that can be readily traveled and foraged through in one day. Orangutans exploiting mainly permanent resources usually traveled shorter distances per day than when concentrating on seasonal foods. Certainly separation in time must serve to minimize competition for food between female units whose extensively overlapping home ranges ensure maximal utilization of the habitat.

The highly asocial nature of orangutan adult males (as compared to females and immature animals) also serves to reduce direct food competition. The resident male whose home range overlaps that of a number of females does not often associate with the females in his area, does not influence their activities, and usually does not compete directly with them for food, but is available nearby to provide his reproductive prowess when necessary. Also, the presence of the resident male may serve to inhibit the forays of other males into a female's home range. Unlike the females who seem to occupy one area permanently, there may be a turnover in resident males.

The fact that male-male competition sometimes reduces direct competition for food between males and females is probably best illustrated by data from Sumatra. In Tanjung Puting rarely were there trees sufficiently productive to attract large aggregations of orangutans. In Sumatra, however, when large figs were in season, aggregations of females and immature animals would gather to feed in one tree. Invariably, there would be only one large male per tree (Rijksen 1974). If a second mature male appeared, he would replace the first, retreat, or be chased away. The females were never in a position where they had to compete directly for food with a number of these huge animals.

There is also evidence from Tanjung Puting that mature males and females were exploiting slightly different ecological niches. Ground foraging seemed to be a normal part of each habituated mature male's daily routine but was much rarer even among the most habituated females.

MacKinnon (1971) reported what amounted to seasonal migrations in an orangutan population studied in Sabah. The Tanjung Puting orang-utans also showed similar movements in concert with the appearance and disappearance of various seasonal food resources. At certain times most orangutans would be concentrated in the inland swamps. At other times units would move independently to the river edge. As far as the females and resident male were concerned, however, these movements seemed entirely within the boundaries of known home ranges. All ranges contained areas of both swamp and dry ground forest, but not all female ranges were contiguous with a river edge. Only the females whose ranges touched a river traveled to the river edge to exploit food sources available there at the appropriate times. The others stayed inland.

It can be hypothesized that the adaptive strategy of wide-ranging males at Tanjung Puting seemed to lie in their success at continually following and exploiting abundant seasonal food crops over the wide areas that constituted their home ranges. In the Sekonyer River system as a whole, major differences in fruit-ripening times were often ob-served between areas only several kilometers apart. One important swamp fruit came into maturity and was being eaten along the Sekon-yer Kanan river edge at least one full month before the same fruit began to ripen in the inland swamps only a few kilometers away. Also, closely adjacent areas varied considerably in the frequencies of various pre-ferred fruit sources. Certainly the propensity of males for more effi-cient ground locomotion (as opposed to arboreal travel) increased their short-term ranging capacity. Males were seen to travel distances of over one kilometer along the forest floor very quickly. Indeed, it became apparent that, as the big males became fully habituated, almost all of their long distance traveling was done on the ground. The most habitua-ted males would sometimes climb down the trunk of their nest-tree in the morning and then climb back into the canopy only after reaching their first food source.

The Tanjung Puting study area data indicate that wide-ranging mature males tended to appear in times of seasonal abundance. For instance, the first resident male's (TP's) return to his originally defined home range after a 2-year-plus absence did not come as a total surprise, since his reappearance coincided with an extraordinary profusion of fruits and flowers. Two common swamp species came into season at about the same time, providing an unusually abundant crop.

Likewise, the central study area seemed devoid of any males during mid-1973, when only minor fruit crops were available and the females were subsisting on unusually large amounts of bark. Perhaps significant-ly, it was during this one time that there was even an absence of resi-dent males. Nick's initial appearance in TP's home range coincided with the first good fruit crops.

ADAPTIVE SIGNIFICANCE OF MATING PATTERNS

Orangutan adaptation poses unique circumstances for a higher primate. Since natural selection has resulted in orangutan populations composed of incomplete breeding units with a relatively low incidence of encounter, simply locating a sexual partner in the forest constitutes the first problem in reproduction. Furthermore, to avoid much direct competition for food, association between males and females is comparatively brief, even in consortship.

According to Trivers (1972), female and male reproductive strategy in the same species is likely to be different, based on the differential amount of parental investment each sex contributes to the offspring. Both Horr (1972) and Rodman (1973a) have pointed out the utility of Trivers' analysis to an understanding of orangutan adaptation in Borneo. In reality, orangutan reproductive strategy is different not only between the sexes, as discussed by Horr (1972), but also between age classes and even classes of males. The mating system observed at Tanjung Puting can be at least partially explained in terms of the adaptive strategies most productive for each class of individuals.

At Tanjung Puting adult females seem to breed, at the maximum, once every 4 or 5 years. This period is correlated with the prolonged dependency of orangutan young. Given this long interval between offspring, receptive fertile females are a rare commodity at any one time, widely scattered throughout the forest. Female reproductive capacity will be a limiting resource for male reproductive success (Trivers 1972). Males must first locate and then compete for this resource (as, in fact, they do). Male-male competition is concommitant with the ecological factors that rendered a solitary existence adaptive for both males and females. The crush on the environment if eight or so mature males followed one female orangutan continuously during her periods of receptivity (as sometimes occurs with chimpanzees; see Goodall 1971) would certainly be nonadaptive for the female.

Both adolescent and adult females seem to prefer fully mature males to the smaller subadults or young adults whom they encounter more frequently. The physical fitness of the big males vis-à-vis other males has been demonstrated—simply by their successful survival to maturity. Full cheekpads, large size, and the ballooning throatpouch represent the badges that presumably are normally associated with breeding success. Mature males advertise their presence over long distances by means of the long call, which serves to attract receptive adolescent females, and, no doubt, adult females as well. Although resident males were capable of very rapid, long distance travel, they sometimes tended to be remarkably sedentary. (The second resident male once moved a circuitous distance of 50 meters during the course of one day.) By means of zeroing in on the resident male's call, which is given an average of twice a day and frequently much more often, a receptive female could probably locate him within a matter of days. This is adaptive behavior for the female, since fertilization as quickly as possible by a genetically fit male enhances her own reproductive success.

Both adolescent males and females are eager to begin reproduction immediately after puberty before full maturity is attained. The initiation of reproduction as early as possible maximizes reproductive success over the whole lifetime for both males and females. Adolescent females in many primate species tend to be sterile, however (for chimpanzees, see Asdell 1946 cited in Goodall 1965). The lack of response by dominant mature males to the estrus of young females has been reported for other primate species (for baboons, see Hall and DeVore 1965). Fully mature orangutan males seemed less than eager to mate with undersize immature females, even when these solicited copulation in a most direct way. The second resident's unresponsive behavior in several instances toward the eager and obviously receptive adolescent (Noisey) can be contrasted with his initiation of mating with an unwilling adult female. Adult male-adolescent female consortships were always initiated and maintained by the female.

Adolescent females generally are not preferred by mature males. Although the evidence is slim, such adolescents may be less selective in terms of their sexual choice than mature females. Although no copulation involving a subadult male was ever female-initiated and was almost invariably without the female's cooperation, certainly the worst struggles were between these males and mature females.

Immature males are in a particularly untenable position regarding reproduction. Receptive females of both age classes seem to prefer the large fully mature males as sexual partners. Extensive association with any adult female—receptive or not—is dangerous. The abrupt appearance of a mature male could mean disaster. Denied easy access to receptive females, subadult males have three alternatives open to then: (a) delay reproduction until maturity and thus avoid violent confrontations with mature males until better equipped to handle such situations; (b) forcibly copulate with unreceptive females on the off-chance that fertilization would occur; (c) locate receptive females before the mature males do.

The first alternative perhaps increases chances of survival but certainly is not viable in terms of maximizing immediate reproductive success. Nonetheless, it must be emphasized that most encounters between females and subadults did not lead to copulation attempts. To attain some reproductive success, subadult or young adult males are forced into the second and third alternatives. Subadult males do forcibly mate with females; mature males generally do not. Almost all copulations initiated by subadult males were resisted by the female.

Perhaps more important, the subadults locate and contact females on a more regular basis than the mature males do. Since they are not capable of attracting receptive females to them by means of the long call, this behavior is essential. Subadults frequently checked the genitals of females encountered as if to ascertain their reproductive state. Having more frequent contact with females as a class, subadult males were able to locate females just as they were coming into receptivity. Three of the four receptive females who later became involved in consort relationships with mature males were seen escorted by subadult males prior to the appearance of the adult male. This is somewhat analogous to baboons—the female at the beginning of estrus mates with subadult males, but arouses the interest of dominant males only at the peak of estrus. The orangutan females showed no interest in the subadult males, however. A receptive orangutan female will move toward a calling male (for example, Noisey towards Nick) and in all probability will soon become monopolized by the resident male with whom she shares her home range. Assuming the female's preference for mature males, simply locating a receptive female just as she comes into receptivity is not enough to ensure success for a subadult. Attempted matings are probably rebuffed and may lead to forcible copulations. The reproductive success of subadult males (as confirmed by subsequent observations of raped females) seems to be rather limited. Female selection and competition with the big adult males seem to exclude subadult males effectively from the reproductive process.

Nonetheless, the tenacity of subadult males regarding sexually receptive females, even in the presence of mature males, was almost incredible. One subadult male doggedly followed a consorting adult pair (TP and Val) from a discreet distance for part of 2 days even though he was periodically driven off by the mature male. He would simply return a short time later. A subadult male attempted seizure of an adolescent female (Noisey) in a brief interval when she was foraging apart from her mature male consort (Nick). There were at least three subadults following this same adolescent-adult pair at one time or another. The adult male's abandonment of the adolescent female to chase away one subadult intruder only resulted in her forcible copulation with two successive subadults who must have been lurking in the background. It is likely, however, that the female was past the peak of estrus since her last copulation with the mature male had taken place the morning of the previous day.

On the basis of data collected in Kutai, Rodman (1973a) has suggested a life history for orangutan males, in which young adult males at physiological maturity (corresponding to the subadults of this study) begin to range widely before permanently settling down to smaller home ranges in maturity. Range size is interpreted as a function of the male's stage of development. Although this hypothesis has merit, the situation at Tanjung Puting was different and much more complicated.

First, some subadult males exhibiting postpubertal characteristics seemed to be present regularly in the study area. The possibility that a few subadult males were ranging widely cannot be excluded since some individual males were very rarely encountered. It is possible that the histories of different males were already beginning to differentiate even at this early stage. Nonetheless, these males were already competing with the mature males for receptive females.

Second, most mature males ranged widely rather than remaining resident in one small home range. The ranges of mature males were not discrete. There was a dichotomy between resident males—those mature males whose location and boundaries of movement over a period of time could be predicted with the same general assurance as that of the females—and the others—nonresidents who vanished and then reappeared in the study area. This dichotomy, although real, was not absolute since individuals could apparently pass from one category to the other and then back again.[3]

The organization of the Tanjung Puting orangutan population into resident and nonresident males might, at first glance, be explained in terms of intense competition among mature males resulting in the exclusion of some males from residential status. This is not adequate, however, to explain all aspects of the data. If the competition was for territories or sections of the forest, one would expect that each area or most areas would support mature males in residence and that these territories would not have much overlap, but this was not the case. Instead, although females and immature animals seemed evenly distributed throughout the study area forest, all areas did not support a resident male. This was true of our most intensively studied and best known area of dry-ground forest, although mature males did pass through it. An immediately adjacent area held two resident males. It might be of some significance that the range boasting two successive resident males (and subsequently both of these individuals at once!) consisted in large part of peat and deep river edge swamps, a habitat that periodically produced large concentrations of food. Perhaps competition among males revolved only about favored home ranges. It must be stressed, nonetheless, that both resident males in turn were observed in this home range when forced to subsist predominantly on permanent resources.

If, as hypothesized, the adaptive success of wide-ranging males depended on their successful exploitation of seasonal fruit crops (the preferred food resource) over large areas, it becomes clear that a resident male who restricts himself to a small home range is limiting himself unnecessarily to the food resources available there and nowhere else. In times of food scarcity, this would be an outright disadvantage vis-à-vis the other mature males. Establishing residency for long periods probably involves a sacrifice. That this sacrifice might not always be worthwhile, especially in seasons of extreme food scarcity, is suggested by the fact no mature males were in residence during mid-1973 when the females were subsisting primarily on bark. It is also suggested by the fact that some areas seemed devoid of resident males.

There must be a reciprocal advantage for the individual mature male to keep him situated in one limited area for long periods of time. The main advantage seemed to involve better access to receptive females. Both resident males were observed in stable consort relationships. Nonresident mature males were never so observed.[4]

A resident male will be regularly in close proximity, although not in direct association, with at least several females. He will be in a position to locate quickly one of these females coming into receptivity, especially since such a female will probably be moving toward him in response to his call. His constant residence in the area probably also serves to familiarize the females with his presence.

It is clear that resident status is an advantage only when receptive females are actually available in the environment. The abrupt disappearance of the first resident male (TP) coincided with a lack of potentially receptive females in his home range. At the time of his disappearance, three females in his range had recently given birth and all others were with dependent young; it is significant that his home range was not immediately reoccupied by a second male (despite continuing visits by other males for several months afterward). Likewise, his sudden reappearance more than two years later was very soon followed by a consortship with an adult female. This consortship was with the same female (Val) who had actively resisted the advances of the second resident male 1½ years earlier. Her resistance was probably a function of her infant's dependency at that time, although individual preferences could have been involved on her part as well. It must be stressed that the consortship of TP and Val existed in spite of the continuing presence of the second resident male (Nick) and at least one wide-ranging male (HH) in the same general area.

We also know that, during the 1½ years that Nick was resident in TP's home range, he entered into only one stable consortship—with the large adolescent (Noisey), a female who probably had no sexual relationship with his predecessor. No adult females appeared available during TP's long absence.

In Tanjung Puting the advantages of residence did not seem to involve access to superior or more abundant food resources. When seasonal abundance occurred in the second male's home range, his presence did not deter the forays of other mature males, although it may have limited them. With the initial intrusion of the two (or more) adult males into his home range, the resident male became obviously agitated and his calling rate increased, but he never managed to make contact with the intruding males. He seemed dominant in that the other males (including TP) avoided him and fled at his approach, but these same males nonetheless refused to leave and continued exploiting the resources of his home range.

It is tempting to hypothesize that resident males, although they have limited contact with their offspring, might, by their mere presence, effect a better environment for them and their mother(s) by guarding the resources of the common range against other animals. Thus, the second resident male did try to contact the other males, albeit unsuccessfully.

We can also speculate, based on the adolescent female's behavior toward the male in periods of nonconsortship, that a particular female might gain access to resources through her relationship to a resident male. No other female was ever seen to enter this male's tree while he was foraging. The adult females avoided him as much as possible although they might ignore his presence if he entered a tree where they were already feeding. The adolescent female, on the other hand, seemed to have no such compunctions and would forage quite close to him in the same tree. The test would be to see if such an attitude on her part continued after she gave birth.

Several other points need comment. It is clear, from observed reactions, that long calls served to space lone mature males in the forest, and furthermore, that such calls attracted receptive females to the calling male. Males consorting with a receptive female called more frequently and sometimes their calls were shorter than usual. The possibility exists that the frequent calls of a mature male consorting with a receptive female may serve to attract the attention of other males to her condition. The apparently somewhat coordinated movement of solitary mature males could be, in part, a response to the presence of a receptive female as well as to seasonal availability of food resources. Mature males, although never in direct association with each other, tend to be encountered in clusters, spatially and temporally. Perhaps it is irrelevant to ask whether the appearance of several males in Nick's home range and the return of TP was effected by seasonal food abundance of unusual proportions or by the presence of a receptive female. Perhaps the two—female sexual receptivity and seasonal abundance— might, to a limited extent, be correlated. Concentrations of food meant congregations of orangutans. It is not surprising that consortships sometimes occurred then. Further, the problem of direct male-female competition for food resources would tend to be ameliorated in these circumstances. The evidence is limited since the sample of consorting pairs involving a mature male is so small (four pairs) but, nonetheless, all were initiated during periods when the orangutans were eating fruit predominantly.

Rather than passing through definite life-stages, the males at Tanjung Puting seemed to be maximizing individual feeding and reproductive strategies. Maintenance and reproductive strategies might not always be complementary; and for some males, the quest for mere physical survival could be paramount. Some mature males could continue to range more widely throughout most of their lives. Such mobility may have its advantages in that a male will not be confined to the reproductive capacity of only the females in a small area. Contacting numerous females should increase the chances of encountering one in estrus, although if another mature male were established in her range, an estrous female probably will have been located and attended. In such a situation, combat between a resident male and a wide-ranging male might take place. It is quite probable that some mature males are effectively excluded from reproduction. In view of the constant competition, forced copulation with adult females would be useful strategy even for mature males.

The sporadic reappearance of the wide-ranging males in the study area does not preclude their establishment in other limited home ranges at other times when this is expedient in terms of access to receptive females. The fact that consortships between specific pairs recurred over a period of months indicates that a wide-ranging male might be forced to limit his wanderings if he wanted to maintain a consort relationship with a periodically receptive female.

Residence in a limited home range can perhaps be best explained as a specific strategy (probably reproductive) adopted under certain conditions by some mature males. Such an explanation would help account for observations of resident males in some locales (Rodman 1973a) and no resident males in other areas (MacKinnon 1971; Horr 1972) as well as the variable situation throughout the Tanjung Puting study area forest. The relationship of residency to dominance is slightly ambiguous. The first resident male won a combat focused on a receptive female, yet this same male avoided other males in his own home range. It could be argued that this male's lack of total dominance was one factor predisposing him to vacate his home range. Yet whatever the eventual status of the first male vis-à-vis the male who later established himself in his range prior to his return, it is clear that he still retained the capacity to compete successfully in terms of reproduction. Abandoning his range did not seem to diminish his mating success. Although this male (when alone) did flee from the second male (Nick) without contact being attained, this fact does not indicate that the second male would necessarily have driven away the first in a consort situation.

Skillful avoidance of confrontation in all but the most critical situations might be useful strategy for all mature males.

It is difficult to assess relative ages of wild primates accurately. The second resident male seemed somewhat younger than the first who could be subjectively characterized as middle-aged. Neither male was particularly large. The nonresident males, on the other hand, ranged all the way from one full-size young adult, whose age could only be discerned by the relative lack of fullness in the cheekpads, to an old, very large, extremely thin, slow-moving individual, whose cheekpads appeared withered and sunken. The old male first appeared in the study area during our third year of fieldwork and then periodically reappeared. He was never observed to call, but he once briefly chased a subadult male, with an astonishing display of vigor. The nonresident class also included several males of approximately the same age as the resident males but larger and seemingly more aggressive. One such large, fierce-looking male was observed actively to avoid contact with both the first and second resident males (after the first returned). Orangutan males may follow certain sets of sequences in their development, but these need not be in a fixed resident-nonresident order of stages, at least in Tanjung Puting. The exigencies of adult male-male competition (combined with maintenance considerations) are probably the most important factors in determining whether an adult male will range more widely than the females or limit himself to a relatively small home range for long periods of time.

Despite some unique characteristics, orangutans are still great apes and capable of complex discriminations and behavior. No matter what reproductive strategy would seem most appropriate in any given situation, it is obvious that personal idiosyncracies and specific experiences will play a part in mate selection and perhaps even appear contradictory to what may seem to be most adaptive. Unfortunately, it was difficult to collect extensive data on individual sexual preferences above and beyond those of age-class. Nonetheless, it appeared that orangutans demonstrated preferences for specific individuals of the other sex. This is not surprising. The subadult who repeatedly raped one female and tried with others was seen accompanying other females with whom he did not attempt copulation. Kinship or peer relationships may have been involved. The interest of the second resident male in one particular unreceptive female has been mentioned also. Individual preferences of this sort certainly must play an important role in orangutan social behavior, but at present their significance to orangutan mating patterns cannot be evaluated. Since consortships between specific pairs were observed to recur over a period of months, however, it would not be too astonishing if the relationship thus established spanned not just one pregnancy but several. Although no one is ascribing monogamy or the like to orangutans, it is possible that a successful male throughout his life could establish and maintain specific relationships with one or more females—perhaps even in different areas—mating with each in turn at the appropriate times in her reproductive cycle.

CONCLUSIONS

Natural selection has structured orangutan populations into incomplete breeding units. Adults of both sexes are essentially solitary, with the females being accompanied only by their dependent young. Contact between two or more independent units at the study area in Tanjung Puting Reserve accounted for 981 hours of observation or 17.9% of all observation time.

The mating system at Tanjung Puting was seen to be a result of differential reproductive strategies between sexes, age groups, and even classes of males. The occurrence of competition among males for females was documented. One combat between two mature males was observed as well as one chase in the presence of females. Aggressive encounters between subadult males and mature males were also observed. The large cheekpadded mature males were preferred as sexual partners by receptive females. These factors, combined with competition between males for females, resulted in the virtual exclusion of subadult males from reproduction, despite their forcible attempts at copulation. Male-male competition and female choice are important components in the evolutionary process of sexual selection. Such processes probably played an important role in orangutan evolution.

Acknowledgments

The research on which this paper is based was funded by the L. S. B. Leakey Foundation, the Wilkie Brothers Foundation, the National Geographic Society, the Herz Foundation, the New York Zoological Society, the Van Tienhoven Foundation of Holland, and the Jane and Justin Dart Foundation. We are most grateful to these foundations and organizations for making our work possible.

In Indonesia, Lembaga Ilmu Pengetahuan Indonesia (Indonesian Institute of Sciences) and Perlindungan dan Pengawetan Alam (The Nature Protection and Wildlife Management arm of the Forestry) served as sponsors and provided much assistance.

My husband and I are very grateful to Mr. Soedjarwo (Director-General of the Forestry); Mr. Siswoyo Sarodja (Assistant to the Director-General); Mr. Prijono Hardjosentono (Head, P. P. A.); Mr. Walman Sinaga (former acting-head P. P. A.); Mr. Soegito Tirtomihardjo (Head Hunting Bureau, P. P. A.); Mr. Eddypranoto Widajat (Head,

Kalimantan Tengah Forestry); Mr. Rombe, (Pangkalan Bun) and Mr. Yusran (Kumai) for the Forestry's support of our work. We are also very grateful to L. I. P. I.: Dr. Bachtiar Rifai (Head); Ms. Sjamsiah Achmad (Head, International Bureau), and Mr. Napitupulu.

The provincial government of Kalimantan Tengah also encouraged and supported our work. We are particularly grateful to Governor Sylvanus and Mr. G. T. Binti (Assistant to the Governor).

We acknowledge our enormous debt of gratitude to the late Dr. L. S. B. Leakey who initially sponsored our work and obtained funding. We would also like to express our deep gratitude to the following individuals who, in one way or another, have provided much help and support: Dr. Rainer Berger, Dr. Joseph Birdsell, Dr. Bernard Campbell, Mr. Robert Gilka, Dr. Jane Goodall, Ms. Mary Griswold Smith, Dr. David Hamburg, Ms. Barbara Harrisson, Ms. Joanne Hess, Dr. Peter Miller, Ms. Nina Sulaiman, Ms. Joan Travis, Trustees and members of the L. S. B. Leakey Foundation, Dr. J. H. Westermann, Mr. Robert, and Mr. Leighton Wilkie.

We would like to thank Drs. Bernard Campbell, Peter Rodman, Irven DeVore, and Robert Hinde for reading the initial version of this paper and for providing their comments. Also I thank Ms. Lita Osmundsen and the Wenner-Gren Foundation staff for their hospitality during the Burg Wartenstein Symposium.

From May 1974 onward, pairs of biology students from Universitas Nasional, Jakarta served for a period of 6 months (each) as research assistants and carried out their own field research on the primates of the study area. We thank Mr. Suharto Djojosudarmo, Mr. Jaumat Dulhajah, Mr. Sugardjito, Mr. Endang Soekara, Mr. Barita Oloan Manulang, and Mr. Yatna Supriatna.

We are very grateful to the Herbarium Borogiense; Head, Dr. Mien Rifai, for identification of botanical specimens and to Dr. J. A. R. Anderson for visiting us in his capacity as forest ecologist. Finally, we thank our local staff: Mr. Ahmad, who aids in the wild orangutan research, Ms. Bahriah, his wife, and Mr. Mujiran.

Notes

1. Immature animals in the process of assuming independence were counted as still belonging to their mother's unit when they met her, even though they might travel much of the time away from her.

2. Identification of gibbons as *H. agilis* was by Dr. J. Marshall. The "Bornean gibbon" *(Himulleri)* is not found in this part of Kalimantan.

3. Observations conducted after this paper was completed indicate that after his return, the first male (TP) resumed resident status in his original home range, sharing it with the second resident male who also remained there.

4. The male Fingers was encountered primarily outside our core 11 square kilometers; it was not definitely known if he was a resident male there.

233

Peter S. Rodman

Individual Activity Patterns
and the Solitary Nature
of Orangutans

The order Primates is fascinating to students of behavior from all disciplines, at least in part because of the nearly universal appearance of a gregarious social system among species of the order. A simple evolutionary hypothesis is that sociality, defined as bisexual group living, is a unitary phenomenon inherited by diverse phyla from a common ancestral species characterized by similar sociality. Such a simple hypothesis fails to take account of, or may even obscure, the variety of possible adaptive bases for structurally similar social patterns, and it specifically fails to account for the solitary nature of orangutans. Although no single hypothesis need explain all exceptions in order to retain some part of the truth, a more powerful hypothesis would take into account conditions that predict exceptions as well as rules. In the present paper I do not presume to advance an alternative to the hypothesis that primates are social because they are primates. I intend to examine elements of the lives of orangutans that appear to be strongly related to their solitary nature and that suggest elements of the lives of any animals that will be related to their sociality. More specifically, I will describe certain activity patterns of individual orangutans and discuss the relationship of these patterns to the solitary nature of orangutans.

THE FIELD STUDY: METHODS AND GENERAL RESULTS

From May 1, 1970 to July 31, 1971, my wife, four Indonesian assistants, and I carried out a study of the behavior and ecology of orangutans and other anthropoid primates of East Kalimantan, Indonesia. The study site covers 3 km² in the northeast corner of the Kutai Nature Reserve at latitude 0° 24′ north and longitude 117° 16′ east where it is part of a large expanse of primary dipterocarp forest. We set up a base camp on the south bank of the Sengata River, which flows east into the Straits of Macassar 160 km north of the Mahakam River delta. Eight species of the suborder Anthropoidea inhabit the study area, including crab-eating macaques, pig-tailed macaques, four species of leaf monkeys (*Presbytis aygula, P. rubicunda, P. frontata,* and *Nasalis larvatus*), Bornean gibbons, and orangutans. While constantly monitoring limited aspects of the behavior and ecology of the five most common species, we concentrated on observation of the behavior and ecology of orangutans.

Procedures

Our general procedure was to search the study area individually or in pairs by walking systematically over a grid of transects cut through the forest. After contact with an orangutan, an observer followed the animal for the rest of the day until it settled for the night. Before dawn the next day the same observer returned to the sleeping site and continued to follow the animal until noon, when he was replaced by a second observer. This system, which was facilitated by use of walkie-talkies, allowed six observers to maintain contact with up to three individuals or groups simultaneously. Normally two pairs of assistants working on alternate shifts maintained contact with or searched for orangutans. My wife (C. F. R.) and I could then move between different animals or search for orangutans as necessary. When we were not present, our assistants kept excellent notes on the animals' behavior. The longest period of contact with a single unit of orangutans was 11 whole days, at the end of which we voluntarily terminated observation in order to search for another orangutan.

FIGURE 1
Adult female Afa (right) and juvenile female Jfa. Note the white spot on Afa's right upper jaw. This photograph records one of two instances of grooming between these two females during more than 300 hours of observation of them together. In both cases the juvenile female groomed the adult, but there was no reciprocation by Afa.

Notes. All observers kept notes in diary fashion with notebook and pencil or with a tape recorder. C. F. R. and I recorded transition times of behavior—the time at which one activity ceased and another began— and described each activity as fully as possible in between. Although this would be a difficult method of recording data for active, group-living animals, it is easy for orangutans because their activities are normally limited to feeding, resting, and traveling, and because each activity takes place in a long block of time. Our assistants recorded transition times as well, but provided less extensive description of activities. They recorded primarily the activity and location of the animal, food type, and local names of food trees.

Individual Recognition of Orangutans. The first orangutan I identified positively (Afa, an adult female) had a round, white mark on her face external to her right maxillary canine (Figure 1). Adult males have wide cheek flanges extending from the sides of the face, and the first adult male I identified positively (BC) had a distinctive notch in his right cheek flange. Other orangutans were easily identified by head shape,

237

height of the hairline, and the length or direction of hair growth on the head. Given the small number of orangutans in the study area, it was not difficult to identify all of them positively.

Habituation. Orangutans reacted to us either with avoidance or with an elaborate agonistic display involving lip-smacking, belching, grunting, howling (males only), tree-shaking, branch-breaking, and vine-rattling. In order to observe their natural behavior it was necessary to observe them from concealment or to habituate them to the presence of an observer. We observed some animals from concealment, but we habituated members of three resident units to observation by persistently sitting in clear view. The success of this effort is indicated by the fact that during the first 10 hours of observation of each orangutan that was eventually habituated, the animals collectively spent 13.7% of their time in some form of display at us, but after the first 20 hours they spent only 2.7% of their time in such display. Although the persistence of some display among habituated animals indicates less than total habituation, the drop in display time provides an objective measure of the degree of habituation.

General Results

The resident population of orangutans consisted of six units of the following composition: two solitary adult males (Figure 2); two adult females carrying infants; one adult female followed by a young juvenile male (Figure 3); and one adult female carrying an infant and followed by a juvenile female. Of these, one adult male (BC), one adult female with infant (Afd), and the adult female with infant (Afa) followed by a juvenile female (Jfa) were habituated. The juvenile female (Jfa) followed her mother closely early in the study, but by the end of the study she left Afa for up to 5 days during which she moved alone through Afa's home range. I thus have records of Jfa's activity when alone and when with her mother.

Altogether the six observers accumulated 1640 hours of observation of habituated and nonhabituated orangutans. In this time we learned that the two resident adult males were almost completely solitary, and that the four adult females, each accompanied by offspring, traveled independently of each other in home ranges of approximately 0.5 km². These six population units combined into larger groups on 13 occasions for a total of 23 hours, or 1.65% of the total observation time. Although one such group involved mating between an adult male (RT) and an adult female (Afc), and although there was some indication of a continuing close relationship between two adult females suggestive of kinship (Rodman 1973a), the majority of these secondary groupings of primary units were chance encounters in single fruit trees with no social interaction among the units.

FIGURE 2
Adult male RT. RT appeared younger than BC, judging by the condition of his facial skin; although younger, he was fully mature.

This brief summary of results that are described in detail elsewhere (Rodman 1973a) is sufficient to demonstrate that orangutans in this study area were hardly more social than is essential for any mammal. To put it succinctly, they were quite solitary. I have previously asserted that the solitary nature of male orangutans is an adaptation to the distribution of female orangutans, but that the solitary habits of females "must" be an adaptation that maximizes their foraging efficiency and their success at raising offspring (Rodman 1973a, p. 206). In the following pages I attempt to explain how the solitary nature of both males and females is related to their size, their activity patterns, and their feeding behavior.

ACTIVITY PATTERNS OF ORANGUTANS

Methods of Analysis

In order to facilitate analysis of the activity patterns of orangutans, I have coded the transcribed notes into the categories feeding, resting, moving, nest-building, and agonistic display. Each record of observations is thus transformed into a series of codes separated by transition times, and it is quite simple to calculate absolute amounts of time devoted to

FIGURE 3

Adult female Afb (top) and juvenile male Jmb. The young juvenile continued to sleep in the nest with his mother at night. During the day he followed her closely, often grasping the hair on her back with one hand as she walked along the tops of large limbs. Note this female's distinctive high forehead.

the various activities. In general, observers did not record times for an activity lasting less than 1 minute. *Feeding* means any or all of the following: gathering, chewing, or swallowing food. *Resting* is essentially inactivity, although there were extremely infrequent and transitory events such as self-grooming, allogrooming, elimination, and play that occurred during activity bouts coded here as resting. For analysis the coded data were keypunched onto cards and then tabulated in various ways as described in the following material.

General Patterns

I have coded behavior of orangutans observed for more than one-half hour at a time, for a total of 1544 hours and 24 minutes. We have observations totaling slightly less than 100 hours in bouts of less than one-half hour each. Observations of habituated animals accounted for 1408 hours and 8 minutes of this time, and the following description and analysis of the behavior of orangutans is based on observations of their behavior. Most of our observation time of BC—the one habituated adult male—was accumulated during March and April 1971, and when I compare his behavior with the other habituated animals, I will use only data from March and April for all animals. It is interesting to look at activities during whole days for habituated animals in order to obtain estimates of variance in activity patterns; we collected 33 whole days of observations—from first movement to last—on individuals, and 9 whole days of observation of Afa, Ima, and Jfa when no distinction was made between the activities of Afa and the juvenile female, Jfa.

Length of Day. The mean time of first movement of habituated orangutans was 0648 hours ± 13 minutes (26 minutes = 95% confidence interval for the mean; N = 76). The mean end of the day (N = 105) was 1804 hours ± 6 minutes. The difference between these two figures gives a mean length of day of 11 hours 16 minutes (± 19 minutes). The mean length of 42 whole days (observed from first movement to last movement) was 11 hours 24 minutes (± 20 minutes); it is satisfying, although not surprising, that these two figures are statistically indistinguishable. The longest whole day observed was 12 hours 45 minutes (Jfa, April 11, 1971) and the shortest day was seven hours long exactly (Jfa, January 30, 1971). It may be significant that the day length was most variable for the juvenile female since she was in the process of separating from her mother, Afa, and would not have established her own pattern of daily activity. Pair-wise comparisons (F-test) show that the variance for Jfa's day length is significantly higher than that of any of the adults, but variance does not differ significantly between any two adults (Rodman 1973b, p. 174).

Diurnal Cycle of Activity. The pattern of distribution of time to the activities of feeding, resting, moving, nest-building, and agonistic display appears in Figure 4. All coded observations of habituated animals after the first 20 hours of observation have been broken down into activities in each of the 52 quarter hours between 0530 and 1830. The interesting features of the pattern produced are two feeding peaks (at 0700-0715 and 1615-1630), a midday rest period, and an increase of moving associated with the afternoon hours. This pattern results from the general sort of day, described as follows.

An orangutan moves into the fruit tree he fed in the night before and feeds heavily for a few hours. He stops for a long rest, may wake and feed, but then moves off for the long jaunt of the day (50-800 meters). This jaunt is associated with intermittent feeding on leaves, bark, or insects, and large adult males are quite likely to travel on the ground (BC spent 20% of his moving time on the ground). The journey terminates at another fruit tree where the orangutan feeds heavily again (afternoon feeding peak). Sometime after 1630 hours he moves abruptly into a neighboring tree and builds a nest where he settles for the night. The next morning the orangutan moves into the tree he fed in the night before, and the cycle repeats itself.

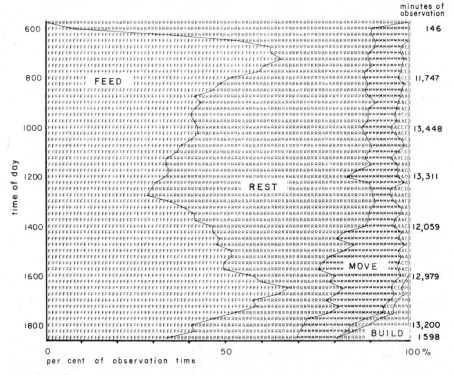

FIGURE 4

The time of day in hours and minutes is shown in the left-hand columns. Each row of letters (F, M, R, B, etc.) represents 100% of the observation time accumulated during one 15-minute period of the day, and one letter in the row represents 1% of that observation time. The numbers in the right-hand column give total minutes of observation during each quarter hour of the day.

Activity Profile. Over all observations of habituated animals after the first 20 hours of contact, orangutans spent 45.9% of their time feeding; 39.2% of their time resting; 11.1% of their time moving; 1.0% of their time building "nests", and 2.7% of their time displaying at the observer. I have described previously the form of agonistic display, and feeding will be discussed in detail later. Resting consisted of sitting quietly, frequently leaning against a tree trunk or sitting enclosed in a cluster of vines. Nonhabituated animals often climbed high into a tree and sat for hours after a period of agitated display. For example, on one day early in our acquaintance with Afa she sat concealed for 7 hours before moving. As I have noted before, some of the brief maintenance activities and transitory interactions between mothers and infants occurred during rest periods. I have counted time observed in the nest before first movement and after completion of the nest as resting time.

Nest-building is a complex activity that has been described in detail by Davenport (1967). Jantschke (1972) described the elements of nest-building among captive orangutans, and his results show that there is probably a large innate component of nest-building. Habituated animals in this study built nests on 75.2% of days when the end of the day was observed (N = 105). The two adult females, Afa and Afd, built new nests on most nights, but the juvenile female, Jfa, frequently used an old nest. The adult male, BC, often slept in the crotch of a large tree without a nest.

The Foraging Index. I have previously used a measure indicating the efficiency of foraging (Rodman 1971), and MacKinnon (1974) has called this a *foraging index.* The index is the ratio of time spent feeding to time spent moving by an animal. For all habituated orangutans taken together the index is 4.1, and this is considerably higher than any estimates for foraging indices of the other primates present in the study area. The tremendous difference in size between orangutans and the other primates suggests, as we might expect, that foraging efficiency is positively correlated with body size.

Feeding and Food. I have analyzed the proportions of time devoted by orangutans to feeding on various foods during March and April 1971, with the following result: 62.1% fruit, 21.7% leaves, and 16.2% other foods (flowers, buds, bark, insects). It is clear that these orangutans were predominantly fruit-eaters, and their pattern of behavior during the day appears to be adapted to careful harvesting of fruit from a few available fruit trees on a day-by-day basis. I assume that they take the suitable fruit from a tree and move to another fruit tree that is frequently of the same species, foraging in between.

The year of the orangutans in our study area consisted of a series of shifts from one key fruiting species to another, punctuated by brief interludes during which they turned principally to the leaves and bark of a few strangling figs *(Ficus* sp.). Table 1 shows the sequence of key fruit species during this study. December, February, the latter part of June, and July were apparently lean months. At these times orangutans fed on various fruits that occurred sporadically through the forest, and they relied heavily on the bark of distal twigs and on leaves of a few individual fig trees.

Although fruit is the principal component of the diet, orangutans eat some leaves or bark each day; this food may serve as roughage or may fill out nutritional needs of the animals. Even in the midst of feeding heavily on a fruiting tree, an orangutan will stop and move into a neighboring tree to feed on leaves intermittently. Some barks—the bark of *Ficus* or of *Artocarpus* sp.—were eaten completely, but only the cambium of others was taken, as in the case of *Uncaria* sp., *Cratoxylon*

TABLE 1
Sequence of key fruit sources August 1970 to July 1971

MONTH	FRUIT SOURCE
August 1970	*Koordersiodendron pinnatum*
September	*Dillenia borneensis*
October	*Dillenia borneensis*
November	*Ficus* sp.
December	No principal source
January 1971	*Dillenia borneensis*
February	No principal source
March	*Dracontomelon mangiferum*
April	*Dracontomelon mangiferum*
	Koordersiodendron pinnatum
May	*Jarandersonia* sp.
June	*Jarandersonia* sp.
	Ficus sp.
July	No principal source

sp., or *Shorea* sp. The cambium might be eaten, or only chewed and discarded. Throughout the year orangutans supplemented their diets with the succulent bases of the leaf stems of several species, including *Pandanus epiphyticus, Asplenius nidus, Aglaonema* sp., and one unidentified palm that grew in relative abundance more than 500 meters from the Sengata River. The local people commonly cut the latter palm and took the heart to eat raw or cooked. Orangutans pulled the leaf-stems from the palm and chewed the base. They only fed on palms short enough that leaves could be reached from the ground, or on palms that grew near a stronger tree from which the orangutans could reach down.

The important conclusion to be drawn from observations of food and feeding is that although a few food sources are common, such as the bark of *Shorea* sp., most *favored* species are extremely rare. Key fruit trees occur in localized clumps or as discrete individuals, and they fruit irregularly. Because of a heavy reliance on fruit as a food source, orangutans must have a highly developed spatio-temporal knowledge of fruit. Major bioenergetic constraints on these ponderous, arboreal animals enforce a highly efficient activity pattern without dependence on random foraging for food, and there is evidence that they make efficient predictions about food locations. I was amazed the day that I watched Afa leave a tree *(D. mangiferum)* where she had been feeding frequently for several days and move due south 750 meters without a stop to arrive at another fruiting tree *(Koordersiodendron pinnatum)*. She fed there for 20 minutes, and then returned along the same path to her original food tree. The fruit of *K. pinnatum* was not yet ripe, although shortly thereafter other orangutans fed in trees of that species in the same area. MacKinnon (1974) reports similar cases of remarkably direct routes to trees previously unseen by him. Obviously individual trees play a major role in the life of an orangutan, and the ability of an individual to predict the locations in time and space must be strongly favored by selection.

Individual Patterns

The activity profile of all habituated animals during March and April is essentially identical to that of the same group of animals over all observations in the study. During March and April orangutans spent 45.3% of their time feeding, 41.1% of their time resting, 11.4% of their time moving, 0.8% of their time building nests, and 1.5% of their time in agonistic display. March and April were unusual months, however, because the four habituated orangutans fed primarily on a single fruiting species *(D. mangiferum)*. We accumulated most of our observation time on BC, the adult male, during these 2 months, and since I am concerned with comparisons between his behavior and that of the three other habituated animals, I will limit analysis to this 2-month period.

Relative Sizes and Sexual Dimorphism. In the field I estimated the relative sizes of the habituated animals and the absolute size of the adult male (relative to my own size when I approached him on the ground). Relative size estimates of the orangutans were possible because all the animals entered the same trees on occasion, and I could judge each against the same tree. Given an estimated weight of 90 kg for BC, the approximate weights of the others are 50 kg, 30 kg, and 40 kg for Afa, Jfa, and Afd respectively. These estimates are consistent with reports that the mean weight of females is approximately one-half the mean weight of males among orangutans (Schultz 1969, p. 201).

TABLE 2
Individual activity profiles of four habituated orangutans (March-April 1971)

	FEED	REST	MOVE	BUILD[a]	DISPLAY[b]	FEED/MOVE
JFA	42.1%	42.0%	12.6%	1.3%	2.0%	3.3
AFD	31.4	55.7	10.6	0.5	1.8	3.0
AFA	47.5	36.3	13.1	2.1	0.9	3.6
BC	57.0	32.6	9.3	0.4	0.8	6.1
ALL	45.3	41.1	11.4	0.8	1.5	4.0

[a] Build = nest-building
[b] Display = agonistic display

TABLE 3

Mean duration of activities on whole days of observation[a] (in minutes, with 95% confidence limits)

	N	DAY LENGTH	FEED	MOVE	FEED/MOVE
JFA	3	728 ± 18	288 ± 135	81 ± 22	4.1 ± 3.1
AFD	7	689 ± 42	245 ± 49	63 ± 22	4.8 ± 2.2
AFA	4	715 ± 10	318 ± 107	105 ± 25	3.6 ± 2.2
ALL FEMALES	14	705 ± 23	275 ± 47	79 ± 16	4.3 ± 1.4
BC	7	688 ± 34	410 ± 88	57 ± 14	7.6 ± 1.8

[a] A whole day is a day on which the first movement of the animal in the morning was observed and the last movement in the evening was observed (and all else in between).

Activity Profiles. Table 2 presents the distribution of time to activities by the four individuals during March and April; there are highly significant variations in activity patterns among the individuals, but it is interesting to note that the adult male, BC, fed the most and moved the least of the four habituated animals. His foraging index is therefore high relative to the females. In order to compare individual patterns of activity, I have examined the day length and absolute distribution of time to feeding and moving on 21 whole days of observation in March and April. Table 3 presents the mean values of these variables for each individual and for all females together.

Foraging Indices. Although there are considerable differences among all individuals in activity profiles, it is interesting that feeding time and moving time vary in the same direction among females; that is, the female who moved the least also fed the least (see Tables 2 and 3). BC, on the other hand, fed more and moved less than any of the others. As a result the foraging index remains relatively constant among the females, but is considerably higher for BC. In order to test the hypothesis that BC's foraging index is the same as the others, and that the difference is only a chance variation, I have compared the mean foraging indices for whole days between individuals and between BC and the females as a group. Comparisons among females show that mean foraging indices do not differ significantly between any two females, but we can reject the hypothesis that the mean for BC is the same as that of any female at the 0.05 level of confidence or better in each case. We can also reject the hypothesis that the mean foraging index for BC is the same as all females taken together at the 0.01 level of confidence. BC is manifestly foraging more efficiently than the females.

The variation in foraging patterns through the day is also interesting. Figure 5 shows the foraging indices for each quarter of the day for BC and for the three females taken together during March and April. The daily feeding patterns of BC and the females are clearly quite divergent, and this divergence is probably a function of differential size; I will discuss the significance of these divergent daily patterns in the following section.

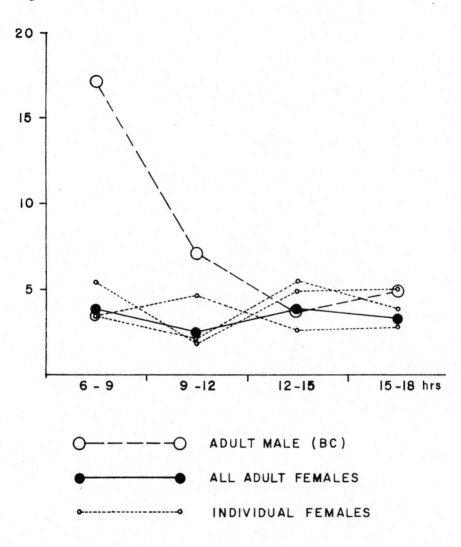

Diurnal Variation in Foraging Indices of

Habituated Orangutans

FIGURE 5

The foraging index is the ratio of feeding time to moving time. The foraging index for each habituated animal and of all females taken together is plotted for each quarter of the day between 0600 and 1800 hrs.

ACTIVITY PATTERNS AND SOCIAL BEHAVIOR

Individual Size and Feeding Habits

Hamilton (1972) has reported that there is a tight positive correlation between size of male rhesus macaques and the amount of food taken in under standard laboratory conditions. This conclusion is not surprising since it has been demonstrated frequently (e.g., Brody 1945) that minimal energy consumption per day measured in kilocalories in a thermoneutral environment (i.e., basal metabolism) is closely related to body weight, with relation:

$$M_{std} = 70 \ W^{0.75}$$

$$M_{std} = \text{basal metabolism (kcal)} \tag{1}$$

$$W = \text{weight (kg)}$$

Laboratories provide high energy food in convenient locations; but animals in the wild normally must forage or hunt for their less nutritious food, and moving in search of food has an additional cost in energy above the basal metabolic rate. Taylor $et\ al.$ (1970) have reported results of a study of the cost of running for a wide variety of quadrupedal animals. The cost of locomotion is a function of velocity and body weight. From data provided by Taylor $et\ al.$ it is possible to estimate the cost of traveling 1 kilometer at velocity V for a quadruped of body weight W. Orangutans of this study traveled a median distance of 0.4 km per day in 1 hour 15 minutes, thus at an average velocity of 0.32 km/hr. For a quadruped the cost M_{hor} of traveling 1 km at 0.32 km/hr is approximately[1]

$$M_{hor} = 17.1 \ W^{0.75}$$

and the total cost of covering the median day journey would be

$$M_{hor} = 17.1 \ DW^{0.75}$$

$$D = \text{day journey in km}$$

M_{hor} estimates the cost to move in a straight line at constant speed for a true quadruped. Arboreal travel is constantly interrupted and requires frequent detours and vertical displacement. Orangutans thus would be less efficient at traveling on the ground in a straight line than a true quadruped; this cost would be increased due to constant acceleration and deceleration in the canopy. It would be possible to model the energetic cost of an arboreal route in more detail by using simple laws of physics, but I will estimate the cost of arboreal travel at twice the cost of horizontal travel on the ground. The vertical component of costs entailed by frequent change of altitude may be estimated as

$$M_{vert} = 2.3\ DW$$

in which I assume that the animal moves up and down about 1 meter for each meter traveled horizontally. The total energetic costs for basal metabolism and travel (ignoring costs of thermoregulation, growth, pregnancy, or lactation) are:

$$M_{tot} = 70\ W^{0.75} + 34\ DW^{0.75} + 2.3\ DW$$

Given an approximate mean daily journey of 0.4 km and the weights estimated above the total energetic costs for BC, Afa, Afd, and Jfa would be roughly 2530 kcal, 1620 kcal, 1370 kcal, and 1100 kcal respectively. Lactating females require more energy per day for their weight than is predicted for basal metabolism, and growth requires additional energy above predicted basal metabolism. The two mature females in this study (Afa and Afd) were carrying suckling young, and the third habituated female (Jfa) was immature and therefore presumably growing. Growth and lactation would reduce the difference between male and female orangutans predicted on the basis of weight alone. On the other hand, males of a species generally operate at higher metabolic rates per unit weight than females of the same weight, and this sex difference would partially offset the effects of growth and lactation (Brody 1945).

Given the estimated weights of the four orangutans and given approximately equal rates of feeding on similar foods, BC should have fed roughly 1.8 times as long as the average of the three females each day. In fact, BC fed only 1.5 times as long as the females (from Table 2). The discrepancy between these figures must be due to the heavy demands of growth and lactation among the females, and to the fact that BC reduced his energetic expenses by traveling less each day than the females and by traveling terrestrially on long jaunts.

Consequences of Large Size. Schoener (1971) noted that large size should increase the variety of items eaten by a predator; i.e., the larger animal should be less selective in its diet. The following data suggest that BC was less selective in his diet than the smaller females. During March and April the mean length of feeding bouts for BC was 50.9 minutes (N = 161). The mean length of feeding bouts for all females was 35.7 minutes (N = 287). BC's feeding bouts were significantly longer than those of the females (t = 1.90 with 161 d.f., variances unequal; $0.05 > p > 0.025$). On the other hand, BC fed in 6.9 bouts per day on whole days (N = 7), and the females fed in 7.2 bouts per day on whole days (N = 14). These means are statistically indistinguishable; thus, BC fed in the same number of bouts per day, but he fed longer at each one. I interpret this to indicate that BC took more food at each feeding location each day, whereas the females took less and could be more selective in their choice of food at each location. In addition females could move farther than BC between food sources and could therefore be more selective of the sources themselves because at each move the females had a wider potential area in which to find food.

Sexual Dimorphism and Social Behavior

Given the anomalous, highly solitary behavior of orangutans, it is interesting to consider the strains on individual orangutans that would be introduced were they social. The total activity pattern of mature male and female orangutans is different at the outset, and the differences are related to body sizes that we must take as fixed. If males and females were to live in a single group, however, the females would have to accommodate the males because the mature male is limited by his size to his pattern of behavior. If he were to maintain contact with a female over a long period of time without adjustment in her pattern of behavior, he would have to move more. In order to maintain contact with her he would have to travel arboreally more than he normally does. Both of these changes would entail increased expenditures of energy, and in order to maintain his new level of activity, the male would have to eat more. Together the increased time of moving and consequent increased feeding would erode his day's resting time considerably. In addition, the daily variation in the male's pattern of feeding (Figure 5) is quite different from that of the females; the females form a homogeneous group. The male feeds heavily in the morning with little moving (high foraging index), and the pattern of females is uniform through the day. There is no established theoretical or empirical basis for interpreting variations in feeding schedules, but it seems logical that a large animal (whose basal metabolism is higher) will awaken hungrier than a smaller animal unless he feeds much more heavily prior to sleeping. If the female did not adjust her pattern of behavior, the morning would be a time of deprivation for the male.

If the female chose to follow a male, her problems would be less acute. Her principal sacrifice would be the opportunity for selectivity in food sources since she could more than fulfill her needs in the time a male spends feeding. She would be moving less as well and therefore would need less to eat than on her own schedule. At each food source, however, the female and her offspring would have less choice of fruit because the male would be taking a large proportion of fruit at each source each day. The female's selectivity *of* sources and *at* sources would be reduced by accommodating the male's pattern of behavior.

Orangutans are bisexually social during mating periods. In December 1970, we observed the behavior of a consort pair, RT and Afc, and the details of their clumsy coordination suggest that the mature male and female patterns of feeding behavior are incompatible. On the evening of December 18, 1970, an adult male (RT) and a female with infant (Afc and Iuc) slept in the same tree. The next morning the three orangutans left the tree at 0610 hours. I followed RT some distance behind, and by noon Afc and Iuc were far ahead of him, out of sight and hearing. (These animals were not habituated, and since neither of the orangutans displayed at me, it is unlikely Afc was aware of my presence). The attempt at coordinated morning feeding ended in separation that was probably enforced by sexually dimorphic feeding habits.

During the conference for which this paper was written, B. Galdikas described a consortship between a mature male and female during which the two animals maintained good coordination of movement. Although this observation appears to contradict my interpretation, Galdikas also remarked that the male of the couple appeared not to feed very much during the consortship. The interruption of feeding during breeding seasons is not uncommon among animals; a particularly striking example is provided by male elephant seals who do not feed at all during the breeding season (LeBoeuf & Peterson 1969). If such a strategy is necessary in order for a male to maintain contact with a female orangutan, it certainly places an upper limit on the duration of consortship. The strategy of starving in order to breed could easily founder for a male orangutan if two resident females were to come into estrus in close succession. The resulting long period of nutritional deprivation would probably weaken a male who chose to maintain close consortship at the expense of normal feeding. Although he would be present to defend access to the female from other males by staying close to her, a weakened condition would leave him vulnerable to aggressive replacement by another mature male at a critical moment (i.e., when a female is ovulating).

It is not a sophisticated theoretical approach to conceive of putting two normally solitary animals together to see what strains would develop were they social. The problems of this solitary, sexually dimorphic species in a hypothetical social situation, however, indicate similar problems for social, sexually dimorphic species such as gorillas and baboons. Hidden beneath the apparently cooperative cohesion of small females and large males in groups of these species is a similar sacrifice on the part of the smaller female; her size alone would allow her to be more selective in her diet if she were not limited to the pattern of movement of males. In this case there must be an overriding element of female interest in proximity to males that outweighs the advantages of selectivity of food. Alternatively, the range of variation in quality of food sources is limited so that selectivity is not an important factor, and females sacrifice nothing by remaining with males or by allowing males to remain with them. This might well be the case for the grazing mountain gorillas who feed predominantly on leaves, leaf-shoots, and bark (Schaller 1963).

Orangutans are not grazers; they feed preferentially on food that is discretely distributed in space and in time. Under these conditions the female has much to gain by exercising her option of selectivity. The quality of her daily diet will be improved by choosing among the fruit at a source, and the number of sources available to her at each moment of her life will be greater because she can afford—energetically—to travel farther to sources. Although there are several possible advantages for females to bisexual social life, such as protection from predators by large males and increased mating efficiency, the conditions favoring such social behavior are absent for orangutans. There are no significant predators (in the area of the present study), and mating is such an infrequent event that females would not profit from constant association with a male orangutan. In addition, given the long birth interval of adult females, an adult male would be favored selectively only be contact with several females, as in the case of gorillas (Schaller 1963; Fossey 1970). Each animal in a larger foraging group would suffer a further decrease in selectivity at each food source, and possibly the amount of fruit on a tree acceptable to *any* orangutan each day would be quickly exceeded if more orangutans were to feed consistently in each fruiting tree.

Another possible advantage of social existence is increased knowledge of food sources among all members of a group over the knowledge of any individual in the group; Kummer (1970) calls this the group's first important function, and Ripley (1970) employs this reasoning to explain aspects of social life among langurs. Given equality of intelligence between males and females, there is no reason why a group of adult females should not be as knowledgeable as the same size group containing a few adult males; an argument based on the advantages of group knowledge cannot explain bisexual groups. Under the feeding conditions of orangutans, one might expect that increased knowledge of the spatiotemporal locations of fruiting trees would be an advantage to a group of females, but several females would also be competing for acceptable fruit on each tree; therefore, each of their ranges of choice at a food source would have to increase unless they were to move more and feed less at each site. The energy budgets of females would limit the increased movement quickly.

It is interesting to note here that female orangutans do share knowledge of fruit trees with their offspring. If my interpretation of female life histories is correct (Rodman 1973a), the female offspring will share part of their mother's range for much of their lives. The knowledge they gain of specific food sources during close early association with their mothers will thus continue to serve them in later life. It is possible to argue that the close association between mother and female offspring will begin to break down when the mother's daily feeding pattern no longer provides both mother and daughter with a sufficient supply of acceptable fruit. As the daughter ages she will eat more, and both mother and daughter will have to increase their range of choice in a tree. At a certain size, the daughter or mother will tend to choose other known fruit trees if possible, thus breaking the close association. If mother and daughter normally use the same food sources on different days, each should find an acceptable crop. A further consequence of this argument is that in trees with an abundance of fruit it will not be unusual to find a mother and a mature daughter in the same place.

My description of sexually dimorphic behavior patterns is highly dependent on observations of a single mature adult male. It can be argued that I have observed an idiosyncratic deviation, but most of that deviation may be explained adequately by size differences between the mature male and the smaller females. Extreme dimorphism in size between mature adult males and females is characteristic of the species as a whole. I therefore feel confident that BC's activity, in as much as it may be explained by his large size, is typical of adult male orangutans. Equally detailed observations of males of the same resident status in the organization of populations should reveal similar differences in activity between those adult males and females. Given the smaller size and consequently different energetic constraints of young adult males, it is not surprising that subadult males have been found in close association with adult females or that analyses including observations of subadult males with those of mature males should reveal smaller differences in activity patterns between males and females than I have observed (MacKinnon 1973b).

Comparisons with Other Great Apes: Two "Test" Cases

Sexual selection has favored great size differences between adult male and female orangutans by one means or another. In light of the foregoing discussion, it can be seen that if females were favored by association with males in a bisexual social existence, there would have been concommitant changes in the diets or considerable reduction in size of both males and females. If males were favored by constant association with females, a much lower limit would have been placed on the size of the males. Comparison with the two other great apes supports both of these predictions and therefore lends some additional credence to the various explanations advanced in this paper. The gorilla is as sexually dimorphic in size as the orangutan, but males live with females in stable social groups. As would be predicted in arguing from the example of orangutans, the diet of gorillas is predominantly *nonfrugivorous;* it seems reasonable to suggest that the foods chosen by the gorilla are of a nature that limits the advantages of selectivity afforded females by their small size (compared with mature male gorillas). The chimpanzee, on the other hand, shows a high degree of association between adult male and adult female individuals while having a predominantly frugivorous diet much like that of the orangutans observed in this study. It is not surprising, given the discussion presented above, that chimpanzees are not very sexually dimorphic in size. This comparison suggests the tentative hypothesis that male-female association in gorillas is advantageous to females in an adaptive sense, and male-female association among chimpanzees is the result of an advantage for males.

CONCLUSION

The distribution and ranging patterns of mature adult male orangutans may be explained as an adaptation to the distribution of ovulating females in space and in time (Rodman 1973a). To summarize the explanation: Given the rarity of ovulating females in space and time, and given the alternatives of constant association with a single adult female or of energetically expensive wandering over large areas of forest in order to contact as many ovulating females as possible, it appears that an adult male will maximize his reproductive potential by defending as large an area of forest as possible from access by other mature males. The size of this territory would be limited at least in part by energetic constraints. Ideally—in a selective sense—the male's range would include the ranges of several adult females to whom he would have exclusive sexual access. Such a pattern of behavior implies strong competition among males, assuming approximately equal numbers of males and females, and the competition among males for desirable territories would be the source of sexual dimorphism in size of orangutans. Strong sexual dimorphism resulting in differential patterns of activity reinforces the segregation of the sexes when combined with a frugivorous diet.

Once female orangutans were observed to be solitary, the solitary and territorial behavior of adult males observed in this study was predictable. The question remaining then is why female orangutans are so solitary. Certain elements of individual activity patterns described above indicate that females benefit from greater selectivity of food by foraging independently of each other as well as independently of adult males. The solitary nature of male and female orangutans in this study area therefore ultimately depends on the relationship of females to the distribution of their food resources.

The solitary nature of these orangutans remains a striking exception to the social rule among the Anthropoidea. Observations reported here argue forcefully, however, that the solitary nature of orangutans is part of a balanced set of complex adaptations, and that the social existence among other primate species may be explained in each case only with reference to the many details of a larger adaptive complex.

Acknowledgment

The National Institute of Mental Health supported the fieldwork described here through grant #MH 13156 to Professor Irven DeVore of Harvard University. The Department of Anthropology and the Hooten Fund of Harvard University provided funds for organization and analysis of the data. I am grateful to David A. Horr and his family, to Irven DeVore, to Robert L. Trivers, to Barbara Harrisson, and to many additional colleagues, students, and other friends for their contributions to this work. The study could not have been completed without the cooperation and support of the Indonesian Government and the government of the Province of East Kalimantan, Indonesia. I am particularly grateful to Walman Sinaga and to many other personnel of the Departments of Forestry and of Nature Conservation and Wildlife Management in Bogor and in Samarinda for their enthusiastic assistance. Carol Ford Rodman assisted in every part of the work described here; the quality and quantity of research carried out in the study would have been severely restricted without her invaluable contributions.

Notes

1. The regression of M on W is obtained as follows: (a) From Taylor *et al.* (1970) estimate the cost per kilometer at 0.32 km/hr for each of seven quadrupeds; (b) Convert the result to kcal/kg (4.7 kcal/10_2); (c) Calculate the regression of M (kcal/kg) on W (kg); (d) multiply the result by W to obtain the estimated cost for the whole animal.

John MacKinnon

Reproductive Behavior in Wild Orangutan Populations

Field studies on the behavior and ecology of wild orangutans, *Pongo pygmaeus,* were run concurrently on both sides of the Segama river, Sabah, for 4 months (July-October 1968) and for 12 months (October 1969-September 1970). Seven months of fieldwork (April-November 1971) was conducted in north Sumatra. Some 1500 hours of direct daytime observation (Borneo c. 1200, Sumatra c. 300) were made on wild orangutans. The period of time over which I was in contact with (i.e., knew the position of) subject animals was much greater than this, particularly in Sumatra, since I preferred to gather synchronous information on the positions and travel routes of several different individuals rather than limit myself to observing single subjects for a full day.

This method seemed most appropriate for investigating the relationships existing among the individual orangutans that ranged over the same areas of forest, particularly as actual encounters between different orangutans were infrequent.

This paper is concerned with one aspect of orangutan behavior about which we are still very ignorant, namely, their reproductive behavior. I am concerned not so much with birth, infant care, and infant development, of which we have some knowledge, but more with courtship, mating behavior, and reproductive strategy. Direct information about this behavior in wild populations is still very scant, and we are forced to look for theoretical models that make sense in evolutionary terms and fit our observed behavior and population statistics.

THE SOLITARY NATURE OF ORANGUTANS

The solitary nature of adult orangutans was obvious in all areas investigated (MacKinnon 1974). Of 191 observed subgroups (animals found traveling together as a single unit), 154 were minimally small, either independent animals traveling alone or single adult females accompanied only by dependent young. The remaining 37 subgroups comprised 16 instances of immature, independent animals tagging onto other population units, and 27 instances of consortship, where a mature (either adult or subadult) male was accompanying a mature female.

Even when different subgroups did meet at food trees, they showed little signs of social interaction; no friendly greetings, no allogrooming and general lack of interest in each other. Only immature animals indulged in social behavior by coming together to play. Granted that adult animals must sometimes come together for reproductive purposes, the orangutan seems to be as solitary as is possible.

The solitary ranging habits of orangutans may seem surprising within the context of the higher primates, which are characterized by their social ways, but they are not so dissimilar from the ranging habits of the orangutan's nearest living relative, the chimpanzee *Pan troglodytes*. Within a rather loosely organized society, chimpanzees range in a very individualistic manner and at some times of year the modal size of foraging parties is one animal (Wrangham, in press).

The orangutan's solitary way of life seems very suitable for such a large, slow-moving, arboreal ape, particularly since fruit, its main food, is usually rare and widely dispersed throughout the forest. Bigger groupings would offer little advantage and several drawbacks:

1. In the absence of serious predation, a large group would bestow little benefit by way of protecting its members.

2. The sum total of experience of all group members would not greatly

exceed that of any one single adult. All are able to retain a good memory of past spatiotemporal food distributions, which is probably the most important knowledge required for their rather simple life pattern.

3. Without greatly increasing daily ranging distance (extremely costly in terms of energy for such large arboreal animals), a group would not discover many more food sources than a solitary orangutan. With more animals to be fed, most individuals would suffer a significant decrease in the quantity and quality of food they could select.

4. In view of the slow reproductive needs of this long-lived species, there is no need for permanent maintenance of complete breeding units. At most, adult females can reproduce only once every 2 or 3 years and the actual mean birth interval calculated for all adult females seen during these studies is only one infant per 5.6 years. The formation of temporary consortships is quite sufficient to ensure enough conception.

ORANGUTAN SOCIETY

In both Borneo and Sumatra, orangutans showed an extremely loose type of society, in which the ranges of several individuals of both sexes overlap. The term *community* seems useful to describe those Sumatran orangutans that ranged over the same area of forest *(community range)*. These animals showed considerable coordination in long-term travel patterns and were frequently clumped around single, large, calling males (MacKinnon 1973b; 1974). Two community ranges in the Ranun area were both greater than 5 square km in size.

The term community may be less appropriate to describe those Bornean orangutans who ranged over the same area since independent animals used the forest space in a more individual manner.

More than 70 different nonresident animals passed briefly through my 5 sq km north Segama research area during 16 months of fieldwork. These were highly mobile animals, and some were possibly individuals that had been displaced by timber-felling operations farther north (see later discussion). The appearance of these nonresident animals was not regular throughout the year. Rather they seemed to arrive in waves, which to some extent corresponded with periods of food abundance in the area. In addition 17 orangutans were seen periodically (on at least three separate visits) throughout the study. They seemed to be resident within individual ranges, but none remained at all times within the study area. All ranges must have been greater than 1 sq km and probably much larger (c.f. Galdikas, this volume).

Orangutan ranges overlap extensively and are not exclusive, except between some adult males (MacKinnon 1974; Rodman 1973a). An orangutan is not *trapped* within its own range in the sense that a gibbon family is confined within its territory. Orangutans can leave their usual ranging areas if these are temporarily low in food or to exploit particularly good fruit crops in neighboring valleys. The fact that orangutans, particularly females with young, remain within extremely small individual ranges in some areas (Horr 1972; Rodman 1973a) tells us much about the local flora and geography, as well as about orangutan behavior.

To understand the apparent dichotomy between resident animals that remain within small, readily definable home ranges and those that do not, we must appreciate that any given section of the forest (area x) is limited in the number of resident animals (y) that it can support at *all* times of the year. In practice, however, most areas are seasonal in food production so that at *some* times of the year they can support more animals than is possible in the leanest periods. If all parts of a forest have lean and rich periods in phase, the total density possible remains only y/x. If, however, different parts of the forest produce different food crops at different times of the year, as is often the case, the overall density can greatly exceed y/x, provided a proportion of the population are mobile between different local food abundances. The actual densities, range sizes, and proportion of resident to mobile members of a given population will depend upon the quantity and quality of food and the spatial and temporal distribution of food that the region produces.

In addition to the movements of more mobile individuals, resident orangutans on both sides of the Segama river in Borneo and also in my Ranun Sumatran area showed similar seasonal shifts in ranging area. The bulk of these populations ranged in lowland riverside parts of the forest during the rainy seasons and early fruit seasons, but moved further back into the hills at the end of the fruit seasons and during periods of dry weather. Fruit availability in both the Segama and Ranun areas was markedly seasonal (MacKinnon 1974), whereas Kutai, where no seasonal shift in orangutan ranges was described (Rodman 1973a), lies much closer to the equator and probably enjoys more year-round stability of climate and thus food supply.

The orangutan's habit of moving down from the hills (or up from swamp forest) to feast on seasonal lowland crops of durians and other fruit has been well known to the human inhabitants of forest-edge *kampongs* for hundreds of years. Mention of such seasonal ranging behavior is made in the accounts of early European travelers to Southeast Asia (Wallace 1869; Hornaday 1885; Beccari 1904). Seasonal movements of orangutans must be accepted as common ranging behavior for this species, not merely the results of abnormal conditions.

The extremely flexible nature of orangutan society, in the sense of spatial distribution, allows the individual orangutans to be spread efficiently to suit differing conditions of food availability. For much of the year, the populations are very dispersed throughout the forest, but different subgroups can congregate to exploit particularly good food sources. Ten orangutans fed in a single fruiting fig tree in my northern Segama study area over a 3-day period. Eight orangutans fed together in two adjacent fruiting *Dialium* trees south of the Segama. In the Ranun area, Sumatra, I watched 14 different orangutans feed in one fruiting fig tree over a 5-day period. Fifteen wild orangutans became clumped in a small area at Ketambe, Sumatra, when a single large fig tree came into fruit (Rijksen, personal communication).

At one time or another, every orangutan must meet all the other individuals whose habitual ranges overlap with its own. We must assume that it is able to recognize each of these individually. It is with these animals that the orangutan makes its limited social contacts and finds suitable breeding partners.

SEXUAL BEHAVIOR

Little sexual behavior between wild orangutans was seen during these studies. In Borneo, I observed seven occasions of copulation between mature animals and one instance of copulation between immature animals. On two other occasions, I heard the distinct rhythmic squeals given by a female orangutan during copulation, but was unable to see the animals involved. I also saw one instance of a male chasing and attempting to catch an adult female, but I was unable to keep up with these animals and did not see the outcome. Once I saw an adult female taking hold of the flaccid penis of her consorting male, as though to initiate sexual play, but he pushed her away.

Of the eight observed copulations, six took place completely in the trees, one started in a tree but ended on the ground, and one took place entirely on the ground. Apart from the immature copulation, males were generally aggressive, seizing apparently frightened females and clasping them round the thighs or waist with their feet. They thrust either ventrally or from the rear. Copulatory bouts lasted for about 10 minutes with males thrusting at about 1-second intervals. In four aggressive attempts at copulation, males struck and bit struggling females. Infants accompanying these females also attempted to fight off the males concerned. It is doubtful whether intromission was achieved on these occasions, and once the male's pink penis could be clearly seen thrusting on the female's back.

Of the seven instances of copulation between mature animals, three occurred during a brief consortship that lasted only 48 hours, and two between animals that were already together when I met them but separated while I observed them. In the other two instances, males encountered females, chased and caught them, then left them again shortly after copulation.

In Sumatra, I observed three instances of copulation between two different pairs of mature animals. In both cases these appeared to be stable consortships. Males approached their female companions and took hold of them. In one case a playful chase occurred before the male caught the female. As in Borneo, males clasped females with their feet while they held onto branches with their hands. Females gave rhythmic squeals in time to the slow thrusting. In two cases, males mounted from the rear, in one case ventrally after an initial playful period of tactile investigation of the genital region of his partner.

Nine other consortships were observed without sexual behavior. Sex-play was observed during the formation of another brief consortship (see discussion following), but this broke up the following day. A male infant was seen masturbating his erect penis with his foot.

Apart from these rather scant observations on wild orangutans, I have observed a variety of imaginative masturbatory and mutually erotic behaviors, as well as some 20 instances of copulation, among captive and semi-wild orangutans (MacKinnon 1974). Both sexes participated enthusiastically in long bouts of sex-play and copulation. These usually occurred in recognizable, regular partnerships, analogous with consort-ships, even when more than two animals were housed together.

In view of the small size of the orangutan penis and the inherent difficulties of suspensory arboreal copulation, it seems that only with female cooperation can intromission be successfully achieved. Such cooperation is usual only within the context of an established consort-ship, in which a female is familiar with and shows little shyness toward her partner. The more aggressive forms of copulation with uncoopera-tive females, which I have referred to as rape (MacKinnon 1971), would seem to be futile reproductively. I saw most of the aggressive sexual behavior north of the Segama River, and, indeed, that population was characterized by a very low birth rate (see discussion following).

CONSORTSHIPS

Sixteen apparently stable consortships were observed in Borneo and another 11 in Sumatra, which involved mature male and female orangu-tans traveling together as a single foraging unit. Such partnerships were usually maintained mutually, the female following the male whenever he initiated travel and vice versa. Consorting partners usually nested in the same tree at night, with the male lower than the female.

Two instances were observed of the formation of consortship, once in Borneo and once in Sumatra. In neither case, however, did these reach a state that could be described as stable, being maintained for only 2 days and 1 day respectively. In the Bornean case, a subadult male encountered an adult female with a small juvenile when their inde-pendent foraging paths crossed. He followed this couple wherever they traveled and three times caught and copulated with the female in an aggressive manner, despite her attempts to fight him off and get away. The consortship broke up on the third day, when instead of following the others, the male moved off in a different direction.

In the Sumatran case, a subadult male encountered a lone adolescent female at a fruit tree. He fed in a neighboring tree and nested close by in the evening, but not until the following morning did he climb up into the same tree and feed beside her. She squealed and tried to move away each time he reached out to her, and she eventually hurried from the tree. He followed, caught her, and drew her onto an old nest. Here he played with her quite gently, but she escaped again and he followed. He paused in a willowy leafless tree and began to sway it to and fro. For over a minute he indulged in a strange courtship display. Chest pushed forward, chin up and limbs outstretched, he swayed gently round in a circular motion, his long hair undulating with the rhythm. He continued to follow her, and when he caught the female, they played again in another old nest. The female still squealed whenever the male tried to examine her genital region, but seemed to be losing her earlier fear of him and joined in the biting, tickling, and grappling of the game. The two animals fed together for the remainder of the day and nested in the same tree that evening, but the female had left him before I reached the tree the next morning and did not return.

Although this was the only occasion on which I saw such a courtship display, I think it is probably common for Sumatran orangutans. This race has much longer hair on the back and arms than do Bornean animals (MacKinnon 1975). Moreover, a type of chest-out, chin-up, bipedal posturing commonly seen in captive Sumatran subadult males is rather similar to the arboreal display I witnessed. I have not seen such bipedal display among captive Bornean animals.

As neither the formation nor the separation of any of the stable consortships was observed, I cannot give any figures concerning the length of such partnerships. Bornean consortships may not usually last longer than a few days, sufficient to ensure a good chance of fertilization but little more. In Sumatra, some consortships lasted several months. One particular consorting pair remained together for at least 5 months.

Female orangutans show a menstrual rather than markedly oestral periodicity, and captive animals show month-round sexual receptivity (Asano 1967), which may well be adaptive in helping to maintain consortships. Consortships probably break up as soon as a female is pregnant. Once she became pregnant, Twiggy, an adult female Bornean orangutan at Regents Park Zoo, became very aggressive toward her partner, Boy, whenever he tried to copulate with her (personal observation). Van Bemmel (quoted in Harrisson 1962) states that captive Sumatran orangutans also cease copulation during pregnancy.

In Borneo only 7 (19%) of 37 observed female-infant groupings were accompanied by males, while in Sumatra 9 (75%) of 12 female-infant groups were seen with males. There are large predators such as tigers and possibly also leopards in Sumatra, which do not occur in Borneo. Also, in Sumatra the orangutan competes for food with aggressive siamangs *(Symphalangus syndactylus),* who will attack young orangutans (MacKinnon 1973b, 1974). We can expect the female orangutan with an infant to be more vulnerable in Sumatra than in Borneo, and this may have led to the prolongation of Sumatran consortships with a protective male to form a complete family unit. The male consorts of females carrying small infants may well have been the sires of such young. On the other hand, it may be that Sumatran females form new consortships sooner after the birth of an infant than do their Bornean counterparts. The only obviously pregnant female seen in Sumatra had no male consort, but she was accompanied by a large infant.

Subadult males were more actively involved in consortship than were adult males. In Borneo 7 (41%) of 17 subadult males were engaged in consortships compared with only 9 (22%) of 41 for adult males. In Sumatra, corresponding figures were 6 (75%) of 8 for subadults compared with only 7 (41%) of 17 for adults.

Those consortships involving adult males were sometimes rather clumsily maintained—females having to wait while males fed and halting their travel intermittently to allow males to rejoin them. Such observations accord closely with expectations arising from Rodman's comparison of activity patterns and foraging strategies of the two sexes (Rodman 1973b; and this volume). Subadult males are only slightly larger than adult females, but fully adult males may be over twice the size. As male orangutans become larger, prolonged consortship with females becomes increasingly difficult. Also, females seem particularly shy of the very large males, moving out of their way or even hiding from them. Possibly mature males can attract to themselves (by calling) those receptive females with whom they have previously enjoyed more prolonged consortships when they were younger. Even in captivity, however, subadult male orangutans are more sexually active than adults (personal observation). Fully adult males are less successful at breeding in captivity (personal communication, Mike Carman, zoo keeper, Regent's Park Zoo Apehouse).

THE DEVELOPMENT OF REPRODUCTIVE BEHAVIOR

As soon as adolescent female orangutans begin ranging independently of their mothers, they are liable to be consorted by mature males. From captive animals, we know that females can reproduce from the age of 7 years, although they do not usually reach full size until they are about 10 years old. From the time of her first pregnancy, a female will be involved in rearing a succession of infants for the rest of her life.

Females can only be regarded as receptive toward males during the period between one infant reaching about 2 years of age and the next pregnancy. I saw only two old females without any accompanying young during the whole study, so it seems that wild female orangutans rarely reach a stage of infertile old age.

We can distinguish four stages in the sexual development of male orangutans:

1. *Immature males,* who are social, playful, and indulge in experimental erotic behavior and juvenile copulation.

2. *Subadult males,* the class most frequently found in consortships with females. They form rather playful, often prolonged consortships and indulge in long bouts of sexual behavior with cooperative females, as well as aggressive sexual behavior with other females. Consortships seem to form from chance encounter, the male taking much of the initiative.

3. *Adult males,* who are characterized by the full development of second-ary sexual adornments including long hair and beards, full throat pouches, high fatty crowns, expanded cheek flanges, and large size. These animals indulge in loud, long-range calling, defending range boundaries against other males. They are less frequently engaged in consortship but do sometimes form brief, sometimes rather clumsily maintained, consortships with females that they meet by chance or that are attracted to their calls. They also exhibit violent, aggressive sexual behavior toward some females they encounter.

4. *Old adult males,* characterized by extreme size, dark color (Borneo), shaggy coats, often bare backs, and terrestrial habits. Such animals still call regular-ly but no longer seem to indulge in any sexual activity. They are extremely unsocial, avoid contact with any other orangutans, and sometimes chase other animals (including human observers!) away.

THE FUNCTION OF CALLING BEHAVIOR

It is not possible to understand the reproductive behavior of orangu-tans fully without considering the function of adult male calling. Adult males give periodic long, loud vocalizations that consist of a series of deep roars and can be heard from more than a kilometer away. I have termed these *long calls* (MacKinnon 1971; 1974).

Two main functions have been proposed for these calls. Firstly they may be used to maintain spacing between different adult males—the call acting as a long-range threat or challenge (MacKinnon 1971; 1974; Horr 1972; Rodman 1973a; Galdikas, this volume). Secondly, these calls may attract sexually receptive females to the caller (MacKinnon 1969; 1974; Horr 1972; Rodman 1973a; Galdikas, this volume).

Adult males were very obviously intolerant of other males and usually managed to keep apart. Females regularly met other females in the forest and females often encountered males, but I witnessed only two instances of adult males coming together during my Bornean studies and both occasions were characterized by violent display with one animal chasing the other away (MacKinnon 1974). In Sumatra, two or more mature males were seen together on several occasions feeding peacefully in the same tree. I observed only one aggressive chase involving two subadult males. None of these animals, however, was a calling male. No other independent male was ever tolerated close to a calling male, and on one occasion three animals, including two mature males, made a hurried exit from a fig tree when a much larger and, presumably, calling male arrived in display and moved into the tree to feed.

Long calls were sometimes given in response to other calls heard, and in both Sumatra and Borneo, individual calling boundaries were recognized (MacKinnon 1974). The behavior prior to and during calls, such as display, branch-shaking and dropping, etc. suggested that animals were aggressively motivated when they called. Calls certainly seemed to demonstrate long-range status rivalry between males and territorial-like defence of space. The individual ranges of several Bornean and two Sumatran calling males showed considerable overlap. Two observations by myself (MacKinnon 1974), at least one by Horr (1972), one by Rodman (1973a), and several by Galdikas (this volume) indicate that adult female orangutans are sometimes attracted to males, presumably by their calls, and that this can result in consortship and/or mating. More frequent reaction of adult females on hearing calls was either to ignore them or, if they were close by, to hide in thick vegetation or even move away from the caller. Possibly these latter females were not receptive. Some carrying large infants might have been expected to be receptive, but it is possible that they were pregnant already.

Although calls may help males to contact females, they are certainly not necessary. A greater proportion of noncalling subadult males than calling adults were found with females. Moreover, calling males accompanied by females still gave long calls, as did several old males who seemed totally inactive sexually.

MALE SEXUAL STRATEGY

Horr (1972) and Rodman (1973a) have proposed essentially identical models for the sexual strategy of male orangutans. Both authors see intense competition between adult males for the limited supply of receptive females. Both see the greatest reproductive success for high-status males with established home ranges that overlap the ranges of several adult females. Such animals call and give vigorous displays to keep other male competitors away and also to attract females that become sexually receptive. The model is theoretically sound and may well represent tactics employed by high-status resident males. It is, however, an oversimplification of the question of male reproductive strategies for orangutans.

The attempts of any male to monopolize several females sexually are hampered by the severe problems inherent to the structure of the rain forest and the dispersed, solitary orangutan society. The high-status male rarely sees the females who share parts of his range, and his calling makes him an easy individual to avoid. Females have almost unparalleled opportunities to exercise sexual selectivity and cuckoldry. Low-status males and passing visitors have excellent chances to consort with or rape resident females without the local dominant male knowing. Many different reproductive strategies are undoubtedly open to males and are employed by different individuals. In a given population, high-status resident males are probably in a minority, and the reign of any given individual as a high-status resident is probably only a small part of his reproductive career. We must consider the sexual strategies employed by other male categories of status and maturity.

One curious point about male breeding strategies is that the stages of greatest sexual activity and of greatest agonistic activity are quite out of phase. It is the subadult males who are most active sexually and who are most frequently found consorting with females, yet these males neither give fully formed long calls nor compete with adults for high status. Not only does male agonistic activity start late in the reproductive career of the individual male, but it appears to continue beyond the period of reproductive activity.

It is not yet possible to prove that old male orangutans in the wild are sexually less potent, although this can be shown in captivity where male orangutans often survive many years after the birth of their last offspring, and show no signs of spermatogenesis on postmortem examination. I infer, based on their observed lack of sexual behavior, a similar stage of impotency in old males in the wild.

A clear contrast was apparent between the two most frequently observed adult males in the Segama study area. Both these males were regular callers. Redbeard was a youthful animal with a neat beard, fine long hair, and well-rounded cheek flanges. He was once found in close consortship with an adult female, was three times found within a hundred meters of adult females, was seen to allow females to feed in his tree on ten occasions, and twice chased and raped transient females he encountered. Harold was a darker colored animal with a bare back, straggly long hair, three stiff broken fingers, and lumpy, scarred cheek flanges. He was never seen in association with any other orangutans and actively avoided an encounter with an adult female who was attracted to his calls (MacKinnon 1974). Redbeard seemed to be sexually vigorous, but Harold, a visibly older animal, seemed to be quite inactive

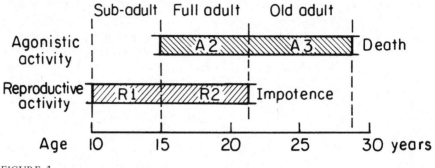

FIGURE 1
Reproductive behavior in wild orangutan populations

sexually. Other old males were seen briefly. None showed any signs of sexual activity, yet they continued to compete for status, giving long calls and aggressive displays. It seems prudent to conclude that sexual vigor and potency decline earlier in the male's lifetime than does agonistic vigor.

We can consequently construct a model for the sexual and aggressive phases exhibited at different stages in the male's maturity (Figure 1). These stages can be allocated on a tentative age scale based on comparison with the rate of development of captive animals. Development of wild animals may prove to be slower than that of their captive counterparts, but at present we have no sufficiently long-term records of wild individuals.

We can distinguish two distinct reproductive phases—subadult sexuality (R1) and full adult sexuality (R2), and also two aggressive phases—full adult aggressiveness (A2) and old adult aggressiveness (A3). The strategy and function of the types of behavior associated with each of these phases requires further discussion.

Rodman (1973a) has inferred that young adult males (he does not use the term subadult) represent a stage of dispersal prior to establishing discrete individual ranges as full adults. There were no subadult males resident within Rodman's Kutai study area. Transient subadult males were observed by me in the Segama study area, but there were also two subadults that I saw regularly. Regularly encountered subadult males have also been reported by Galdikas (this volume) at Tanjung Puting and in Sumatran orangutan populations (MacKinnon 1974; Rijksen, in press). We must therefore consider the reproductive strategies of subadult males within two contexts—when they are resident and when they are more mobile.

It might be imagined that the more mobile members of an orangutan population are less fortunate members, who have been unable to establish resident ranges and who thus have little chance of breeding. This conclusion may be valid under some conditions, but where food and other resources (e.g., minerals) are widely and unevenly distributed, the more mobile members could have a distinct advantage. The chances of transient males breeding with resident females or of resident males breeding with transient females may be slim, but mobile animals can breed with other mobile partners. The transient subgroups observed in the Segama study showed remarkable coordination in their movements (MacKinnon 1974). All age-sex classes were included in these migrations, and several transient consortships were observed passing through the area. Some reproductive opportunities are undoubtedly open to mobile population members.

Subadult males are still too small to challenge the male status establishment and are unable to call females in to themselves, but they are sexually very active. They employ two main strategies; long consortships with relatively cooperative females and aggressive sex or rape with uncooperative females. The former method seems much the most likely to result in reproduction (see preceding discussion), but consortship is not always possible (as between a transient male and a resident female) and rape probably has some reproductive benefit, however slim. Rijksen (in press) has suggested that the rape of resident adult females by resident subadult males at Ketambe in Sumatra represents the establishment of dominance over the females rather than reproductive activity. This theory may be true, but the fact that these attacks are sexual rather than purely aggressive suggests that there is some reproductive benefit in this behavior.

Subadult males spend much of their time in consortship with adult females (Segama, Borneo 41%; Ranun, Sumatra 75%). Some instances of sexual behavior were observed, as stated previously, but some couples were observed for several days at a time without any sexual behavior. It would seem, therefore, that subadult males sometimes consort with females when the latter are not sexually responsive. Presumably the tactic is to befriend and stay with a female until she comes into oestrus, then to copulate with her and try to keep her away from any calling adult males at this critical time (the only time when she is likely to be attracted to another male's calls).

The transformation from subadulthood to full adulthood in male orangutans is sudden and dramatic. In zoos, the filling out of the face, the spurt in hair growth, and the increase in body weight can take place within a few months. In the wild, this metamorphosis is accompanied by the onset of agonistic display and calling (A2). Improperly formed immature long calls were only heard twice, given by subadult males. As a full adult, the male's reproductive behavior (R2) changes. The male's new ability to attract sexually receptive females enables him to show much greater sexual selectivity. Probably this selectivity is necessary, since the full adult is physiologically handicapped by his great weight compared to the subadult and prolonged consortship may be quite difficult (see Rodman, this volume). The adult male can only afford to consort with near certainties. Adult males also sometimes indulge in rape. Presumably again there is some slight gain from such behavior.

The inter-male aggressive behavior (A2) helps to reduce the chances for reproductive activity (R2) of any other adult males within the range of the individual, its effectiveness being dependent on the status and aggressive vigor of the caller. It should be pointed out, however, that aggressive behavior (A2) has little effect in reducing the reproductive activities (R1) of any subadult males, who are easily able to operate within the calling adult's range provided they keep out of his way. Indeed, his calling actually helps them to avoid him.

Whether an individual adult's chances of breeding will be increased or decreased by remaining sedentary within a small range or by wandering over a large area will depend upon that individual's physical condition and status. Low-status males are faced with the alternatives of waiting where they are for the chance of taking over from more dominant animals should these die, become decrepit, or move on; or, they can seek a new part of the forest where the struggle for status may be easier. High-status males may, or may not, achieve the best situation by settling in one place and attempting to monopolize the local females. Any one male's range will overlap the ranges of only a few females, and each of these will only become sexually receptive at intervals of 3 or more years. The situation must often arise when no more females in a given locality are likely to become receptive for perhaps a year or two. The resident male can cautiously maintain his position and wait, or move on to a new location. The supremely dominant male would possibly be better off by trying somewhere else, confidently muscling in against new rivals he encounters.

Perhaps the most intriguing question of all concerns the role of old, sexually inactive males and why they continue to struggle for status. Natural selection would not have favored longevity in adult males beyond the reproductive stage, if these animals did not somehow still contribute to the welfare of their offspring, but were merely competing with them for food. Only if their calling is bestowing advantages on their offspring can this behavior by old, nonreproducing males be regarded as adaptive.

It is possible that the calling of old males conveys information that could be useful to younger, less experienced animals, such as indicating where food might be in different seasons. Probably the most important way in which a calling male can benefit his offspring is in his role as range guardian. By his calling and general aggressiveness (A3), the old adult male may be protecting those parts of the forest that contain his offspring and therefore his genetic contribution to the population. By keeping out the potential influx of strange animals and disrupting the breeding activity (R2) of other adult males in the area, he is in effect maintaining a vacuum, an area where the population density is below the carrying capacity of the forest. In this way he may guarantee that his offspring have sufficient space, and therefore food, to grow up and breed after him. This tactic may be his best way of promoting his own lineage.

Even when his oldest male offspring are already breeding as subadults, the old male may still fulfill a useful function. His sons' reproductive activity (R1) will not be much affected by his aggressive behavior (A3), while they are not yet sufficiently powerful to guard the range themselves. Not until the sons become fully adult will the old male's aggressiveness really start to jeopardize the reproductive success of his own offspring. This situation would not occur before the old male is at least 26 years old. It is not easy to determine the age of wild orangutans, but it is unlikely that many animals reach 30 years of age.

The breeding strategy of the male orangutan can, therefore, be divided into two parts. As a young adult he consorts with as many potentially receptive females as possible to make the maximum genetic contribution to the population. As an older male he consolidates and protects his contribution. The two parts overlap in his period of physical peak as a prime adult. As he grows older he becomes larger and more frightening, which makes consorting more difficult but enhances his effectiveness as range guardian. We do not yet have sufficient information about the paternity of wild infants to assess the relative reproductive success of the different tactics employed at different stages in the career of any male orangutan.

THE NATURAL REGULATION OF NUMBERS

Like many other mammals that have long periods of gestation, the orangutan is a long-lived species in which the whole emphasis of breeding must be on low numerical wastage; the production of only a few young, but maximum care to ensure that these survive. Female orangutans are able to reproduce only once every few years. It is therefore very important for a female orangutan to avoid wasting one of her few chances by breeding at an unsuitable time, thus risking the loss of her baby or, even worse, making a heavy parental investment in an undernourished infant unlikely to reproduce after her. A female's ability to exercise great selectivity over when and with whom she breeds would constitute a great advantage. Her breeding success would probably be maximized by refraining from breeding at unsuitable times such as periods of food shortage, overcrowding, or social disturbance. This selectivity would be similar to that of seasonal breeding animals, which effectively maximize their reproductive chances by refraining from breeding at less suitable times of the year.

Population data from the Segama area in Borneo suggest that the breeding rate in wild orangutans is regulated to suit varying local conditions. My study area B to the south of the river was totally undisturbed by man and showed a high level of reproductive activity. Eighteen adult females seen in this area were accompanied by a total of 15 infants, which is equivalent to a mean birth interval of about 3 years per adult female—about the maximum reproductive rate for this species. There appeared to be no check on breeding in this area. Area A, however, on the north side of the river exhibited a much lower breeding rate. Sixty-seven adult females scored as new visitors (MacKinnon 1974) carried a total of only 22 infants, which is equivalent to a mean birth interval of only 1 infant per 8 years per adult female. The drop in births continued throughout the study; in 1968, 66% of females seen in this area had infants, in 1969 50% did. By 1970, the figure had dropped to 25% and only one of these infants was less than 1 year old. Reproduction seemed virtually to have ceased on this side of the river. This area was not directly disturbed by man, but lay adjacent to the Pin Forest reserve where extensive timber felling operations were underway. The total area of continuous orangutan habitat was being rapidly reduced, and the appearance of large numbers of apparently homeless transient orangutans in area A strongly suggested that the population was socially disturbed and probably in excess of the carrying capacity of the forest. Differences in birth rate between the two areas were closely correlated with differences in the rate of consortship and inversely correlated with male calling rate (Table 1). Birthrate in area B was two-and-a-half times as high as in area A. Males in area B consorted with females two-and-a-half times as frequently, but gave long calls only one-third as often as their counterparts north of the river.

TABLE 1
Habitat, birth rate, call rate, and consort frequency

AREA	LEVEL OF HABITAT DESTRUCTION	INFANTS PER FEMALE	CALLS PER DAY	CALLS PER MALE SEEN	% OF MALES CONSORTING
Borneo, Segama B	None	0.8	0.2	1.3	50%
Segama A	High	0.3	0.7	4.6	21%
Sumatra, West Langkat	Low	0.5	1.3	0.8	100%
Ranun South	Low	0.4	0.1	0.8	45%
Ranun North	Fairly high	0.5	0.5	3.4	20%

Figures for the frequency of male consortships, male calling, and female-infant ratio for three different areas in Sumatra showed less dramatic variation. In general, however, they complied with the picture that forest areas where there had been recent felling are characterized by lower frequencies of infants and consortships and higher calling frequencies (Table 1).

It seems reasonable to conclude that overcrowding, social stress, and/or the presence of strange individuals makes it temporarily more important for males to indulge in calling displays, range protection, and status disputes than to proceed with reproduction. This behavior enhances, rather than reduces, their overall chances of contributing to the next generation. Sexual behavior does not cease in disturbed areas, but, as in Segama area A, is directed in an aggressive and unselective manner toward uncooperative females and is unlikely to result in offspring.

Natural regulation of the recruitment rate seems to occur in a number of animal species (see Lack 1954). Recruitment can be regulated by changes in the mortality of young or larvae, cannibalism, fecundity, clutch or litter size, abortion and embryo resorption rates, territory size, etc. The orangutan, however, may regulate population recruitment rate in a different way—namely by changes in courtship and mating behavior, a method employed culturally by human societies.

Acknowledgment

These field studies were made possible thanks to the support of the James Poulton Fund, the Alexander Allan Paton Trust, the Royal Society and Leverhulme Trust, the Science Research Council, the Department of Zoology at Oxford University, the Sabah Forests Department, and the Lembaga Ilmu Pengetahuan Indonesia (Indonesian Institute for Sciences).

PART FOUR

Theoretical and Laboratory Studies

The earlier field studies of primate behavior were directed toward seeing the animals and accumulating basic information; theoretical analysis was minimal. Today, basic facts are still woefully inadequate, and no one knows how many years of study in how many different locations will be necessary before definitive accounts may be written. As the number of facts accumulate, however, theory becomes increasingly important, for both organization and interpretation of the data.

In his article William Mason takes the reader back more than 40 years, quickly outlining the similarities in the senses between man and ape. Going back to W. Köhler, Mason shows that the old problems are still very much with us, and states his belief that ". . . apes and man have entered a cognitive world which sets them apart from other primates."

R. A. Hinde outlines a view of behavioral description which starts with relationships and builds toward social structure. Principles of organization are considered, and these are used for analysis and finally for the comparison of human behavior with that of other primates.

Joseph Popp and Irven DeVore examine some aspects of primate behavior in the light of the theories advanced by sociobiology. According to the theory as presented, animals act to increase their inclusive fitness. The importance of natural selection has been debated for many years, and kin selection adds an important dimension to the traditional modes of thinking.

Field observations and theoretical treatments are supplemented by R. K. Davenport's account of the behavioral disturbances of captive apes. The behavior of the disturbed animals gives many clues to the nature of learning, with wide theoretical and practical applications.

Emil Menzel's carefully controlled experiments show how chimpanzee behavior may be communicated and how the behavior of one animal may influence that of another. Clearly, experiments augment understandings coming from the field work.

Mother toque macaque eating peanuts with her infant. Photograph by P. Dolhinow.

William A. Mason

Environmental Models
and Mental Modes:
Representational Processes in the Great Apes

As confirmed evolutionists we can agree that what really counts, in the long run, is not how a creature experiences the world—its perceptions, thoughts, and feelings—but how it behaves. Nature is a ruthless pragmatist and does not care a fig for elaborate plans or refined sensibilities that do not materialize in some useful act. In the evolutionary game it is performance that wins the day. On this issue at least, the behaviorist and the classical ethologist join hands, for both assume—and rightly— that behavior is the primary means of coping with the environment for the vast majority of animals.

Is there any need, then, to consider representational processes in the great apes, or for that matter, in any other animal? The question is not whether representational processes exist, but whether we must treat them as a serious object of study. All behavior that is guided by sensory information—that is, most of the behavior that interests us—implies some type of schema or functional image of the environment. Behavior is the endpoint of an information-processing sequence. And were it not guided in some fashion by knowledge of the contingencies and constraints of the real world, its utility would be severely limited.

Nevertheless, with some species we can predict behavior with reasonable confidence when we restrict our attention only to what goes into the organism and what comes out, to physical events and overt reactions to them, to *stimuli* and *responses*. We seem to be able to account for a great deal of the behavior of insects, fish, birds, and some mammals without having to face the problem of how the world is represented to them. We do, of course, require some knowledge of the *effective stimulus*. We cannot go very far in predicting a rat's behavior until we have knowledge of its sensitivity to odors, its visual capabilities, its hearing, and the like. We characteristically describe these stimuli in the language of the physicist, however, and are seldom forced to try to reconstruct the rat's perceptual world from such fragments of knowledge about its sensory systems as the laboratory provides.

For most purposes in the study of rat behavior, the fiction of the *empty organism* is, at least, defensible. We do not have to be greatly gifted with the capacity for *indwelling*, as Polanyi (1959) calls it, in order to provide a reasonable account of how the rat's behavior is organized. If this were the case with every species, then behavioral analysis would be greatly simplified and the task could be neatly and exhaustively divided between the classical ethologists and the radical behaviorists—both of whom stress the sufficiency of objective behavioral data.

If the fiction of the empty organism works, there are no strong scientific reasons to go beyond it. To be sure, it might be of some interest to explore the experiential world of, let's say, the pigeon, but if such excursions do not lead to more accurate predictions of pigeon behavior, there are no compelling scientific arguments for pursuing them. My position is that an exclusive focus on physical input and behavioral output leads to better scientific results with some species than with others, and that when this approach does not work well in studying certain organisms, our predictions of behavior will be improved in proportion to the amount and quality of our knowledge of (i.e., our inferences about) the organization of their perceptual world. In this paper I will advocate the view that the great apes fall clearly within this last group—in which the empty organism is no longer a useful fiction—and that a satisfactory explanation of their behavior requires that we give careful and explicit attention to the elusive details of how they construct their world and order their experiences within it.

278

My aim is to characterize this constructive process, based on what is now known. For more than 40 years, scientists have been investigating the sensory thresholds of nonhuman primates, examining their ability to respond to differences in color or size or form, and asking whether they can deal with relations such as same, different, or intermediate. In addition to these sharply focused studies we have a long tradition, beginning with Köhler and Yerkes, in which the problems posed to the animal are loosely structured, and the ape's success in solving them is of less interest than how it goes about the job. Studies of home-reared chimpanzees are another rich source of information on representational processes. Finally, we can draw upon the recent efforts to teach language to great apes, a procedure that David Premack, one of the leading contributors to this enterprise, believes is largely a matter of mapping knowledge that the organism already possesses (Premack 1971b).

The present emphasis will be on description, rather than explanation; on phenomena, rather than mechanism. Although I will have little to say about the cumulative effects of experience in shaping representational activity, it scarcely needs to be noted that it plays a fundamental part at all levels—from the development of elementary functions of the sensory apparatus to the emergence of the highest order of problem-solving skills. The reader interested in more causally oriented reviews may consult other sources (e.g., Mason et al., 1968; Riesen 1966; Rumbaugh 1970). Here the concern is to describe how apes order and structure their perceptual world, rather than to suggest how such organizational tendencies are brought about.

SENSORY PROCESSES

The firmest ground for our entry into the perceptual world of the apes is through the study of sensory processes, although it will not carry us very far. Audition and particularly vision have received most attention. As in most research we will consider, the apes most often studied were chimpanzees. The data indicate substantial similarities between ape and man. (For reviews see Nissen 1946; Prestrude 1970; Riesen 1970.)

The limits of sensitivity to light are about the same in chimpanzee and man, and in most regions within the visible spectrum the ape discriminates between different wave lengths as accurately as we do. There is some indication, however, that the chimpanzee's ability to discriminate hues in the red-yellow region of the spectrum is not as good as our own. With respect to visual acuity, the thresholds for chimpanzee and human subjects are essentially similar, as is the ability to detect differences in size. Menzel compared responses of an adult chimpanzee and human subjects to food pieces of various sizes and found correlations between chimpanzee and human judgments were 0.86 or better (Menzel 1960). The threshold for perception of movement is also comparable in ape and man (Carpenter & Carpenter 1958). The ape's ability to discriminate simple forms is highly developed, a skill that is taken for granted in many of the situations we will consider later. Fantz's work suggests that capabilities for form and pattern discrimination are present in ape and human soon after birth (Fantz 1965).

The ape's sense of hearing is similar to ours in terms of thresholds and upper and lower frequency limits, although their sensitivity to high frequency sounds may be somewhat better than our own. They also localize sounds readily and distinguish variations in tempo. There have been no detailed investigations of the perception of patterned sounds, such as that involved in human speech or music.

Systematic information on olfaction and taste are lacking. Those who have worked closely with chimpanzees generally have the impression that the ape is more responsive to odors than man, but owing to strong cultural sanctions surrounding odor in western society, this suggestion cannot be accepted uncritically. The data are somewhat better for taste. Kellogg and Kellogg (1933) compared the responses of a young chimpanzee and a human child to sour, salt, sweet, and bitter substances. They concluded that the order of preference for the child was sweet, sour, salt, and bitter, whereas for the chimpanzee it was sour, sweet, salt, and bitter.

With respect to other sensory modalities, evidence is meager. Most people who have had close acquaintance with chimpanzees conclude that they are less sensitive to pain than human beings, an impression that is based largely on the fact that apes seem far less hesitant than man to undertake actions that would result in pain for us and show little overt response to injuries that would be extremely painful to a person. They respond to tickling as humans do, and on the basis of the limited data available, their proprioceptive and labyrinthine senses are on a par with man's. For example, chimpanzee thresholds for discrimination of lifted weights are about the same as human thresholds established under comparable test conditions.

In summary, for those modalities where systematic data are available, and this is chiefly for vision and audition, the basic receptor activities of the ape are essentially similar to our own. Moreover, although I have not reviewed comparable data on Old World monkeys, I believe the evidence would show that the sensory systems of these primates correspond closely to those of the great apes and man. Thus, what gets into the organism as reflected in basic sensory functions does not seem likely to provide significant information on the psychological contrasts among these species. Rather it would appear that if differences do exist in how the environment is ordered and perceived they will be found on some other level of organization, such as attentional factors, the inter-relations between different modalities, the development of object relations, and the effects of context and meaning.

The following quotation from Gibson is to the point:

> The input of the sensory nerve is not the basis of perception as we have been taught for centuries, but only half of it. It is only the basis for passive sense impressions. These are not the data of perception, not the raw material out of which perception is fashioned by the brain. The active senses cannot simply be the initiator of *signals* in nerve fibers or *messages* to the brain; instead they are analogous to tentacles and feelers. And the function of the brain when looped with its perceptual organ is not to decode signals, not to interpret messages, nor to accept images. These old analogies no longer apply. The function of the brain is not even to *organize* the sensory input or to *process* the data, in modern terminology. The perceptual systems, including the nerve centers at various levels up to the brain, are ways of seeking and extracting information about the environment from the flowing array of ambient energy (1966, p. 5).

PATTERNS AND STRUCTURES

In real-life situations the organism is required to go far beyond the simple tasks of detecting the presence or absence of stimulation within a particular modality, or of discriminating between discrete stimuli arranged along a single dimension such as hue or pitch. As Gibson emphasizes, environmental energies are not so neatly packaged. On logical grounds we can identify two major organizational strategies for dealing with the problems posed by the complexity and variability of ambient energy. One strategy is for the organism to respond to only a few recurrent and highly specific patterns or configurations of energy—which hold some obvious biological significance for the species—and to treat all other elements in the situation as though they were extraneous and inconsequential. Selectivity can be accomplished at any level or combination of levels, from the receptor to the central nervous system. This type of *fixed tuning* organization is exemplified in the concepts of the *key stimulus* and the *innate releasing* mechanism. The other strategy is to respond to many elements of a complex situation, but to order the elements according to certain general perceptual rules, such as the Gestalt laws of proximity, similarity, continuity, common fate, and closure (Haber & Hershenson 1973; Hinde 1970; Marler 1961). Actually, it is most unlikely that many organisms operate exclusively within one or the other of these organizational possibilities, and a mixed strategy will probably be the rule in most situations. Nevertheless, the distinction will be useful here because there are large differences among species in the nature of the mix; and the data indicate that great apes are prominent examples of animals that have the ability, if not a strong propensity, to respond to relations among elements, as well as to their specific attributes.

This tendency is clearly seen in responses to pictures and other two-dimensional arrays. Both the Kelloggs and the Hayeses noted in their home-reared chimpanzees an early spontaneous interest in photographs, and unequivocal signs that the objects represented were recognized. For example, the Hayeses' Viki placed her ear against a photograph of a wrist watch, and on another occasion made food barks while tapping a picture of a candy bar. In formal tests she demonstrated her ability to discriminate between pictures of a wide range of familiar objects, presented in color and black-white photographs and in line drawings, and to match pictures to their corresponding objects (Hayes & Hayes 1953).

A matching procedure was also used to investigate Viki's ability to perceive the number of elements in a stimulus. For example, she was shown a sample card bearing from one to six spots, and two choice cards, one bearing the same number of spots as the sample (but of different sizes and different spatial arrangements), and the other bearing a different number of spots. Viki's task was to select the card that matched her sample when the only clue to successful matching was the number of spots. Her performance on such problems was very similar to that of precounting children, but deteriorated when numbers were large and differences between stimuli were slight (Hayes & Nissen 1971).

With respect to other perceptual preferences and organizational tendencies, the data indicate that apes can readily be taught to discriminate stimulus arrays on the basis of internal relations such as presence of an odd element, middleness, on-top-of, and sameness-difference. They prefer to look through a clear window rather than a window that produces a distorted image, and show a tendency to fill in spaces when presented with the opportunity to draw upon a paper displaying various stimulus elements. (For details see Rumbaugh 1970.) Contrary to Schiller's (1951) original conclusion, however, they show no tendency in these productions to balance figures, nor to complete an incomplete array (Smith 1973).

Two rather obvious conclusions are suggested by the material reviewed in this section. The first is that, in spite of the obvious importance of elucidating the perceptual proclivities of the ape, the major impetus behind the many careful studies that have been completed has been to uncover higher-order processes involved in concept formation and problem-solving behavior. The traditional interests of perceptual psychology in illusions, constancies, and structuring tendencies have received comparatively little attention in studies with the great apes. Investigating these processes in a subject who is unable to report directly to the experimenter poses formidable practical problems, and this may account in part for the paucity of data on some questions.

The second conclusion is that the traditional methods of training and testing animal subjects in this kind of research need to be combined with other procedures. One of the by-products of recent efforts to teach language to the great apes is the development of a flexible system of communication between experimenter and subject that could serve as a very powerful adjunct to more conventional approaches. The content of existing chimpanzee vocabularies is certainly broad enough to permit an animal to be instructed in the relevant features of a task, as human subjects are, and at no greater risk to objectivity. It is easy to imagine how a few apes possessing the communicative skills of the Gardners' Washoe, Premack's Sarah, or Rumbaugh's Lana could provide in a few years more systematic and extensive data on perceptual processes than we have had in the entire period since the study of great ape perception began.

THINGS AND ATTRIBUTES OF THINGS

Thus far we have considered the ape's ability to impose patterns or to create structures based on relations among the elements making up a complex stimulus array. Here we go beyond the question of the ape's ability to order the immediately given, and ask whether it is also capable of organizing its experiences along more abstract lines. To what extent is it able to respond to a constant aspect of a variety of stimuli (concept formation), to integrate information coming through different sensory modalities (intermodal transfer), and to differentiate the physical attributes of the sign for an object from those of the object itself?

There is compelling evidence that apes have little difficulty classifying stimuli according to gross differences in physical attributes. In the light of evidence we have already considered, it is not surprising that an attentive and cooperative subject can do very well when required to respond to color or form or size as abstract dimensions. Of greater interest are those problems in which more subtle differentiating characteristics must be employed. For example, are apes able to distinguish photographs of adults from photographs of children? To classify objects according to function? Or to classify the same objects in several different ways? Such questions have been most thoroughly explored with Viki (Hayes & Nissen 1971). Working with pictures that she had encountered for the first time, Viki discriminated animate from inanimate objects, children from adults, and complete from incomplete pictures—for example, a hand with three fingers missing vs. an intact hand—at levels of accuracy between 79 and 89%, comparing favorably with human beings of her age.

Viki also showed a spontaneous interest in sorting objects into groups, a characteristic that Kohts also noted in her young chimpanzee.

This behavior was exploited by the Hayeses to investigate Viki's conceptual world. With little or no prompting, she would classify objects into homogeneous groups, such as large nails vs. small nails, black buttons vs. white buttons, small red blocks vs. small red balls, blue wooden diamonds vs. blue wooden triangles, spoons vs. forks, and keys vs. joined nuts and bolts. She was also quite consistent in classifying tools for eating vs. tools for writing, except for a pair of chopsticks, which she placed with the writing implements.

This is an intriguing error. Since Viki had often seen people eating with chopsticks, the Hayeses suggest that she was sorting objects according to material properties, rather than function. This idea was supported by similar results from another test. Viki was given different kinds of hardware (e.g., nails, bolts, and brads) and 25 buttons, all made of plastic, glass, or bone, except five that were made of metal. She placed all the buttons together except for the metal ones, which she put with the hardware, again suggesting that she was sorting by material properties rather than use. Likewise, when she sorted buttons vs. money, she placed buttons with holes in the middle in one group, but placed buttons without holes with the coins. Such findings do not rule out the possibility that the chimpanzee can classify objects according to function, of course, but they clearly suggest that physical characteristics are prepotent dimensions.

Sorting tests were also used with Viki to demonstrate that she was able to classify objects according to more than one common property. For example, she could sort red and blue poker chips and marbles, or stainless steel and green plastic forks and spoons, according to form or color. In one case Viki demonstrated that she could use several bases for classifying objects by sorting buttons into two colors, then into two forms, and then into two sizes. She was also able to make use of differential cues in order to select color or form as the correct basis for classification. For example, when the objects were placed on a blue test tray, the appropriate response was to match objects according to color, whereas when the test tray was white the appropriate response was to match according to form.

There is no reason to believe that Viki's achievements were unusual for chimpanzees, nor indeed beyond the reach of rhesus monkeys. One may suppose, however, that the ape shows a stronger tendency than the monkey to respond spontaneously to abstract dimensions and that it is capable of managing a somewhat larger number of them. One of the most convincing demonstrations of the ape's ability to adopt an abstract attitude is provided by Premack (1971b), who demonstrated that a chimpanzee distinguished between the physical properties of the sign of an object and the object itself. The ape, who had already been trained to communicate using plastic tokens as words, was first asked a series of questions about the properties of a red apple, with the actual apple present during the interrogation. Her responses indicated that she perceived the apple much as we do: red rather than green, round rather than square, and having a stem-like projection rather than a plain contour. Next she was required to answer the same questions when the apple was replaced by its *word,* a triangular piece of blue plastic, and she assigned to the plastic the same properties she had previously assigned to the apple. She was, thus, clearly able to disregard the physical properties of the word that was before her and describe the object to which it referred.

Apples, oranges, and the myriad other things that we deal with every day obviously comprise multiple attributes that we perceive through different sensory modalities. Oranges have a distinctive color, shape, texture, odor, and taste; when touched they feel different from apples or peaches. The harmonious and seemingly automatic synthesis of these heterogeneous inputs as properties of one and the same object is so obvious a fact of human experience that we are likely to take for granted that a similar process is at work in other primates. But this is by no means the case. In fact, apart from man, the only primates for which evidence of intermodal integration is compelling and unambiguous are the great apes. In a series of carefully controlled experiments, Davenport and Rogers were able to demonstrate that apes can accurately identify by touch alone objects that have only been seen or, conversely, can recognize on sight an object that they have previously experienced only by touch (Davenport & Rogers 1970; Davenport & Rogers 1971; Davenport, Rogers & Russell 1973). Gallup's demonstration that chimpanzees provided with an opportunity to view their mirror images form a *self-concept,* implies, at the very least, a high degree of intermodal integration, and here too, Old World monkeys are sharply different from the apes (Gallup 1970).

These findings suggest that apes spontaneously classify their experiences with objects according to salient stimulus attributes, or at least, can readily be trained to do so. Their ability to abstract such dimensions as size, form, color, brightness, and the like, certainly seems to be on a par with that of a human child of 2 or 3, and probably exceeds that of monkeys, although this remains to be established. Complementing this tendency to analyze objects according to abstract dimensions is the ability to synthesize heterogeneous attributes as the properties of a single object. In this regard, ape and man would seem to be sharply separated from the other primates.

In apparent contrast to man, however, the ape makes little use of functional criteria in classifying objects, a point that is stressed by Kohts and supported by the Hayeses. It should be emphasized, however, that this important question has not been investigated systematically. Tool use offers a promising avenue for further exploration of the ape's ability to conceptualize objects according to function, rather than physical appearance. For example, in a situation requiring a hammer, would an ape choose stones, clubs, and wooden mallets over rubber balls, hoses, and whisk brooms, even though it had no previous opportunity to use such objects in solving problems requiring a hammer? Given experience with various devices for pounding, digging, reaching, and climbing, would it show any tendency to group them according to use? Köhler's (1925) observations were not encouraging. Recently, however, a test-wise chimpanzee showed impressive skill in selecting from a collection of tools the one that was appropriate to the task (Döhl 1968; 1969). Moreover, the fact that apes are apparently capable of a generalized understanding of certain actions implying some reference to environmental objects—take, give, insert, bring—suggests some ability to work with functional categories (e.g., Premack 1971).

DEALING WITH SPACE AND TIME

The world as we experience it is extended in space and time, as indeed it is for many other animals. We are here concerned with how that experience is organized and represented in the great apes. Traditionally, this question has been treated as the problem of memory, of insight, and of foresight. I assume that these are related facets of a single constructive process, which draws upon previous experience and present circumstances.

This modest assumption seems fully consistent with the evidence we have considered, and we will encounter nothing new in principle in this section, although the phenomena will be somewhat different from those we have touched upon previously. I doubt that any reader, having come this far, would take exception were I to assert at this point that apes have good memories, that they remember selectively, that they anticipate the consequences of their own actions (and to a lesser extent the actions of others), and give every indication of perceiving their surroundings as structured and constrained. All this has been so often demonstrated and in so many different contexts, that it is difficult to appreciate that it leads to what is probably the most lively and enduring issue in all of primate behavioral research.

The controversy centers on the problem of *ideation* or *thought,* or in Hebb's terms ". . . some sort of process that is not fully controlled by environmental stimulation and yet cooperates closely with that stimulation" (Hebb 1949, page xvi). The issue has several aspects: First, how autonomous is this process? To what extent is it emancipated from immediate environmental control? (This is not to be confused with the philosophical question of determinism vs. free will.) Second, how should this process be characterized? What is the basis of thought— images, symbols, prior associations, covert behavior? Third, is there any fundamental *discontinuity* between man and all other animals in the nature of this process? (No one would claim that there were not differences in degree, of course.)

I doubt that any of these questions (at least as I have phrased them here) will ever receive a final and decisive answer, but they have served to focus attention on problems that might otherwise have been ignored, and they will probably remain heuristically useful for some years to come. To appreciate their influence on the character of primate research and on the current climate of opinion, it will be helpful to turn back a few years to a time when the issues were more sharply drawn.

The battle was joined with the publication of Köhler's *Mentality of Apes* (1925), in which he asserted that chimpanzees were capable of insight, which he defined as ". . . the appearance of a complete solution with reference to the whole layout of the field." What seemed to rankle Köhler's critics most was his rejection of the principle of association as a sufficient explanation of the chimpanzee's intellectual achievements. Consider the following comment by Pavlov, who, at the time he made it (1934), was himself studying higher nervous activity in apes:

> The tendency to draw a psychological difference between ape and dog based on the process of association is nothing more than the secret desire of the psychologists to evade a clear solution of the problem and to render it mysterious and extraordinary. In this pernicious, I should say, disgusting tendency to depart from the truth, psychologists like Yerkes and Koehler fall back on such barren notions, as, for example, that the ape went away, "meditated at leisure" in a human-like way, "has found a solution," etc. (1957, p. 556).

Much has gone forward since Pavlov wrote and, as is so often the case history has discovered some merit on both sides of the argument. We are left with a clear demonstration that the chimpanzee is capable of sizing up a situation and of selecting a course of action that will take him unerringly to the goal on the very first attempt, and with the certain knowledge that his ability to perform with such facility is dependent upon prior experience. In short, Köhler was right; insight as he defined it does occur. The associationists were likewise correct, to the extent that prior experience is a demonstrably important factor in such achievements.

We can consider only a few of the findings in support of such a conclusion. Rensch and Döhl tested a 5-year-old female chimpanzee on a graded series of maze tasks. Each maze pattern was mounted on a board and covered with glass. To solve the problem the ape used a magnet to draw an iron ring (subsequently exchanged for food) toward the exit. The chimpanzee eventually became quite proficient in solving these tasks, even when the specific pattern was new to her and rather complicated. For example, on the last 100 problems she chose the correct path in 86% of the trials. It is significant that on the more complex patterns she spent as long as 75 seconds after sitting down at the maze—ostensibly figuring out the correct pathway—before making her first move.

Human subjects tested on some of these same tasks also took time to select the correct path, but they required, on average, less than half as much as the chimpanzee, although they moved the ring along the chosen route only slightly faster than the ape did (Rensch & Döhl 1968).

Menzel's recent investigation of spatial memory in chimpanzees confirms the apes' ability to perceive the layout of the task, and provides compelling evidence that they recall important specific details within the situation and are able to work with such information in a constructive or creative way. In his research, young chimpanzees were carried by the experimenter around a large outdoor enclosure and shown up to 18 randomly placed hidden foods. They were then taken back to a starting point and released. The animals not only remembered most of these hiding places and the type of food that each contained, but they recovered the food by routes more or less in accord with a least-distance principle, regardless of the pathway along which they were carried by the experimenter (Menzel 1973a).

Although the experiment did not examine the role of experience in the development of this achievement, it seems likely that it was important. The chimps had lived in the cage for more than a year, and it is reasonable to assume that their familiarity with it was a factor in their success. The role of experience is more clearly indicated in Menzel's research on leadership with the same animals and in the same enclosure. He found that followers came to anticipate the direction of travel of the leader (who knew the location of hidden food) and would sometimes run ahead of him and search for the food in clumps of grass, behind logs, or in other likely places (Menzel 1973b).

There is little doubt, then, that chimps perceive their surroundings as more or less articulated structures—as wholes and parts within wholes —that they are capable of selective retention and constructive, creative application of information about the environment, and that they are even capable of a certain degree of foresight. In those few cases in which direct comparisons have been made between ape and human, the ape generally finishes second best, as one would expect. It seems likely that the similarity between ape and man is greatest in the spatial realm, particularly in those situations in which all the essential information is immediately available (although it may be far from obvious) or only a few elements are missing. Detours, roundabouts and the like—which require the animal to select a route that deviates sharply from its normal predilection for a straight-line approach—are plainly difficult for the ape, as anyone who reads Köhler will appreciate, but they are by no means impossible. A reasonable estimate is that the ape's ability to *project* or map some action onto an existing situation is about on a par with that of a child between 4 and 7 years of age.

The more remote the action is from present circumstances, the greater the discrepancy between ape and man. Apes possess the ability to anticipate the outcome of events in progress, of course, as do a vast number of other animals (although strangely enough, we have virtually no systematic data on the nature of limits of this ability in apes). They appear to be severely handicapped, as compared to man, however, in any situation that requires them to plan ahead for more than a matter of minutes, or at most, hours. Tools seemingly are fashioned in response to the need of the moment and abandoned when the job is done; nests are built anew each night, even though perfectly serviceable ones are available from another time; the remains of today's feast are not laid by against the possibility of a lean tomorrow, even though apes do not seem to lack the motivation or the simple instrumental skills required to establish a small private cache.

THE CRITICAL DISTINCTION BETWEEN APE AND MAN

To many writers the ape's limited foresight, its feeble imagination with respect to future events, marks the crucial and definitive difference between ape and man, and is the key to our cognitive supremacy. Köhler put it in these terms:

> ... though one can prove some effects of recognition and reproduction after considerable lapses of time... this is not the same as "life for a longer space of time." A great many years spent with chimpanzees lead me to venture the opinion that besides the lack of speech, it is in the extremely narrow limits in *this* direction that the chief difference is to be found between anthropoids and even the most primitive human beings. The lack of an invaluable technical aid (speech) and a great limitation of those very important components of thought, so-called "images," would thus constitute the causes that prevent the chimpanzee from attaining even the smallest beginnings of cultural development (Köhler 1925, p. 267).

Köhler's conclusion has become the leitmotif running through all subsequent discussions of the psychological distinctions between ape and man: Apes are deficient in the ability to symbolize (e.g., Cassirer 1944; Langer 1942; White 1942; Whyte 1950), to impose arbitrary forms, functions, or rules on the environment (e.g., Bastian & Bermant 1973; Holloway 1969), or to reify complex functions (e.g., Bronowski & Bellugi 1970). Human cognitive ability is species-specific, a unique constellation of functions peculiar to man (Lenneberg 1967). There is no question, of course, that these attempts to distinguish man from ape are in the right general direction. After all, the record is plain. One can be certain that apes in nature lack anything like language in the human form; that they have, at best, the merest rudiments of material culture; that they are not given to extended reflections on the past, nor elaborate speculations about the future.

Are we to conclude, then, that we have reached the point where investigation of representational processes in the great apes can go no further, have arrived at the boundary that unequivocally separates ape from man? One does not have to be on the side of the apes to reject this conclusion as premature. In fact, it is precisely here that we know least about the relevant dimensions of comparison; that detailed evidence is most wanting; where the methodology being developed by the students of chimpanzee language has most to offer. Attributes such as richness of associations, reflective thought, symbolization, images, and the like, so often cited as differentiating man from ape, remain elusive and ill-defined in the human case and can hardly provide a firm basis for interspecies comparisons. To foreclose the issue now is to say, in effect, that the ape utterly lacks abilities and functions that have yet to be described adequately in ourselves.

At stake is the question of human cognition, which we cannot hope to resolve completely except by reference to other species. The humanist can with strong justification assert that man is unique and must be approached on his own terms, that human experience provides an ample and demanding program of study. He cannot, however, claim that an exclusive focus on man will ever lead to a full understanding of why human cognition is special in a biological sense, or how man came by his unusual perceptual abilities. To accomplish this a comparative framework is required, in which the great apes will necessarily be given a prominent place.

In terms of a concern with representational processes, the preoccupation with learning and *intelligence*—construed as the ability to solve set problems—which has dominated investigations of anthropoid cognition, has led in an unfortunate direction. The roots of the difficulty can be traced to Köhler's *Mentality of Apes.* Köhler clearly recognized that his research told something about how the world was represented to the chimpanzee, a question that obviously interested him keenly. He also emphasized, however, that his approach shed light on chimpanzee intelligence, which is not entirely the same thing.

The medium he chose to investigate both questions was the *roundabout* problem—food behind a barrier, or buried in the ground, or suspended out of reach. This was a good start—at least it is difficult to imagine how one might begin with either sort of question without relying on lures and physical devices, and the measurement of successful performance—and Köhler carried out his enterprise with great ingenuity, there was in this approach an inherent source of ambiguity. Köhler was sharply critical of Thorndike's studies of animal intelligence on the grounds that all relevant features of the problem were not available to the animal. To avoid this difficulty, Köhler designed his tasks according to his view of how an intelligent creature such as the ape ought to behave. He evaluated performance not merely on the basis of objective criteria of failure or success in attaining rewards, but also on whether the errors were good or bad, based largely on his impressions of whether the ape took the essential features of the problem into account. Thus, *success* (which could occur accidentally) and *failure* (which could be either intelligent or a crude stupidity) were diagnostic only when combined with the experimenter's judgment of whether the ape correctly perceived the structure of the problem.

Subsequent developments have tried to resolve this ambiguity by emphasizing specific skills and concrete achievements. In the general push toward objectivity, the distinction between methodological and philosophical behaviorism became increasingly blurred; successful performance became the focus of scientific attention, rather than the covert representational processes upon which such accomplishments partly depend. In Gilbert Ryle's terms, a concern with *knowing how,* or the development and exercise of skill—which can be demonstrated fairly directly—all but eclipsed the interest in *knowing that,* or the acquisition and organization of information—which can only be approached indirectly and inferentially.

The metaphor that best describes this last approach is *mapping.* In the realm of skillful performance, the individual experiment has a kind of self-sufficiency. When dealing with representational processes, however, the isolated result adds only a few details, at best, which gain their principal significance in relation to a larger composite picture. The aim of this picture is a reconstruction, so to speak, of the ape's models of reality.

To be sure, there is no reason to suppose that the ape's construction of the world is precisely the same as our own. The principle that each species occupies its own unique perceptual domain was expressed long ago by Jacob von Uexküll in his concept of the *Umwelt,* and nothing we have learned since has altered the essential validity of this idea. Ape and man, after all, are quite different species and they have followed different evolutionary paths for countless generations. To assume that representational processes in the ape are merely simpler versions of those in man, that the workings of its mind are but feeble reflections of our own, is to commit a gratuitous anthropocentrism. Nevertheless, it is clear that the *Umwelten* of ape and man are indeed similar in many respects. Time and again, we have imposed our own categories of experience on the apes and have found solid evidence of correspondence, in spite of the millennia that separate us in evolutionary time.

True, we are not brothers under the skin; neither can our kinship be denied. Here, we have moved from simple sensory systems, to more complex perceptual processes, to certain aspects of concept formation, arriving finally at questions of imagination and thought—from the world as it is immediately presented through the action of the receptors, to those mental operations that are the subject of our strongest proprietary concern. At each succeeding level the correspondence between ape and man has been somewhat less than at the one before. To suppose that this increasing divergence is not real is naive. At the same time, it would be tendentious to overlook the fact that with the increasing divergence there has been a corresponding reduction in the sharpness of the questions, the precision of the methods, and the amount and quality of the data that have been obtained. The essential terms of our uniqueness have yet to be defined.

CONCLUSION

A primary justification for investigating representational activities in the great apes is that knowledge of such processes leads to better predictions of their behavior, psychology's traditional shibboleth and goal. The ape structures the world in which it lives, giving order and meaning to its environment, which is clearly reflected in its actions. It is not very illuminating, perhaps, to describe a chimpanzee as figuring out how to proceed, while it sits and stares at the problem before it. Certainly such an assertion lacks originality, as well as precision. We cannot escape the inference that some such process is at work, however, and that it has a significant effect on the ape's performance. At the descriptive level that has mainly concerned us here it would seem better to be vaguely correct than positively wrong.

A second major source of concern with representational processes in the great apes is our desire to gain a comparative perspective on human mental activity. There is a strong consensus that the great apes are our closest living kin, with the orangutan occupying a somewhat more remote position than the chimpanzee and gorilla. That leaves a great deal unsaid. Evolutionary sequences and degrees of phyletic relatedness are by no means agreed upon by the experts. Whatever the final outcome of their deliberations, however, we can be certain that the great apes provide the best comparative material available to us, and we must use it as judiciously as we know how. Unfortunately, Köhler's remarks, made more than 50 years ago, apply with equal force to the contemporary scene:

> Since the rise of evolutionary ideas, highly emotional questions, wishes, and assertions have been directed to the nonhuman primates. Even the simplest observations, which can be made almost daily, easily meet with suspicion under these conditions, because lively prejudice is much more common than experience in this area. And, indeed, this danger threatens just about as much from those who wish to find the highest possible characteristics in apes as from those who would at all costs deny them anything human (1971, p. 211).

On the basis of findings such as those reviewed in this paper, I am persuaded that the apes and man have entered a cognitive domain that sets them apart from all other primates. The question of whether the differences between man and ape are in degree or kind cannot be answered simply on the basis of present evidence, and may well prove to be beside the point. How the issue will be resolved in a given instance will very much depend upon the level of organization that is being considered and the data language appropriate to it. For example, at the level of speech—in the sense of vocalized language—the contrasts between man and ape are manifestly qualitative. On a different level, such as the ability to form concepts or to combine various acts or subroutines into larger functional units, the differences appear to be matters of degree. Differences in the neural correlates of mental activity are even more probably of this sort.

In any event, notwithstanding the impressive contrasts between human and anthropoid achievements, there is no unequivocal evidence at this point of a grand abyss separating the cognitive processes of ape from man, nor for that matter, monkey from ape. Perhaps the evidence will be forthcoming, although recent experience gives no encouragement to this belief. Clearly, nothing will be gained from foreclosing the issue before all the data are in. In the meantime, we are only at the beginning of discovering how the environment is represented in the great apes. To guide our thinking we desperately need more comparative data on monkeys, on the apes themselves, and on humans, particularly the preschool child. The methods are available, the enterprise is feasible, and whatever the outcome, we are bound to see some of our most prized and distinctive attributes in a different light.

Acknowledgment

In addition to the participants in the Wenner-Gren conference on the Great Apes, where the original version of this paper was presented, I want to thank the following friends for their helpful comments and suggestions: J. Bastian, D. Cubicciotti, A. Elms, D. Fragaszy, G. Mitchell, and R. Sommer. Support was provided by NIH Grants HD06367 and RR00169 and the Wenner-Gren Foundation.

Two gorilla brothers (aged 68 mo. and 27 mo.) play-wrestling

R. A. Hinde

The Nature of Social Structure

Chimpanzees have been reported as living in loose communities (Goodall 1968; Itani & Suzuki 1967; Nishida 1968; but see Wrangham, this volume), gorilla in small bisexual groups (Schaller 1963, Fossey 1974; this volume), and orangutans are usually solitary (Galdikas, Horr, MacKinnon, Rodman, this volume). Among monkeys, an even greater range of social behavior is to be found. This diversity of social structure is of concern to primatologists for two main reasons. First, attempts have been made to relate different types of social structure to the ecological conditions where they are found (e.g., Crook & Gartlan 1966). These have met with little success, in part because the environmental conditions have been described too crudely (Struhsaker 1969; Crook 1970;

Clutton-Brock 1972), and in part because the characters of social structure on which attention has been focused, namely group size and sex ratio, may not be the consequences through which natural selection molds the behavior of individuals (Hinde 1974). Second, the diversity of social structure raises the question of the relation between the structures observed and the behavior of the individuals that gives rise to it. Although some progress has been made with this problem (e.g., Yamada 1971; Mason 1973), many workers have underestimated the complexity of the problem (see discussions by Crook 1970; 1974), and again the main focus has often been limited to the parameters of group size and sex ratio.

Although for many issues the proportion of males to females and the size of troops may be important, they are clearly not the only ways in which troops differ. Indeed the question of just what the concept of *social structure* should embrace would seem to merit further consideration by primatologists. This paper contains an examination of certain aspects of that concept. Although referring primarily to studies of nonhuman primates, it attempts also to indicate how they relate to studies of social structure in human societies.

THE CONCEPTUAL FRAMEWORK

The discussion here depends in part on a conceptual framework outlined elsewhere (Hinde, in press a, b), which attempts to relate the concept of social structure to the empirical data actually observed in particular instances. This framework involves three levels—interactions, relationships, and social structure. In terms of behavior, *interactions* can be characterized in such terms as "A does X to B," or "A does X to B and B responds with Y." Description of X or Y involves reference to the sort of behavior involved, and also to its quality—e.g., whether the act was rough or gentle, whether the goals of the interactants were related or conflicting, and so on. *Relationships* involve a series of interactions between two individuals known to each other, and can be characterized by the content, quality, and patterning of those interactions. *Social structure* in turn can be characterized by the content, quality, and patterning of relationships.

At each of these levels it is necessary to proceed in two directions (Figure 1). First, from data about particular interactions, relationships, or structures, it is necessary to seek generalizations that describe the same phenomena over a wider range of occasions, individuals, circumstances, cultures, or species. Of necessity, the broader the applicability of a generalization, the more information must be added to it to relate it to a particular instance. This sequence of generalizations of increasing breadth is illustrated by the successive rectangles from left to right in Figure 1.

Second, at each level it is necessary to search for principles that will explain the empirical data: these are represented by the discontinuous circles in Figure 1. At each level more than one set of principles will be required, and the principles may interact: those required at any one level will enhance those required at the preceding level.

Each of the three levels in this scheme demands rather different techniques and concepts for its analysis. The study of interactions must be based on some understanding of the behavior of individuals in non-social situations—an understanding that may be based on physiological or behavioral-psychological analysis according to the nature of the problem, the stage of the analysis, and the predilections of the investigator. In addition, the study of interactions must involve analysis of the additional complexities that arise because two or more individuals are involved, including processes of inter-individual communication, verbal and nonverbal (e.g., Goffman 1969; Hinde 1972a), and analysis of how such communication contributes to, or mitigates against, coordination between the behavior of the participants. The study of relationships requires, in addition, understanding of the way in which the interactions between two individuals known to each other are patterned in time: this is discussed elsewhere (Hinde, in press c). The study of social structure is discussed in the rest of this paper.

THE CONCEPT OF STRUCTURE

Surface Structure and Structure

As will be apparent from Figure 1, the study of structure must be based on that of relationships, and thus ultimately on that of interactions between individuals. This statement does not imply that social structure could be predicted from the study of the behavior of individuals or even pairs of individuals, but that it must be deduced from the way in which individuals interact in a full social situation. Exactly how the concept of structure is to be related to the empirical data on the interactions of individuals, however, is a matter of some controversy. Sometimes, especially in sociological studies of particular groups, it is used for something quite close to empirical reality. Perhaps more usually, especially among anthropologists, it refers to something derived from it. Fortes (1949) wrote ". . . structure is not immediately visible in the 'concrete reality.' It is discovered by comparison, induction and analysis based on a sample of actual social happenings in which the institution, organization, usage, etc. with which we are concerned appears in a variety of contexts" (p. 56). Lévi-Strauss (1953) goes even further, stating categorically that "the term 'social structure' has nothing to do with empirical reality but with models which are built up after it."

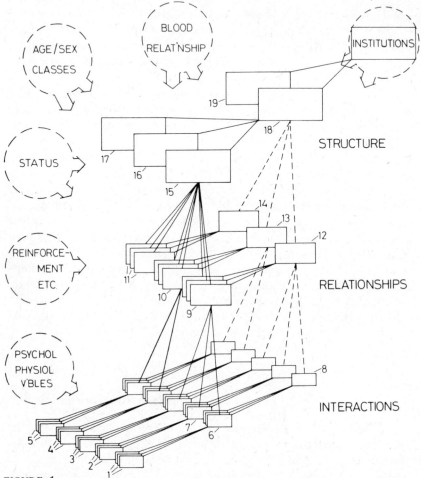

FIGURE 1

Diagrammatic representation of the relations between interactions, relationships, and social structure. Interactions, relationships, and social structure are shown as rectangles on three levels, with successive stages of abstraction from left to right. The discontinuous circles represent independent or intervening variables operating at each level. Institutions, having a dual role, are shown in both a rectangle and a circle.

In the specific instance of a nonhuman primate, the rectangles might represent:

1. Instances of grooming interactions between a mother A and her infant B.
2. Instances of nursing interactions between A and B.
3. Instances of play between A and B.
4. Instances of grooming between female A and male C.
5. Instances of copulation between A and C.
6. First stage abstraction—schematical grooming interactions between A and B. Abstractions of grooming interactions between other mother-infant pairs are shown behind, but the specific instances from which they were abstracted are not shown.
7. First stage abstraction—schematized nursing interactions between A and B. Abstractions of nursing interactions of other mother-infant pairs are shown behind.
8. Second stage abstraction—schematized grooming interactions between all mother-infant pairs in troop.

Here the term *surface structure* is used to refer to the pattern of relationships actually observed in a particular instance, and *structure* for something of greater generality derived from the study of one or more particular instances. It is considered that the degree of abstraction, the distance from surface structure at which it is desirable to seek for structure, will vary with the problem in hand, however, and the rules of abstraction must be defined in such a way as to preserve the features relevant to that problem. Three features of surface structure that must always be taken into account in seeking for generalizations are considered in the next section.

Three Characteristics of Surface Structure

The Patterning and Nature of Relationships. Each level of the conceptual scheme in Figure 1 depends both on the nature and on the patterning of the elements in the level below it. Relationships involve properties that depend on how the component interactions are patterned (Hinde & Simpson 1975; Hinde, in press c). Patterning is equally important at the level of structure. As a simple example, a group of four individuals could be sustained by three relationships, but would have a very different structure according to whether those relationships all focused on one individual, or linked the four individuals into a chain. A description of patterning may of course be hierarchical, involving subunits within larger entities (e.g., Kummer 1968).

Although patterning is crucial, it is not all: for most purposes an adequate description of structure requires reference also to the nature of the relationships involved. Where each of three individuals has relationships with each of the others, the structure can be described as

---

9. Mother-infant relationship between A and B. Mother-infant relationships of other mother-infant pairs are shown behind (but connections to grooming, nursing, etc., interactions are not shown).

10. Consort relationship between A and C. Other consort relationships shown behind.

11. Specific relationship of another type (e.g., peer-peer).

12, 13, 14. Abstraction of mother-infant, consort, and peer-peer relationships. These may depend on abstractions of the contributing interactions.

15. Surface structure of troop containing A, B, C, etc.

16, 17. Surface structures of other troops (contributing relationships not shown).

18. Abstraction of structure of troops including that containing A, B, C, etc. This may depend on abstractions of mother-infant, etc., relationships.

19. Abstraction of structure of a different set of troops (from another environment, species, etc.).

This scheme is of course schematic only, and could be expanded in a number of directions. For example, at the level of structure, further abstractions from 18 and 19 could provide a picture of structure characteristic of species but independent of environment. Or further levels could be added—at the lower end, referring to the mechanism underlying interactions, and at the top, to represent population structure, the patterning of the relationships between troops in an area.

triangular, but this tells us little unless we also know which of the relationships are hostile and which friendly. The degree to which the nature of the relationships needs to be specified will of course depend on the problem: for instance, if we wish to compare the rate of information flow through structures of different types, the description could be quite crude. Insofar as the nature of the relationships do need to be specified, however, social structures differ from physical ones: a cube is a cube whether made of wood or stone.

The Dynamic Nature of Surface Structure. In everyday language, the term structure tends to have static implications. Such implications are quite foreign to the concept of social structure.

In the first place, the description of social structure must involve a time dimension. Since relationships involve a patterning over time of interactions between two individuals, relationships themselves must always be described with reference to a span of time. Likewise surface structure, involving a patterning of relationships, must also contain reference to a time dimension.

Second, the relationships within a group are likely to affect each other. Even in the simple case of the spatial relationships between three individuals, it is unlikely that each can sit at his preferred distances from the others: the distance that any one sits from the other two is affected by their preferred distances from him and from each other. Some degree of tension in relationships is thus inseparable from group living. To cite some examples from more complex relationships in non-human primates, a mother can affect the relationships of her infant with its peers, and other females affect the relationship between a mother and her infant (Rowell et al. 1964; Hinde & Spencer-Booth 1967); coalitions may affect agonistic interactions (Kawai 1958; 1965; Struhsaker 1967); the relationship between the leader and the adult females in a troop of Japanese macaques may have far-reaching effects on the pattern of relationships in the group (Yamada 1971); and each relationship within a small captive group of rhesus macaques may affect each other (Hinde 1972b). If one individual is removed from a small group, or indeed if one relationship changes in nature, there are likely to be widespread repercussions on the social structure. The surface structure is essentially dynamic.

Third, in nearly every group the relationships of particular individuals change with time. In a macaque troop, for example, infant males become juveniles, subadults, and may eventually join the central hierarchy, thereby showing a constantly changing pattern of relationships. For the troop as a whole, however, the surface structure may stay more or less constant from year to year, showing only relatively minor changes between birth season and mating season as the infants grow older and the relationships between the adults fluctuate with their endocrine state.

Finally, the relationships between individuals may be affected by environmental factors, with consequences upon group structure. For example, food availability affects the incidence of aggression in captive rhesus monkeys (Bernstein 1969), the size of black spider monkey troops (Durham 1971) and chimpanzee groups (Wrangham 1974), and the extent to which herds of gelada baboons split into one-male groups (Crook 1966). Anubis baboons also split up into small foraging parties when food is scarce (Aldrich-Blake et al. 1971). The availability of sleeping sites also affects the size of baboon troops (Crook and Aldrich-Blake 1968; Kummer 1968). In general, insofar as environmental factors affect either the social behavior of individuals, or other aspects of their behavior with consequent repercussions on their social behavior, it must affect group structure. Again, surface structure is to be considered not as static, but the consequence of continual interaction with environmental factors.

Stability. Although social structure must be considered in dynamic terms, both the surface structure of particular troops, and the more abstract structure common to different troops of the same species, have considerable stability in that the dynamic changes that occur do not usually exceed certain limits. This stability can be considered at three levels.

First, there are consistent differences between species in their troops' social structures. (Sometimes, indeed, there are consistent differences between groups of closely related species—for instance, macaques and baboons tend to form multi-male groups, and *Cercopithecus* monkeys tend to form one-male groups [Struhsaker 1969]—but at this level there are also many exceptions.) Species differences in social structure must depend ultimately on genetic differences, although direct evidence is available in only one case. Hamadryas baboon troops consist of units, each containing one male and several females, but in anubis baboons, the smallest unit includes least several males. A population of hybrids showed an intermediate type of social organization, with one-male groups that were small and unstable (Nagel 1971).

Second, within any particular species, surface structure tends to fluctuate in a cyclical fashion. In baboon and macaque troops, as we have seen, the structure may change annually as the young are born in a limited birth season and then gradually mature, returning to its initial form with the next crop. In other species the cycle may be longer. Thus, Eisenberg et al. (1972) have suggested that most arboreal monkey troops return periodically to a one-male state. They suggest that the one-male troop grows by recruitment to yield a subgroup of juveniles, with the more senior males forming another subgroup. As this *age-graded* male troop develops, the younger males may form yet another peripheral subgroup of their own. As the troop increases in size, it becomes unstable and may split. Indeed, under favorable conditions, the multi-male troops of macaques grow and split from time to time (Furaya 1969; Nishimura 1973) so that their structure also can be said to be stable only over long time spans.

This long-term stability implies that the group must provide an environment for the developing young, such that they develop into individuals who form relationships similar to those formed by their elders. If one considers the structure of natural troops in the light of recent studies on the development of social behavior, it is clear that the balance between the various developmental forces necessary to achieve this is a delicate one. In many macaque troops, for instance, the balance is such that subadult males become peripheral for a while but not permanently: from what we now know about how the early social environment can affect later aggressiveness (Sackett 1968; Harlow & Harlow 1969), it is easy to imagine that a slight shift in genetic or social factors could produce either males who attempted too vigorously to remain central, with resultant excessive discord, or males who left and never returned.

If the long-term structural stability is to be taken into account in our generalizations, it is clear that, in many species at least, description of social structure must not stop at the individual troop. For example, long-term stability may depend on a *reservoir* of potential troop leaders who spend a large part of their lives as solitary individuals or in bachelor bands (e.g., Sugiyama 1967; Fossey 1974). Description of the social structure in such a case must include the (temporarily) nonbreeding individuals as well as the reproductive groups.

Third, we have seen that, superimposed on the species norm and such cyclic fluctuations as occur, environmental factors can bring considerable changes in the pattern of relationships that individuals form. Some examples were considered in the preceding section. In addition, differences between troops resulting from cultural variations or the personalities of the members may affect the social environment of the growing young (e.g., Itani 1959). Furthermore, chance variations in the sex ratio of births may lead to considerable fluctuations in troop composition (Rowell 1966a). Such facts also have ontogenetic implications. First, the propensity of an individual to form a pattern of relationships characteristic of his species and demographic category, although labile in some degree, must also have considerable stability. Second, although affected by social experience and perhaps, more specifically, by observational learning (Crook 1970; 1974), the effect cannot be so great that social experience within the range of the species norm produces a propensity to form a pattern of relationships that is outside the species norm.

This relative stability in the structure of nonhuman primate social systems implies a degree of "equifinality" (Bertalanffy 1960): in spite of small perturbations, the structure regains its initial nature. This, however, does not necessarily imply self-regulating mechanisms other than those involved in the behavior of individuals: the structure can be seen as merely a consequence of the pattern of relationships that individuals form.

Although the social structures of nonhuman primates show considerable stability, it is now clear that, contrary to earlier views (e.g., Parsons & Shils 1951), the social structures of most human groups change with time (Mead 1934). This does not mean that mechanisms promoting stability are absent, but that they are not fully effective. Processes of socialization, pressures towards conformity, and institutions, in addition to the constancies in the behavioral tendencies of individuals, can act at least to slow down the rate of change. Change itself could arise in three ways:

1. Extrinsically caused environmental changes, including long-term climatic and ecological changes.

2. Intrinsically caused environmental changes, including the depletion of food resources, demolition of natural vegetation, agriculture, etc.

3. Intrinsic changes in social structure independent of environmental change.

The first of these changes occurs in animals, however, the resulting changes in social structure are long-term and primarily the result of natural selection acting on individuals. The second kind of change is at least rare in nonhuman forms, in part because the reproductive potential of resources is usually sufficient to permit recovery from the level at which it ceases to be used as an economical resource. In man, this kind of change occurs frequently. It seems likely that changes in social structure brought about thereby involve positive feedback, a relatively small initial deviation produced by environmental change resulting in increased divergence either directly or as a consequence of further environmental change engendered by the initial social change (Buckley 1967). Whether the third kind of change—intrinsic changes in a previously stable social structure—ever arise *spontaneously*, without extrinsic cause, is a circular issue, depending on how stable is defined.

Conclusion

In this section I have emphasized the need to distinguish surface structure, the empirical pattern of relationships derived from data on particular troops, from structure abstracted from it. I have also considered briefly some aspects of structure that are crucial to its nature and should thus be preserved in attempts to move from surface structure to structure. We have seen that information about both the patterning and the nature of relationships is crucial, and that the inter-related aspects of the dynamic nature of social structure, and the extent to which it is stable, must be preserved.

THE PROCESS OF ABSTRACTION

In moving from the surface structure of particular groups to the abstract level of structure, we must attempt to specify aspects of the content, quality, and patterning of relationships that are independent of the particular individuals concerned and show regularities across categories of individuals and across societies. Primatologists have so far given little thought to the precise methods whereby this is to be achieved. In that they have been mostly concerned with characteristics of group size and age-sex composition; in making generalizations they have presented average (mean or median) data from several groups. Means, at any rate, are likely to be affected by aberrant instances, however, and thus include individual idiosyncracies. It could be argued that a mode would be a better guide.

Primatologists have come nearer to abstracting elements of an essential or *normal* pattern of relationships in their qualitative statements, when these take such a form as "a central group of adult males, with females and young nearby, and sub-adult males on the periphery." Such statements, however, have two shortcomings. First, the way in which they have been derived from hard data on specific instances too often remains unspecified. Second, such statements give little indication of the nature of the relationships between individuals, and imply an essentially static view of social structure: they are thus adequate only insofar as the dynamic aspects, discussed in the previous section, can be neglected.

How are we to escape from mummifying structure as we describe it? Consideration of four cases, in which hard data that will present a more dynamic view of structure are becoming available, may help to point the way.

1. Both the general characteristics of, and the differences between, the (surface) structures of troops of Japanese macaques have been described (qualitatively) with great insight by a number of workers (e.g. Sugiyama 1960; Imanishi 1960; Yamada 1971).

2. Kummer (1968) has used sociograms to portray aspects of the relationships of the one-male groups of hamadryas baboons.

3. Bygott has used both dendrograms (Figure 1, this volume) and spanning trees to show the degree of association between individuals in a community of chimpanzees.

4. Deag (1974) has produced networks, derived from studies of spatial proximity and of social interactions, for a troop of Barbary macaques.

In all four cases, data about the nature and patterning of relationships in particular groups are available, and permit generalizations about
the relationships formed by particular classes of individuals in those
troops. If no further material was presented, the group structure would
appear to be static. Each of these studies, however, presents material
that reveals the essentially dynamic nature and the mechanisms underlying the stability of the group. Growth and fission of troops of Japanese macaques have been described in some detail (e.g., Furaya 1969)
and the determinants of inter-troop differences discussed (e.g., Yamada
1971); Kummer himself (1967; 1968) has described the development of
hamadryas one-male groups; a fair amount is known about the social
development of chimpanzees (Goodall 1968b; Itani and Suzuki 1967;
Nishida 1970), and Bygott (this volume) has described changes in the
adult male hierarchy; and Deag (1974) has described many aspects of
the social development of Barbary macaques.

What is needed, however, is a means of integrating the ontogenetic
material with that on structure to show the dynamic nature of the
structure and the cyclical changes that it undergoes. Some attempts in
this direction have been made in each of these groups of field studies.
Perhaps the most extensive concern the hamadryas baboon, where
Kummer (1968; 1971) has provided sociograms and descriptions of
one-male groups in successive stages of development, and indicated the
nature of the changes in the dyadic relationships.

But even this is not all. The patterning of relationships results in
part from the dynamic effect of relationships upon relationships. Two
studies discuss this effect: Kummer (1967) discusses *tripartite* relations
in hamadryas baboons, and Deag (1974) describes how the relations
between adults are affected by the birth of infants.

Furthermore, if we are to understand how structures remain stable
or change with time, we must understand not only the dynamics of the
relationships within the group, but also how they are affected by interaction with the environment. This understanding requires knowledge of
the variability and plasticity in the constituent elements of the structure. So far we know practically nothing even about the potential
variability in the structure of similarly constituted groups of the same
species living in similar natural environments. There is limited data on
troop differences in structure in the Japanese macaque (Yamada 1971),
and on local differences in the structure of chimpanzee populations
(references cited previously). Hanby (1975) has some challenging data
on captive rhesus macaques. One of the difficulties here is that research
workers have often tended to regard differences between troops living
in the same environment as an embarrassment, to be swept under the
carpet, rather than treating knowledge of those differences as an
essential element for the understanding of social dynamics.

305

The difficulties with abstracting structure will now be apparent. Although the term structure seems to imply something capable of diagrammatic representation, the dynamic nature of social structure mitigates against it. What we must aim for, therefore, is not a picture of structure, but an account—an account that may utilize pictures of elements or temporary states (Hanby 1975), but is essentially an account of the processes underlying the dynamic structure. Our research for generalizable structure thus links with our second aim—the search for principles of organization.

PRINCIPLES OF ORGANIZATION

At the two upper levels in Figure 1, relationships and social structure, the data can be *described* in terms of the content, quality, and patterning of the next lower level (interactions and relationships respectively). These properties of content, quality, and patterning remain through successive stages of abstraction. Although the patterning of elements at each level gives rise to emergent qualities at the next, description can remain in basically the same language as that used in the initial description of the interactions. If we wish to *understand* the patterning at any level, however, we must employ new concepts that are not intrinsic to the initial data. These concepts can be embodied in groups of principles each of which concerns the patterning at the level in question.

The groups of principles primarily relevant to the patterning at each level are indicated by the discontinuous circles in Figure 1. In practice, principles determining the patterning at one level may affect also qualities at the next lower level. Status, for instance, may affect both how interactions are patterned in a relationship, and thus the quality of that relationship, and also the patterning and quality of responses in each interaction. For example, how much each of two grooming partners groom the other, and the quality of the grooming, may both be affected by the partners' mutual status (Simpson 1973a). The groups of principles primarily relevant to the patterning of interactions in a relationship are discussed in more detail elsewhere (Hinde, in press c) but, since those necessary at any one level embrace those necessary at the next lower level, they must be mentioned briefly here.

Principles of Organization at the Relationship Level

Two groups of principles appear to be of particular importance in nonhuman primates:

Status. The labeling of the two partners in a relationship as dominant and subordinate usually has primary relevance to behavior in agonistic situations. If it is based on data concerning only one aspect of behavior

(e.g., A attacks B more than B attacks A), it is merely descriptive. If it has relevance to several aspects that are correlated with each other across relationships (e.g., A attacks B more than B attacks A; A threatens B more than B threatens A; B grooms A more than A grooms B; and similarly for pairs C and D, E and F), it begins to have some explanatory power concerning the ways in which the different interactions in the relationships are patterned. In some contexts age appears to be a better predictor of the direction of nonagonistic interactions (e.g., grooming) than is success in agonistic encounters (e.g., Kummer 1968; Bygott, in press).

Kummer (in press) has provided evidence that dominance status has strong predictive power over the course of the formation of relationships between gelada baboons. His observations show that a newly acquainted pair of baboons usually go through four stages—reached with the first fighting, first presenting, first mounting, and first grooming, respectively. A stage can be omitted, but changes in the ordering of the sequence are rare. The speed with which animals progress through the sequence *(compatibility)* seems to be faster the greater the difference in dominance status between the two individuals, and slower the greater their absolute levels.

Reinforcement. Given the proven utility of *reinforcement* as an explanatory concept in nonsocial situations, it seems reasonable to try it out in the context of social relationships. Such an approach has been worked out in some detail in the human case by Homans (1961) and Thibaud and Kelly (1959). Starting from the basic assumption that individuals continue to perform activities in the context of a relationship insofar as those activities produce positive reinforcement from others and insofar as they do not involve excessive cost, Homans has produced a system that throws considerable light on human interpersonal relationships. Although essentially reductionist and not beyond criticism (e.g., Davis 1973), it may have important implications for studies of nonhuman primates. In particular it emphasizes the fact that relationships involve long series of interactions in time, with the course of any one interaction being influenced by events in the fairly remote past and possibly even by expectations for the future. Too often, hitherto, primatologists have focused on the immediate outcome of interactions, to the neglect of their long-term consequences (but see e.g., Rowell 1966b; Simpson 1973b).

Although the diversity of reinforcing events is much smaller in nonhuman primates than in man, one of the most important categories for both can conveniently be subsumed under the title *social approval.* It has been argued elsewhere (Hinde, in press a and b) that submissive gestures, presenting, mounting, grooming, and greeting behavior can perhaps best be understood in these terms.

307

It is not, of course, suggested that these principles are alone sufficient to predict the patterning of interactions within a relationship—others concerned with, for instance, familiarity per se remain to be worked out. Furthermore the sets of principles will interact: for example the extent to which grooming is reinforcing may vary with the dominance relationships of the individuals concerned.

Principles of Organization at the Level of Group Structure

In many nonprimate vertebrates, individuals congregate together because they are attracted to other members of the species *in general.* A similar generalized attraction no doubt occurs also in primates, but of far greater importance for the structuring of groups are the attractions of each individual to particular others. That these attractions result in relationships that differ in content and quality of interactions will already be apparent: the point to be emphasized here is that the group *is* a network of such relationships rather than a number of individuals drawn mutually together. This point is particularly well illustrated by Kummer's (in press) study of group formation among gelada baboons previously unknown to each other: each individual formed a series of bonds (see above) with others in succession until each was bonded with all others in varying degrees. In this section we are concerned with principles that provide understanding of the patterning of relationships that individuals form.

In nonhuman primates, at least three groups of principles are known to be important, namely those concerned with:

1. Relationships within or between age-sex classes (e.g., peer-peer; adult male-adult male; male-consort female).

2. Blood relationships or some correlate consequent upon early experience.

3. The influence of status.

In each case the principles concern the exclusion of individuals from basically possible relationships, as well as the various possible degrees and qualities of relationships that may be formed. Within each of these groups there may be parallel principles (e.g., those relating to different age-sex classes), or the principles may be hierarchically organized. As an example of the latter, a high level principle of wide applicability (e.g., the possibility of understanding a number of facts about the patterning of relationships in terms of a concept of status) may be superordinated to principles of more restricted scope (e.g., those concerned with the differences in the relationships between individuals far apart and close together in the hierarchy).

The surface structure of any particular group is to be regarded as an expression of such principles and of the interactions between them. Thus in a triad A, B, and C, the relationships may reflect the facts that

A and B are adult males, C is A's (half) sister, and the dominance order is A, C, B. Only in rare cases, such as in a group of previously unacquainted 3-year-old males, will one of the principles find relatively pure expression.

We could search for these organizational principles at any level of abstraction from the surface structure of a particular group to those of more generalized structures. We could be satisfied with principles from which the structures to be found in a particular species in a particular environment can be derived: these will comprehend the natures of the species and of the environment. We could seek for principles of wider generality, from which structures in diverse environments could be predicted: these would still comprehend the nature of the species. We could seek for principles at still more abstract levels. In general, the more abstract the level to which the principle applies, the more empirical facts it is relevant to and the less precisely does it predict them (i.e., the more additional factors must be considered).

Although principles of organization have not been worked out in detail at any level, they are beginning to take shape in the material available. It is not my aim to review this material extensively, but some aspects may be considered.

Relationships Within and Between Age-Sex Categories. Every species of nonhuman primates provides evidence of propensities for preferential relationships within or between age-sex categories, but the nature of those propensities varies widely, even among the anthropoid apes. Gorillas live in small groups led by one or more silverbacks (fully adult males). Within those groups, adult females and immatures appear to seek the proximity of silverbacks, but blackbacks usually spend more time on their own (Fossey, this volume; Harcourt, this volume). Chimpanzees live in more open communities, each consisting of several dozen individuals acquainted with each other, but not continuously associating together. They forage either alone, in small family (mother and offspring) parties, or in small groups. Such groups may contain a consort pair (McGinnis 1973), or a number of males (Bygott, in press), but practically any other combination is possible. When family parties meet, the young are especially likely to associate together. The orangutan, by contrast with both of these species, is largely solitary (Galdikas, MacKinnon, Rodman, this volume). Among monkeys, the diversity is even greater: although generalizations across species are prone to many exceptions, within any one, propensities for particular types of association are always clear.

Blood Relationships. With the possible exceptions of solitary species, such as the orangutan (as discussed previously) and those species that travel in nuclear families (e.g., titi monkey, Mason 1973), blood relationships probably play a crucial role in determining social structure. As yet, however, longitudinal studies have been made on very few species, and the data are fragmentary. Among chimpanzees, long-term preferential associations along matrilines are well established (Goodall, this volume), and at least one case of a long-term association between (half) brothers is known (Bygott, this volume). The most detailed studies concern the Japanese macaque, in some troops of which the matrifocal families form a dominance hierarchy with all members of one being dominant to all members of the next(e.g., Kawai 1958; Kawamura 1958; Oki and Maeda 1973), and rhesus macaques (Sade 1972a). In the latter species, also, the rank of daughters is often determined (and that of sons sometimes influenced) by that of the mother, and younger daughters come to rank above their elder sisters when about 3-4 years old. Comparative studies indicate differences in the importance of blood relationship even between closely related species kept under comparable conditions (Rosenblum 1971).

Dominance and Status. In the study of relationships between pairs of individuals, the terms *dominance* or *status* can be merely descriptive of the direction of agonistic interactions. Used as intervening variables, however, they can also have an explanatory function if the relevant dependent variables are correlated (see preceding discussion). In the group situation, if A is dominant to B and B is dominant to C, it usually happens that A is dominant to C. The individuals can thus be arranged in a dominance hierarchy. In reference to a relationship, dominance-subordinance concerns the patterning of interactions, but in reference to a group, it concerns the patterning of relationships. Here too, however, it may be merely descriptive, if it refers to only one aspect of the relationships, or explanatory, if it refers to a number of correlated ones.

Although dominance-subordinance in nonhuman primate groups usually depends on aggression or on potential aggressiveness, dominance and aggression are not to be confused. Dominance-subordinance relationships between pairs of individuals, or dominance hierarchies in groups, often succeed periods of aggression or aggressive display and are accompanied by a reduction in aggression. In a hierarchy, agonistic interactions between two individuals are usually less frequent, the greater the difference in rank between them (Bernstein 1970). In some species, at least, the ratio of the amount of agonistic behavior received to that given is a better predictor of basic rank than are the actual frequencies of either (Deag 1974). Furthermore, the hierarchy is often maintained by the avoidance behavior of the subordinates as much as by the overt aggression of the dominants (Rowell 1966b).

Evidence is in fact accumulating that dominance rank may be predictive of the nature and direction of interactions other than those associated with agonistic behavior or competition. For instance, among the adult male chimpanzees in one community in the Gombe Stream National Park, the more dominant individuals tended to be involved in more grooming bouts, and to be groomed for longer, than their inferiors (Simpson 1973a). Similarly Sade (1972b) found a relationship between grooming and dominance position in female rhesus monkeys: those high in the dominance order tended to have a higher *grooming status* (defined here as the sum of the number of animals grooming the focal animal, the number grooming them, and so on). Deag (1974) found that the frequency with which individual Barbary macaques occurred as *nearest neighbors* to others was correlated with their basic rank.

There are, however, many studies that show little or no relation between dominance rank and the direction of grooming and other gestures of social approval. For instance, Rowell (1966b) found little relationship between dominance rank (assessed subjectively) and *friendly* gestures in captive baboons, and Bernstein (1970) found no significant correlation with grooming behavior in captive groups of several species. Such findings, however, are to be expected for two reasons. First, individuals may address gestures of social approval to each other for reasons quite unrelated to the dominance hierarchy: the most popular individual is not necessarily the most dominant (e.g., Virgo & Waterhouse 1969). Second, such gestures may be used both by superiors to inferiors and by inferiors to superiors (e.g., Blurton Jones & Trollope 1968; van Hooff 1972; Kummer, in press): their direction or frequency is thus not necessarily related to rank. In addition, as we have seen, aggressive potential is not always of primary importance in determining the direction of interactions: age, for instance, may be even more important.

In this context it is profitable to refer again to Homans' (1961) use of the reinforcement concept in interpreting human relationships. He points out that, in any human interaction, each participant will expect his rewards to increase with his costs. The more costly to one partner is an activity (and thus probably the more valuable to the other), the more valuable to him must be the rewards that he receives from the other (and thus probably the more costly the interaction to the other). If each is being reinforced by a third party, each will expect to receive rewards commensurate with his costs. Beside exchanging rewards with each other, people appraise the rewards and the costs they incur in relation to those received and incurred by others.

Homans suggests that *distributive justice* is realized when the profit (i.e., the reward less the cost) of each man is directly proportional to his *investments* (i.e., that with which he has been invested and/or which he has acquired). What counts as an investment is to a large degree culture-specific, but such things as age, wealth, maleness, strength, wisdom, and acquired skills are recognized in many societies. The concept of investment is clearly related to that of *status*, which is determined by similar independent variables, and may be symbolized in the form of address.

The extent to which individuals of nonhuman species compare themselves with each other in respect to the rewards they receive and costs they incur is quite uncertain. There is increasing evidence that characteristics other than potential for aggression may count as an investment and contribute to status. For example, in a study carried out after Simpson's, Bygott (1974) found that age was correlated more significantly than any agonistic measure with the frequency of grooming and of being groomed. Furthermore, the frequency with which males displayed, although highly correlated with agonistic measures, was also more correlated with a measure of how often individuals were chosen as grooming partners than with age. Future studies will probably reveal that analyses of social structure in nonhuman primates have hitherto focused too much on status as related to agonistic potential, to the neglect of its other determinants. In addition, although it often seems that the dominance hierarchy is constant and independent of the context, it may well be that, as in human societies, there is not always only one hierarchy but a number, with status not absolute but context-dependent. In any case, the hierarchical principle is only one of the principles determining the patterning of relationships, and its expression is usually to some extent obscured by others. Sade (1972b) found a relationship between dominance and grooming status in female rhesus monkeys, but that in males was obscured by the occurrence of cliques.

We have already referred to Kummer's study of the manner in which the formation of a bond between individual gelada baboons was affected by their dominance relationships. He also studied the formation of relationships between triads (two recently unacquainted individuals, both male or both female, and a third of opposite sex strange to both) and in small groups. Although the dyadic relationships were formed in the same way as when only two individuals were present, bond formation tended to take longer and not to proceed as far along the sequence of fighting, presenting, mounting, and grooming (see previous discussion). Indeed, a dyad that formed a bond might regress to an earlier stage of bond formation when a third individual was introduced. Although recognizing that other types of explanation are possible, Kummer disregarded the variables of sex and previous acquaintance and interpreted the order in which the various dyadic relations formed in terms of dominance-subordinance, elaborating a series of rules that appeared to describe the observed events most economically. In the triadic experiments (see previous discussion), he found that dyads of high and almost equal compatibility were incompatible in one individual, the one with the lower sum of dominance statuses being delayed or even suppressed.

Kummer found the greater complexity of the group situation as compared with the dyadic one to be due primarily to two processes—inhibition and intervention. Once a dyad was formed, the dyad-forming capacity of an onlooker with either member of the dyad or with others was partially or completely, but often only temporarily, inhibited: males were more affected than females. Intervention in an interaction between two members of a potential dyad was more likely the greater the compatibility of the intervening animal with one of the members, the less his compatibility with the other, and the greater the compatibility between the two members. An animal was especially likely to intervene in an interaction between the two individuals on either side of him in the dominance hierarchy. Within a group, interventions fade with time, permitting group solidarity.

Conclusion

In the previous sections, I have reviewed briefly three groups of principles that help to explain the patterning of relationships in groups of nonhuman primates. These principles are to be regarded as akin to natural laws—derived from particular instances, but employing concepts not necessary for the description of those instances. It must be emphasized that such principles have not yet been worked out in any formal sense: the only attempt in that direction is Kummer's delineation of rules determining the formation of relationships in gelada baboons in terms of dominance-subordinance, and even there it is not yet clear whether principles of other types (e.g., age-sex classes) are derivable from those of dominance-subordinance or are needed to supplement them.

THE COMPARATIVE STUDY OF HUMAN AND ANIMAL GROUPS

In preceding sections, certain organizational principles throwing light on the structure of nonhuman primate groups were discussed. That they are applicable to the patterning of relationships also on the human level is immediately apparent. Of overriding importance in the human case, however, is a further set of principles concerned with institutions. *Institution* is used here to refer to one or more recognized positions in a society that constrain the behavior of the incumbents, and thus covers unique positions (e.g., the king), recognized relationships (e.g., marriage), and large-scale systems (e.g., the National Health Service). In that the common feature is one or more positions with recognized rights and duties, it is linked to the concept of *role*. The usage approximates that of Parsons (1951). An institutionalized relationship or structure is thus one in which the participants have recognized rights and duties.

Institutions have occupied a dual role in sociological studies. Insofar as they represent a sought-after pattern of relationships, a goal of one or more of the participants, institutions have the same status as organizational principles: just like status, blood relationships, and age-sex class relationships, they provide insight into the causal mechanisms that shape the actual pattern of relationships that we observe. Some institutions, like marriage or some all-male groups, are based on the organizational principles found in nonhuman societies, but others depart more or less radically from them. Systems of sociological kinship do not necessarily follow blood relationships (Fox 1972; Leach 1966), and it is not easy to discern the biological roots of a democratically elected parliament. In such cases, the surface structure we observe will be a product of interaction between the organizational principles of the institution and other organizational principles that affect the incumbents' relationships, as well as of individual idiosyncracies. The factors that distort the sought-after institutional structure will vary from case to case, and it may well be that the best general description we can abstract from the social structures we observe is the institution. Thus in sociological studies, institutions sometimes have the nature of organizational principles, and are sometimes the product of processes of abstraction from the raw data—although it will be noted that the processes of abstraction differ somewhat from those used in studies of nonhuman primates.

The crucial importance of institutions in human societies provides an important point of difference between animal and human social structure. The regularities of structure in animal societies are statistical regularities, consequent upon similarities in the relationships individuals form. The relationships that they form are not, *so far as we know,* guided by rules or by communally accepted ideals. For this reason, the concept of role is of little help in understanding the determinants of individual behavior, although it may contribute to an understanding of the consequences of individual behavior for the group (Hinde 1974; in press, a, b). This does not necessarily mean that the cultural transmission of structural norms is totally absent in nonhuman forms: the issue of the extent to which a young macaque male uses as models those individuals he sees behaving as peripheral or dominant individuals is still undecided (see Crook 1970). Animals receive no verbal instructions on how to behave, on the successive positions in society they are likely to occupy, or on abstract subgroups with specific codes that they may one day enter. The fact that monkeys and apes do not seek for relationships whose characteristics have been verbally defined must be a major factor limiting the complexity of their social structure; however, this does not mean that studies of nonhuman species are irrelevant to those of man. As discussed elsewhere, a comparative approach, in which the differences between human and animal groups were used to throw light on the special properties of the former, may well be fertile (Hinde, in press b).

CONCLUSION AND SUMMARY

Our approach to the study of group structure is, thus, by no means straightforward, even for nonhuman primates. It involves first, from a study of particular interactions in particular groups, description of the content, quality, and patterning of relationships within those groups. Abstraction from the pattern of relationships in particular groups will provide generalizations of social structure valid for a range of groups. At each stage, such descriptions or generalizations must necessarily contain a reference to the time dimension, for structure is dynamic, and its characteristics may fluctuate or change with time. Understanding of how relationships are patterned within a limited time interval and, more important, how they fluctuate or change with time demands knowledge of groups of organizational principles that explain the patterning of interactions within relationships and of relationships within structure. A description of group structure must be not merely one of the relationships between individuals at one time or even of their changing relationships over time, therefore, but rather must be in terms of the processes in time that determine those relationships. In seeking for abstractions or generalizations valid for a number of groups, we must attempt to make generalizations that retain the dynamic nature of structure, and thus concern the processes of group dynamics rather than solely the end products of their action. Although the study of social structure in nonhuman primates has not yet reached a stage of sophistication that makes a general systems approach possible, that would seem to be the direction in which it must aim. Of course this does not mean that description of more limited parameters of structure are valueless—only that their limitations must be recognized.

Acknowledgment

This work was supported by the Medical Research Council and the Royal Society. I am grateful to a number of colleagues for discussion, including especially M. Thorndahl, M. Fortes, J. Hanby, and M. J. A. Simpson.

Joseph L. Popp and Irven DeVore

Aggressive Competition and Social Dominance Theory: Synopsis

Discussions of the origin and modification of aggressive behaviors abound in the biological and social sciences, where aggression is most often attributed to proximate causes such as frustration. Here we consider aggression in the context of natural selection; specifically we ask under what circumstances will aggressive behavior increase the representation of the aggressive actor's genes in future generations? In answering this question we provide a brief outline of a model of optimal competitive strategies, in terms of animal behavior and morphology, for the maximization of an individual's reproductive success.[1]

Human beings consider premeditated murder the most reprehensible of crimes, and intragroup aggression of any sort is cause for concern in all societies. It is therefore not surprising that many observers of animal aggression have offered explanations that suggest such behavior is non-functional or maladaptive. Most often these arguments assume that undisturbed groups of social species in natural settings maintain a high level of pacific stability. Accordingly, aggression, and especially the killing of conspecifics, is viewed as a breakdown of orderly social relations, a form of social pathology—often the result of outside pressure such as human interference, overcrowding, and the like. Although such extraneous influences may foster higher levels of aggression in some circumstances, such explanations are wholly inadequate for most instances of intraspecific conflict and lethal aggression.

Although evolutionary explanations of aggressive behavior have been advanced before, they have often been flawed by species-advantage arguments. Such arguments suggest that aggressive behaviors are perpetuated because the *species* benefits when the *strongest* (hence, by implication, the *fittest)* individuals reproduce. Such explanations characterize a number of popular and influential works that attempt to explain the evolution of fighting (e.g., Eibl-Eibesfeldt 1961; Lorenz 1966). By contrast, we adopt Darwin's view (1859) that selection at the individual level is the evolutionary principle of overriding importance, and support recent advances by Geist (1966; 1971), Parker (1974), Maynard Smith (1972, 1974), and Maynard Smith and Price (1973) which have made an outstanding contribution to the understanding of aggressive behavior in terms of natural selection. In the following sections we will attempt to demonstrate that aggressive behaviors, when understood in an evolutionary framework, become predictable.

THEORY OF AGGRESSIVE COMPETITION

Consider a simple cost-benefit analysis of aggressive competition in a model of optimal behavior, where costs and benefits are measured in units of reproductive success, and aggression is viewed as a strategy that, under a given set of environmental and genetic constraints, may maximize an individual's contribution of genes to succeeding generations. In discussing the principles of natural selection that determine the optimal rate and form of aggressive competition, it is logical to ask first what costs of competition can be adaptively incurred by an aggressive actor. In calculating the maximum costs that an actor should be willing to accept in an aggressive encounter (C_a), there are four groups of variables of primary significance: (1) the probabilities of access to the disputed resource through aggression or through alternative strategies, (2) the benefit of access to the object of competition, (3) the effects of the competition on relatives of the actor, and (4) the intrinsic competitive abilities of the actor and his competitor.

In determining an actor's maximum adaptive expenditure in a competitive encounter, the following algorithm can be used. Multiply the probability of the actor's access to the disputed object if he competes aggressively (p_{a_1}) times the benefit of access to the disputed object to the actor if he competes aggressively (B_{a_1}); subtract from this quantity the probability of the actor's access to the disputed object if he does not compete aggressively (p_{a_2}) times the benefit of access to the disputed object to the actor if he does not compete aggressively (B_{a_2}). Hence, $C_a < p_{a_1} B_{a_1} - p_{a_2} B_{a_2}$.

The above inequality takes into consideration only the potential costs and benefits of aggressive competition to an actor, and assumes that the competitor is unrelated. Following Hamilton (1964; 1972) Maynard Smith (1964), and others, however, it is clear that the actor must also take into account the potential costs and benefits of the competition to his competitor if that competitor is a relative—a not unlikely condition in a social group of primates. Thus, the probability that a related competitor will gain access to a resource if the actor is or is not aggressive (p_{c_1} and p_{c_2}, respectively) and the potential benefits to the competitor under both of these conditions (B_{c_1}, B_{c_2}), as well as the costs of the aggressive competition to the competitor (C_c), must be multiplied by Hamilton's (1972) regression coefficient of relatedness (b_{ac}), where b_{ac} is defined in the absence of inbreeding as the probability that a gene is present in the genotype of the actor that is identical by descent with a gene in a randomly chosen gamete of the competitor. We would expect, therefore, that natural selection will favor an individual that competes aggressively with a conspecific when the following inequality is true:

Inequality I: $C_a < p_{a_1} B_{a_1} - p_{a_2} B_{a_2} + b_{ac}(p_{c_1} B_{c_1} - p_{c_2} B_{c_2} - C_c)$

The actor's maximum adaptive expenditure for an aggressive encounter is represented by C_a. This includes such components as the caloric expenditure during a fight, the cost of repairing injuries, the risk of permanent injury or death, and so forth, all expressed in terms of their effects on reproductive success. Likewise, B_{a_1} represents the total benefit of access to the object of competition. For example, in competition over a food item, B_{a_1} includes components such as the net nutritional value of the item relative to other food items in the environment, measured by its effects on reproduction, as well as other consequences of the competition (such as a higher rate of access to similar resources in the future, if such action means that the same competitor is more likely to avoid future competition for those resources).

According to Inequality I, aggressive competition is most adaptive when: (1) the cost of aggressive competition to the actor is small; (2) the benefit of the disputed object in terms of the reproductive success it will confer upon the actor is large; (3) aggressive competition confers a probability of access to the disputed resource by the actor that is large compared to the probability of access if aggressive competition does not occur; (4) the competitor (i.e., the recipient of the aggressive act) is genetically unrelated; and (5) the cost of aggressive competition to the competitor is small if the competitor is genetically related ($b_{ac} > 0$). In fact, with exceptions to be noted in a following section, if the competitor is unrelated, the actor will be selected to discount the costs of aggressive competition to the competitor (C_c) entirely.

When Inequality I is true for one or more actors, aggressive competition is expected to result and continue over time until the inequality is no longer true for any of the actors. Individual a is predicted to win in an encounter with individual β when the following inequality is true:

$$\text{Inequality II: } K_a C_a > K_\beta C_\beta$$

where,

$K_a = $ a's intrinsic competitive ability, i.e., a's ability to inflict costs of competition on β per unit time;

$K_\beta = $ β's intrinsic competitive ability;

$C_a = $ the maximum adaptive expenditure by a for aggressive competition, as determined by Inequality I;

$C_\beta = $ the maximum adaptive expenditure by β (considering β as the aggressive actor) for aggressive competition, as determined by substitution in Inequality I.

More generally, if a winner in a dyadic aggressive encounter is defined as the individual that takes possession of the disputed resource when the aggressive interaction is terminated, then the winner can be predicted to be the actor with the higher value for the quantity $K_a C_a$. In theoretical terms, aggressive competition may be viewed as an attempt by an individual to modify the cost-benefit function of its competitor in such a way that continued attempts by the competitor to gain access to a resource are unprofitable and maladaptive. Injury, or the threat of probable injury, to a competitor can elevate the costs of continued competition until these are no longer offset by the potential benefits of access to the disputed resource. When such a point is reached (i.e., when Inequality I becomes false), natural selection will favor the competitor that terminates the aggressive interaction; in a dyadic encounter with β, the amount of β's losses are predicted by the value of C_β. The last individual to terminate its aggressive behavior will normally gain access to the resource, and therefore be defined as the winner. If a wins in an encounter with β, its cost of competition equals the quantity $K_\beta C_\beta / K_a$, which is less than or equal to its maximum adaptive expenditure (C_a) for that encounter.

It follows that in aggressive encounters an individual will be selected to act in ways that will increase the rate of expenditure of an opponent's competitive effort while reducing its own rate. In populations where this competitive strategy includes attempts to inflict physical damage on opponents, natural selection will favor those individuals who develop defense strategies (assuming the defense strategy is less costly than the injuries it prevents). As a result, an aggressive actor need not necessarily inflict physical injury on an opponent for the actor's behavior to be adaptive; it is sufficient that the actor force his opponent to adopt a strategy of defense that is costly in time and energy. As Maynard Smith (1972) noted: even "ritualized fights" (which may involve little chance of injury to the participants) can thus be settled.

Asymmetry in the cost-benefit functions of competitors is the key to the termination of aggressive interactions. If two competitors knew the precise value of all variables in Inequalities I and II prior to an aggressive encounter, we would not expect such an encounter to occur; the losing competitor would be known in advance, and would be selected to avoid the competition. For an excellent discussion of assessment strategies see Parker (1974).

Even when the values for the relevant variables are not precisely known, the disparity between the values of $K_a C_a$ for two potential competitors may be so large that the inferior competitor actively avoids the interaction; thus, some potential conflicts may be selected with little or no costs of competition. In other cases, competitors may be so closely matched in value of $K_a C_a$ that aggressive conflict will occur. In this circumstance even the winner will be forced to expend most of his C_a before the outcome of the encounter is determined.

It follows from Inequalities I and II that when two individuals are equal in other respects, the competitor that can derive a greater benefit from the object of competition is expected to win the encounter. Similarly, a competitor with the lower costs of competition per unit time of interaction has a competitive advantage when other variables are equal (see Figures 1 and 2). The cost of aggressive competition per unit time relates directly to the concept of intrinsic competitive ability that was introduced in Inequality II; it is not the individual with the greater C_a that wins the encounter, but rather the individual that expends the smaller relative portion of C_a per unit time. Hence morphological structures used in fighting evolve in populations, where such variations arise

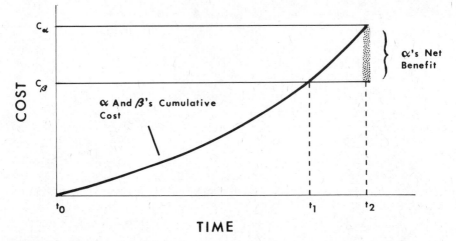

FIGURE 1

A hypothetical aggressive interaction is depicted, in which the two competitors differ only in that a can derive a greater benefit from the disputed resource than β. Accordingly, a's maximum adaptive expenditure in aggressive competition (Ca) is predicted by Inequality I to be greater than the maximum adaptive expenditure by $\beta(C\beta)$. Since both a and β incur costs at an equal rate, it can be seen that β will reach his maximum expenditure at t_1 but a will not do so until t_2; a is therefore expected to attain access to the resource and will be defined as winner.

by mutation, because the costs of competition per unit time are higher to individuals who lack such structures than to individuals who possess them.

THREAT DISPLAYS AND SUBMISSIVE GESTURES

The role of ritualization in animal conflicts has perplexed many ethologists. The fact that animal conflicts often appear to be settled without the use of lethal weapons or serious injury to the participants has led to the widespread assumption that self-restraint during aggressive encounters is practiced for the good of the species (Eibl-Eibesfeldt 1961; Lorenz 1966). We have already argued that species-advantage arguments are largely inappropriate models for the evolution of aggressive behavior (see Lack 1966; Williams 1971; and E. O. Wilson 1975 for a discussion of group selection theory and its relation to social behavior).

Yet it is still pertinent to ask why competitors often give agonistic displays prior to aggressive encounters rather than opportunistically attacking an unsuspecting opponent from behind or broadside. And why should the victor in a competitive encounter acknowledge the submissive gestures of the loser instead of killing it? These questions and other theoretical aspects of aggressive behavior have been discussed by Geist (1966; 1971), Maynard Smith (1972; 1974), and Maynard Smith and Price (1973); their arguments are important for primate behaviorists, and we include some of their points when we apply our model of aggressive competition to the questions raised.

It is easy to appreciate that agonistic displays may benefit the *recipients* of such displays, since advance knowledge of an impending attack enables the individual to prepare to fight or flee. It is much less obvious, however, why an individual would be selected to provide unambiguous signals of aggressive intent for a potential victim; it would seem that greater damage could be inflicted on the opponent through surprise attacks. It is important to recognize that individuals are not selected to maximize the losses of their *opponents,* but to maximize their own gains in reproductive success; these two operations are not necessarily equivalent. Inflicting costs on an opponent beyond those required to win a given encounter does not usually result in added benefits for the aggressor. In fact, when the cost of seriously injuring the opponent exceeds the cost of winning the encounter, or when b_{ac} is greater than 0, such behavior will be disadvantageous to the actor. Agonistic displays will be favored by natural selection when they alleviate the need for forms of aggressive competition that would be more costly to the individual. Thus, if agonistic display alone is sufficient to drive a competitor away from a disputed resource, and if display is on the average less costly to the individual than aggressive behavior, and if the benefit of agonistic display in preventing needless aggressive escalation is greater than the cost of putting one's potential opponent on guard, then agonistic display will be an adaptive strategy.

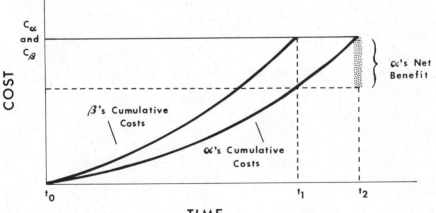

FIGURE 2
Two individuals are depicted who differ only in their intrinsic competitive ability (K_a), the actor with the higher value for K_a is expected to win the encounter. Although in this example a and β have the same value for C_a, a will win the encounter because he suffers a lower cost of competition per unit time.

The ability to predict the *outcome* of an aggressive encounter before it occurs is also advantageous to a potential aggressor. In Inequality I, it was shown that as the probability of winning an encounter approaches 0, the maximum costs that an individual is willing to suffer also approaches 0. Predicting the outcome of a conflict permits an individual to avoid the conflict completely or terminate it early when it is apparent that he would lose. As a result, individuals who are sensitive to cues that are useful in predicting the outcome of an aggressive encounter can be favored by natural selection. Individuals are expected to modify their behavior according to such characteristics of their opponents as body mass, horn size, and stage of dental eruption when these characters relate to their opponent's intrinsic competitive ability, that is, when they potentially affect the outcome of an aggressive encounter. Bighorn sheep have been shown to modify their aggressive competition as a function of their opponent's horn size. Poorly endowed male bighorns either avoid conflicts with males having larger horns, or compete in ways that reduce the probability of retaliation, such as mimicking female behavior (Geist 1971). Body weight and horn size also accurately predict the outcome of agonistic encounters in cattle (Bouissou 1972). Red jungle fowl show postural variations that permit the human observer to predict the outcome of an encounter by the fowl's preliminary display behavior (R. H. Wilson 1974), and it would seem likely that natural selection has also favored such predictive abilities in the birds themselves. When agonistic display elicits a reciprocal display by the opponent, the initial actor will benefit to the degree that the elicited display permits him to optimally modify his competitive behavior. In other words, if one can estimate the cost of competing with an opponent by viewing his weaponry, then inducing a display of that weaponry prior to combat may be advantageous.

The Evolution of Exaggerations of the Variables in Inequalities I and II and Its Relevance to Displacement Activities

In populations in which individuals use the behavioral and/or morphological characters of potential competitors to determine whether or not aggression will be adaptive, it is widely recognized that those characters on which intrinsic competitive ability is estimated will tend to be emphasized or exaggerated by natural selection. For example, in species in which body size is a determinant of the outcome of aggressive encounters and individuals modify their behavior as a function of their opponents' body size, then individuals that exaggerate their body size during display will have a competitive advantage. Hence, chimpanzees show piloerection during displays, anubis and hamadryas baboon males have thick mantles of hair, and orangutan males, in addition to fleshy facial pads, produce impressive vocalizations with inflatable laryngeal pouches. In species in which the horn, antler, or tusk size of an actor influences the behavior of a competitor, it is often noted that the prominent display of these organs to competitors is a common pattern in agonistic encounters. Furthermore, the evolution of exaggerations of such morphological features through automimicry has been hypothesized (Guthrie and Petocz 1970).

Wallace (1973), in a discussion of the evolution of misinformation in animal communication, argues convincingly that, where exaggerations evolve, large exaggerations produce a disproportionately large increase in fitness of the actor compared to small exaggerations; *small lies* have about a 50:50 chance of improving a competitor's understanding of a situation, whereas *big lies* deceive competitors with near certainty. Similarly, it is our belief that selection for increasing exaggeration of characters relating to intrinsic competitive ability (K_a) is expected to proceed until the specific character is so exaggerated that a positive correlation between its appearance and the intrinsic competitive ability of the individuals within the population no longer exists (or until such characters reach the stage of development where counterbalancing selection through other pathways inhibits further change). One such counterbalancing selection force is the evolution of compensatory adjustment capabilities in the conspecifics that observe these displays, i.e., the ability of competitors to devalue the exaggeration by the appropriate amount (Parker 1974). R. L. Trivers has suggested to us that in populations in which actors devalue the exaggerations of competitors, natural selection can act to maintain both the exaggeration and the ability to devalue the exaggeration simultaneously; even though exaggerations are devalued appropriately by competitors, those competitors who have been selected to devalue the exaggerations of conspecifics have not necessarily been selected to recognize those individuals in the population who do *not* exaggerate. Hence, selection can occur against individuals who do not exaggerate because other members of the population, through not discriminating them from exaggerators, devalue the nonexaggerators' intrinsic competitive ability below its true magnitude —with negative effects on the fitness of the nonexaggerators. For example, in a population of individuals who exaggerate body size by piloerection, and also devalue the apparent size of potential competitors by a fixed degree, individuals who exaggerate will be assessed by potential competitors to be near their true size, but mutants who do not show piloerection will be assessed by potential competitors as smaller than their true size, and will suffer a decrease in fitness through an increased rate of challenges by competitors smaller than themselves.

Trends in the evolution of exaggerated body size and fighting structures have been generally appreciated since Darwin; however, some additional principles follow from the model of aggressive competition presented here. Inequality I includes a number of variables relevant to the outcome of aggressive encounters. Exaggerations by individuals need not be limited to morphological structures that are related to intrinsic competitive ability (K_a), but may theoretically include any of the variables in the cost-benefit functions for aggressive competition. Thus, an actor who can exaggerate the value of the disputed resource (B_{a_1}), so that the opponent overestimates the benefit that the actor can derive from it, will have a competitive advantage. In functional terms, such exaggerations might be made by an individual who mimicked higher levels of motivation, to gain access to a disputed resource, than he in fact felt. Or an individual may exaggerate the value of winning an encounter (B_{a_1}) by misrepresenting the circumstances under which the encounter occurs. Cormorants display sham incubation behavior between bouts of aggressive competition (Kortlandt 1940). Numerous other species, including African cichlid fish (Heiligenberg 1965), the three-spined stickleback (Tinbergen and van Iersel 1947), the killdeer (Phillips 1972), the herring gull (Howard 1935), and others cited by Armstrong (1965) show real and sham nest-building and nest-tending behaviors during agonistic encounters. Simmons (1970) has noted that sham nesting in brown boobies differs from real nest-building behavior, in that building materials are manipulated in a conspicuous rather than functional manner and that the movements are directed toward the rival rather than toward the nest. In the current ethological perspective, these behaviors are *displacement activities,* i.e., nonadaptive behavioral overflow resulting from high levels of motivation or conflicts between drives. In contrast, we suggest that these behaviors may have a straightforward adaptive significance; the presence of a nest or eggs would usually imply higher benefits from winning the encounter for their owner than for the intruder, and individuals who advertise such asymmetry when it exists, or delude an opponent into believing that such asymmetry exists through sham behavior, will have a competitive advantage. Natural selection would favor individuals that mimic incubation behavior because in the past opponents have been selected to recognize individuals incubating eggs and avoid competition with them for nesting sites or territories (where other territories are unoccupied or are occupied by birds that have not reached the egg-laying stage), since the presence of eggs implies a high level of investment by the resident (and therefore a high level of *motivation* to protect the investment), and indicates that he (she) will be very difficult to defeat and supplant. Accordingly, the usual assumption that *displacement activity,* such as the behavior described, occurs outside appropriate contexts may be unfounded, and perpetuated only by failure to recognize that the presence or purported presence of a nest (or offspring) can modify the behavior of a competitor and thus be directly related to the outcome of the encounter. Indeed all behaviors classified as displacement activities are so labeled not because they have been empirically or theoretically demonstrated to meet certain criteria, but rather because no other explanation for them is known. (See Walther 1974 for a more detailed criticism of the failure of the concept of displacement activity to meet functional criteria.) One need not object to the ethological model of *displacement activity* in principle, however, to realize that some behaviors so described will eventually be explained in functional and evolutionary terms.

Another consequence suggested by this model is that actors who exaggerate their own probability of winning the encounter (p_{a_1}) have an advantage; an effect that can be achieved by displays that project a confident, self-assured manner. On a related subject Morris (1957) and Maynard Smith (1972) have noted that, regardless of the motivation of an individual, or his probability of winning a fight, agonistic displays are often characterized by typical intensity, i.e., display behavior of known probable losers that makes it difficult or impossible to distinguish them from probable winners by the display alone. Likewise, concealing the costs of competition (C_a) that one is suffering (at least as long as there is some probability greater than 0 of winning the encounter), would lead an opponent to underestimate the rate of cost he is inflicting upon you, and to overestimate the length of time you can compete adaptively—thereby increasing the probability that the opponent will terminate the encounter early. It is common knowledge that among humans the full magnitude of an injury sustained in a fight is often not consciously appreciated until the fight is over and injuries that in other contexts would be perceived as painful may go unfelt during aggressive interactions. To avoid the external display normally associated with the symptoms of injury may be an effective way to conceal the presence of an injury; the best way not to show a response to a painful stimulus is not to feel it.[2]

Likewise, under conditions in which the costs of competition include the loss of time or energy available for feeding efforts (i.e., when foraging time or costs are limiting), sham feeding behavior during an agonistic encounter can lead an opponent to underestimate the costs of competition to an actor. Among Thompson's gazelles, *displacement feeding* is so elaborate and persistent in encounters between males at territorial boundaries that it may actually reduce the cost of competition to the participants through caloric intake from feeding during the agonistic encounter (Walther 1974). Displacement feeding in other species such as the Great Tit, although elaborate to the degree of showing

327

specific movements that are appropriate to the bird's location (i.e., pecking in trees and leaf-turning on the ground), probably does not provide significant caloric intake (Hinde 1952). We suggest that many such observations of displacement feeding may be susceptible to reinterpretation in terms of functional exaggeration of C_a.

An individual who can convince its opponent that an unrealistically high regression coefficient of relatedness exists between them would also have an advantage. Opportunities for such deception are probably rare in nature since an individual would be strongly selected to avoid confusing an unrelated individual for a close relative. We do note, however, that in human cultures a common behavior used to increase group solidarity is the inaccurate evocation of the kinship idiom in terms of address. Also, among a number of species, new immigrants adopt the scent of resident conspecifics by rubbing themselves at local sites of pheromone deposits, with the possible effect of convincing the residents that the immigrant is a familiar (and related) individual.

Finally, an individual can gain a competitive advantage in an aggressive encounter with a competitor that is known to be a relative by exaggerating the costs of competition to himself, i.e., the actor gains an advantage by suggesting to the related opponent that the competition is inflicting very high costs upon the actor. Thus, under some circumstances related individuals may be expected to feign serious injury, or higher rates of costs of conflicts to themselves, during a fight. Among humans, for example, we would expect siblings to have a lower threshold for vocal complaints and to cry more often and more intensely when roughhousing among themselves than when doing so with unrelated children. There are yet subtler strategies that probably find their greatest expression among humans (e.g., convincing the opponent that the disputed resource is valueless to him). Such strategies are not discussed here, but see Trivers (1974) on related issues in parent-offspring conflicts.

Degree of Ritualization in Aggressive Competition

If submissive actions serve to prevent or reduce injury, their adaptive advantage for an individual who is losing a competitive encounter is obvious; but why should a victor acknowledge such signals and allow the loser to back down, instead of pressing the advantage and killing the submissive individual? As already stated, the outcome of an aggressive encounter is dictated by the cost-benefit functions of the participating individuals. If an individual in an aggressive interaction terminates its aggressive behavior at or shortly after the time that its opponent gestures submissively, it will gain access to the disputed resource and on the average gain a net benefit for the entire interaction (as shown in Figures 1 and 2). If the winner continues to act aggressively toward the already submissive opponent, however, the situation changes substantially. If we assume that the victor continues the aggression with the *intent* to kill or seriously injure his submissive opponent, a new set of cost-benefit functions rapidly develop. Since no costs of competition could ordinarily exceed the costs of a fatal injury or, alternatively, since the benefit of saving one's life is considerably higher than the benefit that could be derived from a disputed resource, the individual whose submissive behavior has failed to terminate its opponent's aggression, will under most circumstances (i.e., where high b_{ac} is not involved), fight desperately in all-out self-defense. By contrast, the only benefit for the potential assassin would be the elimination of just one of many competitors.

The costs that a potential assassin is willing to incur, therefore, would usually be lower than the costs the victim is willing to suffer in self-defense. If we assume that there is no great disparity in the value of K_a for the two opponents, the aggressor will find it adaptive to terminate the aggressive encounter prior to killing an opponent. We can conclude that winning an aggressive encounter over a disputed resource that is of trivial value, compared to the value of continued survival, is a poor indicator to the victor that he could kill an opponent in continued combat.[3] In addition, under natural conditions the submissive animal often has the opportunity to escape, and this further reduces the mortality directly attributable to aggressive competition (although it does not necessarily reduce the number of competitors who are eliminated because they have been excluded from essential resources; see, e.g., Watson & Moss [1970] on the direct and indirect results of aggressive competition over food sources in animal populations). Note that the preceding argument is not at all equivalent to the frequent assertion that organisms possess an innate inhibition against killing conspecifics: whenever differences between two competitors in intrinsic competitive ability times the maximum adaptive expenditure for aggressive competition, as represented in Inequality II, are sufficiently large, murderers can be favored by natural selection. Although there is a number of noteworthy examples of strategies favored by natural selection that lead to the killing of conspecifics, the cost-benefit functions do not often meet such criteria. Exceptions include some primate species (e.g., Blaffer Hrdy 1977) in which infants are killed by presumably unrelated or distantly related adult males, who are competing for access to a female's reproductive effort.[4] Additional data on mortality resulting from aggressive competition are provided by Geist (1971) on ruminants, Bergerud (1974) on caribou, Schaller (1972) on lions, LeBoeuf, et al. (1972) on elephant seals, Mykytowycz and Dudzinski (1972) on rabbits, and Meyburg (1974) on eagles.[5] In numerous species mortality arises from cannibalism; such behavior has been well documented and is not discussed here.

SOCIAL DOMINANCE THEORY AND AGGRESSIVE COMPETITION

The Analysis of Dominance

Since it is unlikely that the competitive abilities, i.e., the values of K_aC_a, will be identical for any pair of contestants at a particular moment in time, one expects that aggressive competition will lead to differential rates of access to disputed resources. It is in this way that our previous considerations of aggressive competition relate to social dominance, since social dominance has traditionally been considered to involve priority of access to a disputed item (ordinarily a resource such as a food item or estrous female); for the purposes of this discussion, the dominant individual is the one who attains access to the resource (or a larger fraction of it), and the subordinate individual is the one who fails to gain access (or gains access to a smaller fraction of the resource). Thus, the terms *dominant* and *subordinate* are used here synonymously with *winner* and *loser* in the previous discussion of aggressive competition. Accordingly, Inequality II can be used to predict, within narrow limits, the rank ordering of individuals with respect to their relative dominance. In paired encounters, when individuals are acting optimally in their own evolutionary interests, the individual with the relatively higher value for K_aC_a will win the encounter and be defined as dominant; the individual with the lower value for K_aC_a will lose and be defined as subordinate. When all possible pairings between an individual and all other conspecifics with which it interacts are made, its rank in a hierarchy is defined as the number of individuals within the group minus the number of individuals over which it is dominant. Ordinarily, empirical tests and/or naturalistic observations determine only an individual's ordinal position in dominance rank, and not the absolute distance between such positions. The values of K_aC_a, however, may be plotted as a continuous function over an individual's entire life so that, when compared with the K_aC_a values of his competitors, it would predict both the ordinal rank and absolute distance between the ranks of competitors at all times.

Certain qualifications and omissions in our definition of dominance require explanation. We have deliberately restricted this discussion to dyadic encounters, not because aggressive and dominance interactions are to be understood solely in dyadic terms, but because they provide the only circumstances under which dominance status can be measured empirically. Furthermore, we restrict the definition of "dominance" to priority of access to disputed items, because in the literature of the social and biological sciences the term has so often been employed as if it were synonymous with such terms as *group leader* or *troop defender*, that it is no longer a useful scientific term without this restriction. We also assume that aggressive competition in dominance interactions will involve disputes over resources of adaptive value to the individuals involved—i.e., that these interactions ultimately affect the reproductive success of the individuals. The evolutionary importance of dominance behavior may not be obvious under some laboratory conditions, but we assume that in nature a higher position in a rank order will typically lead to higher rates of access to a limiting resource.

We have already discussed how the value of a resource is relevant in determining the amount of aggressive competition that is adaptive in attempts to attain access to it. Further, one should *not* assume a priori that: (1) two different resources are of equal value to an individual, or (2) a particular resource is of equal value at all times to an individual, or, (3) a particular resource is of equal value to two different individuals. Perhaps the single most important conclusion from a consideration of social dominance in terms of aggressive competition theory is that *dominance hierarchies are expected to be time- and resource-specific.*[6] This conclusion suggests that a number of widely accepted concepts and methods associated with earlier paradigms of social dominance are invalid. The notion that hierarchies produced by measuring rates of access to a single resource may be generalized to speak of social behaviors in *other contexts* may frequently lead to error. Empirical descriptions of dominance hierarchies must include data on the disputed resource for which rates of access have been measured. We do not imply that an individual's dominance status with respect to access to a particular resource will never be correlated with its dominance status in competition for access to other resources; indeed such a correlation may occur frequently. Our point is that such similarities between resource-specific dominance hierarchies are, in fact, correlations—a particular dominance status with respect to access to a food item does not necessarily imply that an individual will have the same dominance status with respect to other disputed resources. Positive correlations between the dominance rankings of an individual in competition for two different resources are expected to arise from similarities in the value $K_a C_a$ in competition for both resources, relative to other members of the social group. Obviously, under some circumstances differences in the intrinsic competitive ability (K_a) among the members of a social group may be so large that short-term fluctuations in dominance status with respect to commonly disputed resources are rare or nonexistent. Such a consistency in dominance status, however, must be demonstrated, not assumed.

The literature on social dominance contains numerous assumptions that lead to confusion. In some cases the dominance status of an individual is considered to be solely a function of the intrinsic competitive ability (K_a) of that individual. Based on this definition, social rankings are viewed not as the outcome of competitive interactions, but as the result of purely physical attributes such as body size, strength, sex, and age—factors that we recognize as important, but not sole, determinants of the outcome of aggressive encounters. Others, who for the most part have ignored the evolutionary implications of aggressive competition, suggest that social dominance is a role assumed by individuals, rather than the result of competition for limiting resources. Commonly, the term social dominance is applied only to those groups in which hierarchies are stable for long periods of time, thereby confusing the question of temporal stability of dominance hierarchies with the question of their existence at any single time. Perhaps the most common view is that of *general* dominance, i.e., the assumption that the members of a social group can be ranked in a single hierarchy that describes the status of each individual regardless of the context of the social interaction. This assumption is implicitly made by all field workers who, when collecting data on dominance, lump the outcome of encounters with respect to competition for mates, food items, sleeping places, etc. into a single hierarchy. For example, Sade states:

> My criteria for ranking individuals in a hierarchy of aggressive dominance derives from the aggressive interaction itself, rather than from the causes, context, or consequences of that interaction. These are the most unambiguous criteria, since the behaviors that mark the winning monkey usually are not also used by the losing monkey in the same interaction, and are independent of the context of the interaction, whether it be competition over an infant, food, or a female, or simply an act of malevolence (1967, pp. 99-100).

Although convenient in collecting data for statistical analysis, such operational definitions obscure points of biological significance. Even if Sade's hypothesis on the resource independence of agonistic gestures is true (and we would expect it to have exceptions), he has failed to address a hidden assumption, and has therefore not provided an appropriate justification for his model of aggressive dominance: the real issue is whether the *outcome* of an agonistic encounter (not the *gestures* employed by *pairs* of individuals) is independent of the context of the interaction. The shortcomings of criteria such as Sade's are best demonstrated by example. Imagine a female rhesus monkey in two types of aggressive encounters: in one type, the object of competition is a food item of trivial value in terms of reproductive success; in the other, the female's dependent infant is threatened with possible injury or death. According to Sade's criteria and the model of general dominance no distinction is made between these two types of encounters, and the female's status in each is presumed the same. In contrast, we argue that the female is expected to win more encounters in defense of her infant than in defense of a trivial food item under normal conditions (considering the relative benefits that would be obtained from winning each). We suggest that, if considered in terms of two hierarchies, the female's status measured by all encounters won in defense of her offspring may be quite high even if she ranks low in dominance as measured in competition for food. Lumping them into a single category obscures the biological causes for the outcome of each type of encounter, and, depending upon what fraction of the total sample consists of each, the final status assigned to the female in a general dominance ranking will vary widely. Similar arguments can be made against lumping the outcomes of encounters over food with those over competition for mates and for other important resources. In short, the concept of *general* dominance, as it is frequently applied in primate studies, leads to methodological error when it is employed to determine dominance rankings and individual dominance status. More seriously, it obscures the issues that underlie the outcomes of single encounters, and precludes interpretations of social dominance and aggressive competition in terms of life history strategies, reproductive success, and inclusive fitness.

We reemphasize that the statements above are not meant to imply that dominance rankings of individuals with respect to many or all common objects of competition will never be highly correlated. Indeed, in both natural and captive groups of primates it is often obvious that a single animal has priority of access to all disputed objects. This is especially true when groups are naturally composed of members with great disparities in K_a, and in captive populations where dominance with respect to access to food quickly generates large differences in K_a values for group members, which in turn lead to a stable and broadly applicable hierarchy of access to all disputed items. What we do refute is the belief that general dominance always applies, and the view expressed by Richards (1974) and others that it *must* apply for social dominance to be a useful concept.

SOCIAL DOMINANCE AND SOCIAL EXPERIENCE

As previously discussed, it is adaptive for an individual to be able to predict the outcome of an aggressive encounter; such an ability permits him to reduce costs by avoiding conflicts that will be lost and to increase benefits by competing to the end in encounters that he can win. In species that form long-term associations among a small set of individuals, one method of predicting the outcome of a competitive encounter is by the recollection of past encounters with a specific opponent. Past competitive experience with a known opponent under circumstances similar to the present competitive interaction can be useful in estimating the cost-benefit function for the opponent in aggressive competition. For example, if individual β has lost in all previous aggressive encounters over food items with individual a, it will be to β's advantage to avoid such competition in the future until the variables in Inequality

II change in his favor. The best strategy for a subordinate individual who knows from past experience that it cannot win an encounter (i.e., pa_1 is near 0) is to avoid the competition. It is this principle that is responsible for the often observed decline in the frequency of aggressive behavior when the members of a social group have had sufficient time to form dominance hierarchies.

Considering the same example from a's viewpoint, it may be to its advantage to seek an increase in the rate of competitive interactions with β, since it can dominate β with relative certainty. In the extreme form, it will be to β's advantage to show submissive behavior immediately upon being challenged by a, since any expenditure in competitive effort always represents a net loss. Appropriate behavior by subordinates that are acting in their own best interests may, therefore, involve active avoidance of the dominant individual or active avoidance of situations with the potential for conflict, e.g., the avoidance of foraging activity near the dominant individual. Behavior such as this has led to a great deal of confusion in the social dominance literature: to the unwary observer, the illusion is created that dominant individuals are passively dominant in a role that has been conferred upon them by subordinates for reasons unspecified. In contrast, we emphasize that both dominant and subordinate individuals must be viewed as actors that have been selected to display behaviors appropriate to the natural social environment for the maximization of their reproductive success. Dominance hierarchies do not exist because they bring harmony and stability to the social group, but as the consequence of self-interested actions, in the evolutionary sense, by each group member.

ALTERNATIVE STRATEGIES

Among the great apes, male chimpanzees employ several reproductive strategies, some of which are alternatives to direct aggressive competition. Both Lawick-Goodall (1968) and McGinnis (this volume) report male-male aggression in competition for estrous females, and consistent with these findings Nishida (this volume) reports that a single male was involved in 46% of all copulations observed during his study. Thus, under some conditions dominant male chimpanzees may be able to monopolize sexually receptive females. Alternatively, Lawick Goodall (1971) and McGinnis (this volume) report highly promiscuous breeding patterns in some circumstances, where males that are unable to exclude male competitors from access to estrous females form large groups attending a single female and copulate successively, or opportunistically steal copulations when more dominant males are temporarily distracted. Under such conditions, one expects that sexual selection will favor those males that deposit the largest quantity of sperm in the female reproductive tract—hence, the extremely high relative testicular weight among male chimpanzees as compared to the other great apes and man. Thus, in addition to aggressive competition, male chimpanzees employ sperm competition as a reproductive strategy. Schultz (1938) and Wislocki (1942) report the following body and testes weight for the great apes: chimpanzees weighing an average of 44.34 kg had an average combined testicular weight of 118.8 gm, or 0.268% of their body weight; gibbons weighing 5.54 kg had 4.6 gm or 0.083%; orangutans weighing 74.64 kg had 35.3 gm or 0.047%; gorillas weighing approximately 200 kg had 36 gm or 0.018%; men weighing 63.54 kg had 50.2 gms or 0.079%. Chimpanzees clearly have testes that are significantly heavier, both relative to body weight and absolutely, than the testes of all other Hominidea. Such divergence cannot be attributed to allometric growth, but relates directly to differences in the breeding strategies of chimpanzees from the other species mentioned. A detailed analysis of sperm competition as an alternative mating strategy in the primates will be presented elsewhere.

Prostitution behavior by subordinates, in the attempt to attain access to limiting resources, is another alternative to aggressive competition. The fact that estrous female chimpanzees have differentially high access to food items in the presence of sexually mature males is well known in both captive and wild populations (Yerkes 1940; Teleki 1973). Since proximity to a sexually receptive female is conducive to sexual access, driving away an estrous female in competition over food items confers fewer net benefits than driving away a nonreceptive female. Hence, males show high rates of sharing with estrous females, and females are expected to recognize such tendencies and respond in a way that is most efficient in exploiting them, e.g., by unhesitatingly approaching males that are ordinarily dominant to them with respect to food or by approaching with sexual presentation. A similar analysis can be made of grooming behavior shown by subordinates in their attempts to gain access to a desired object that is in the possession of a more dominant individual.

THE MAINTENANCE OF DOMINANCE STATUS

Previously we have discussed how the avoidance of conflicts with individuals known to be dominant can be a subordinate's best strategy. Subordinates that act as though the status of dominant individuals is immutable, however, will not necessarily be acting optimally. Subordinates that recognize and capitalize upon opportunities to reverse the dominance status between themselves and higher ranking individuals will be favored by natural selection. Such behavior is analogous to the evolutionarily stable strategy of prober-retaliator discussed by Maynard Smith and Price (1973).

Since conspecifics will estimate the rank of an individual and its ability to maintain that rank by its behavior and appearance, dominant individuals are selected to provide unambiguous signals of their high status and ability to maintain it. Strutting behavior, tail-erect postures, and open-mouth threats are well-known gestures in primates that are displayed primarily by dominant individuals. In a number of species, dominant individuals show scent-marking behavior at a higher rate than subordinates, e.g., in duikers (Ralls 1974) and the white rhinocerous (Owen-Smith 1974); in mice the increased size of the pheromone-producing preputial glands among dominant animals has been well documented (Bronson & Marsden 1973). Such differences in marking behavior can be attributed to the advantages dominant individuals derive from making a prominent and unambiguous display of their presence, and the advantages to subordinate individuals of remaining inconspicuous when in the territory or range of conspecifics with higher competitive abilities.

High-ranking individuals past their prime are expected to attempt to conceal their declining competitive abilities. Hence, we have observed that aging anubis baboon males play a subtle strategy of avoiding direct confrontation with younger males in their prime (DeVore 1966). In competition for highly localized food resources, an old dominant male will often take early possession and rapidly consume a large portion of the food items present. When the challenging, prime males begin intense harassment, however, the old male, rather than engaging in a conflict that he would likely lose, strolls away giving the impression of complete disinterest in the resource. Thus, by asserting dominance at critical points, but avoiding aggressive encounters when they would result in reduced status, an old individual may be able to prolong his period of high reproductive success. This argument applies in particular to competition for estrous females, since the benefit associated with access to a single food item decreases during the period of senescence as a function of declining reproductive value, but the benefit associated with copulations does not. Old males then are expected to have a lower dominance ranking in competition for food items than in competition for access to estrous females. For example, DeVore (1965; Hall & DeVore 1965) reported that an old male baboon, Kovu, who ranked last in priority of access to food incentives, ranked highest in attempted copulations, and second highest in the number of completed copulations during the same time interval, among the six adult males in the SR troop at Nairobi Park.

CONCLUSION

Considering aggressive competition and social dominance as the consequences of strategic behavior by individual actors who are attempting to maximize their genetic contribution to succeeding generations helps explain the origins, frequency and functions of such behavior. We have modeled how an individual would be expected to act in its own evolutionary interests with respect to aggressive behavior and other competitive strategies, and believe that such a model is consistent with naturalistic observations of animal behavior. In addition, the analysis of agonistic behavior in terms of the principles of natural selection leads to novel predictions and provides interpretations for a number of behaviors that have previously been considered obscure or paradoxical.

Acknowledgment

We thank the symposium participants for their comments on this manuscript, especially Robert A. Hinde. Among the many others who have made suggestions we gratefully acknowledge the help of Martin Etter, Steven Gaulin, Sarah Blaffer Hrdy, Lewis Hurxthal, Jeffrey Kurland, James Malcolm, John Pickering, Jon Seger, John Tooby, Madeline Wagner, Pat Whitten and especially Norma Foltz and Valerius Geist. For help in editing and preparation of the manuscript we thank Nancy DeVore, Steven P. Stepak, and Bonnie Nelson; for preparation of the figures, Nancy Hurxthal. Both for comments on the manuscript and for lectures and discussions relating to many topics treated in this paper, we are most grateful to Robert L. Trivers. This paper is affectionately dedicated to Sherwood L. Washburn. The support of NSF and H. F. Guggenheim Fellowships to the authors, NIMH Grants MH-13156, MH-27934, and Training Grant GM-01938, is gratefully acknowledged.

Notes

1. The original version of this paper, entitled "Aggressive Competition and Social Dominance Theory," was prepared for the 1974 Wenner-Gren Foundation symposium "The Behavior of Great Apes." This synoptic version, prepared at the request of the publisher in July 1978, includes only one-third of the material presented in the original version. The full text of the original will be published in *Sociobiology and Social Science*, I. DeVore, editor. We have drawn from the longer work examples of animal behavior and selected aspects of the modes of analysis that seem particularly relevant to understanding primate behavior. However, six major sections have been entirely deleted in this synopsis, including: (1) the life historical perspective, (2) ecology and social organization, (3) troop defense, peacekeeping, and agonistic buffering, (4) reversals in dominance status, (5) dominance triads—a kinship hypothesis, and (6) coalitional behavior. Although not included here, these six topics include issues of fundamental importance in research on aggression and dominance, and the reader is referred to the longer version.

2. We appreciate that two simpler hypotheses must be demonstrated as insufficient to explain this phenomenon before the above can be accepted. Reduced sensitivity to pain during aggressive interactions could be a physiological mechanism facilitating high values of K_a, or sensitivity could be adjusted as a

function of C_a so that the threshold for terminating the aggressive encounter is reached at the appropriate time.

3. When the value of a disputed resource approaches the reproductive value (Fisher 1958) of a competitor, such a competitor is expected to risk fighting to the death wherever Inequality I is true. Under such circumstances, winning the initial encounter with such a competitor is therefore an indication that it could be killed in continued conflict. In general, this implies a relationship between C_a and Fisher's reproductive value: when other variables are equal the competitor with the lower reproductive value has a competitive advantage in an aggressive encounter. However, because Fisher's reproductive value (F. R. V.) ignores some kin effects on reproduction while including others, it is not an entirely appropriate measure of an actor's willingness to assume the risk of permanent injury or death in competition. We suggest that a more appropriate formulation is "extended reproductive value" (E. R. V.), which includes not only the expected future offspring of an individual at a particular age, but also its net future effect on the reproduction of its relatives. It is obvious that among human populations F. R. V. somewhat overestimates the value of young members, in terms of their genetic contribution to future generations (thus underestimating their willingness to assume risk in competition), but underestimates the value of older members of the population (thus overestimating their willingness to assume risk). Women past the age of 50, e.g., have an F. R. V. of zero, but a more realistic appraisal of their contribution to future generations through E. R. V. grants the possibility that, as mothers of still immature children, as aunts, as grandmothers, etc., they may still affect kin significantly.

4. We believe that our model of aggressive competition has predictive value. From the available data on ecology and social behavior we suspect that patterns of infanticide by adult male patas monkeys *(Erythrocebus patas)* will be discovered under natural conditions when a new male replaces the resident male of troop. To our knowledge, such behavior has not yet been reported in this species.

5. Among eagles, like some species of hymenoptera, natural selection favors *fratricide,* i.e., patterns of aggressive competition that lead to the killing of full-sibs. Note that our model of aggressive competition predicts even the murder of close relatives when the object of competition is sufficiently valuable, the reproductive success attainable through alternative behaviors is low, and/or the disparity in K_a between the relatives is large.

6. Territoriality may be viewed as the consequence of aggressive interactions that are location-specific, see Brown (1963, 1964) on territoriality.

A chimpanzee exhibiting disturbed behavior

R. K. Davenport

Some Behavioral Disturbances
of Great Apes in Captivity

Robert Yerkes in his studies of great ape behavior was fortunate in that the animals in his original colony were wild-born and, as far as can be determined, were behaviorally normal. Second, the animals that were available to him and in which he was most interested were chimpanzees, animals that are extremely adaptable to captive conditions. He was thus provided with a behaviorally normal stock to which other animals coming from different environments, especially in the later days of the laboratory, could be compared. Except for temper tantrums, negativism in certain situations, etc., there are almost no records of gross and persistent behavioral abnormalities until experimental manipulations of the early environments were begun in the 1940s and 1950s. The vast majority of abnormal behaviors, behaviors not seen in the natural habitat and/or maladaptive insofar as the human observer can determine, have occurred in animals that experienced peculiar early environments. Between 1925 and 1973, the Yerkes laboratories acquired about 170 chimpanzees and approximately 239 were born in captivity. Approximately 36 orangutans were wild-born and captured as youngsters before being brought into the colony; about 29 infants have been born to these animals. Between 1961 and 1973, there have been 20 gorillas in the Yerkes collection; 19 of these were wild-born and one infant has been born into the colony.

SPONTANEOUS NEUROSIS

Perhaps two of the best known instances of spontaneous *neurosis* or *psychosis* in the chimpanzee were described by Hebb (1947). One of these cases was Alpha, who was separated from the mother very early and raised by human beings until the age of about 1 year. She then lived for the most part with three other females. This group was frequently handled by the staff and, according to laboratory records, her early history was unremarkable. At the age of 12 years she developed a gross behavioral disturbance. Suddenly, Alpha refused all solid food that was given to her in large chunks, but she would accept foods that had been cut into small pieces. A number of tests were given to determine conclusively that she was afraid of large pieces of food. She would take pains to avoid passing near these pieces. Over the next 4 months, there was a gradual improvement in this condition until she was no longer fearful of food, but developed a sudden and marked avoidance of the attendant who fed her. This lasted for a period of 7 days, at which time the fear of food returned and the avoidance of the care-taker disappeared. For 3 years, this illness of alternating fear of food with fear and avoidance of the caretaker continued. Hebb was unable to determine the cause for this behavioral abnormality and likened it to a human neurotic condition.

A second case was Kambi, a wild-born animal who was raised in the lab with other chimpanzees of her own age during the developmental period. Her behavior at maturity was noted to be unpredictable, emotionally unstable, and introverted. At postadolescence she appeared to have bouts of periodic depression. She seemed to be fairly well adjusted for several months at a time and then would enter a severe depression for periods as long as 6-8 months. She would sit for hours with her back to the wall, staring at the floor of her cage, and was unresponsive to other animals and attendants who passed by her cage and attempted to interact with her. She was so socially unresponsive that she had to be caged alone in order to obtain enough food for survival. During her periods of apparent *normality,* her frustration tolerance was described as low and she had frequent temper tantrums.

In neither case was Hebb able to attribute the disturbances to any specific experiences of the animals, such as being bitten by insects or maltreated by caretakers.

Subsequent to her illness, Alpha returned to being a laboratory-normal chimpanzee, was sociable both with other chimpanzees and humans, and became the mother of many offspring. Kambi died, apparently from dysentery, while she was still behaviorally abnormal. Laboratory records and recollections of long-time caretakers indicate that no other similar spontaneous neuroses have occurred.

All other behavioral abnormalities observed at Yerkes appear to be related to special conditions of rearing; these will be described in this paper.

REARING GROUPS AT YERKES

In general, apes at the Yerkes Primate Research Center fall into certain groups based on their birth and rearing. The first group is wild-born animals. These are animals who were born in the natural habitat, presumably captured at about 1-2 years of age, brought into the laboratory shortly thereafter, lived in peer social groups, and were subjected to normal laboratory stimulation including playing with other animals, interaction with experimenters and caretakers, exposure to solid food and other manipulanda, etc. A second group is lab-born and mother-deprived. These were animals born within the laboratory and separated from the mother early in life, usually within the first few weeks or months, and reared artificially in the laboratory nursery where they had visual and auditory stimulation and some play access to other animals as well as interaction with the caretaking staff and investigators. Third was a group of animals born in the laboratory and separated from their mothers within 12 hours postpartum and reared in extremely deprived environments for the first 2-2½ years of life. They lived during this period in small enclosed cubicles. The environment was highly controlled as well as impoverished. All caretaking procedures were done through sphincters on the front of the cubicles and interaction of any kind with human beings was limited to about 15 minutes total per day, none of which involved direct body-body contact or visual observation of the caretakers. These animals wore diapers and received only liquid food during the time in restriction. There were 16 animals so treated. Although all the environments restricted stimulation and experience, there were several different conditions in the type and degree of restriction provided. It was assumed that the degree and type of restriction would result in differences in learning ability, social behavior, etc.; however, extensive observational studies and experiments spanning several years have failed to detect any consistent differences among these rearing groups. For the purpose of this paper, therefore, they will be referred to as restricted environment animals.

A fourth group was composed of animals born in the laboratory and allowed to remain with their mothers for periods of 6 months to several years. They were subsequently separated and reared in peer groups in the normal laboratory situation.

A very few other animals received special care, such as separation from the mother early in life, and were raised by human beings in an atmosphere like that of a human child.

GENERAL HEALTH

A number of studies have emphasized the adverse effects of the absence or deprivation of maternal care on the physical development of human children (Bakwin 1943; 1949; Ribble 1943; Spitz 1945; 1956; Spitz & Wolfe 1946). These studies have had a profound and pervasive influence on child-rearing practices and theories. Because of the affinity of the child and the chimpanzee, and the superior experimental controls that are possible with the latter, it was felt that the present series of experiments could add to our knowledge of the physical results of maternal deprivation, some of which have been questioned on methodological grounds (Pinneau 1955). The health and physical development of restricted environment animals, therefore, were compared to two groups of chimpanzees that were reared during infancy in grossly different conditions (Davenport et al. 1961). One group was composed of 24 animals reared in the Yerkes colony by the natural mother for periods ranging from 1 month to about 2 years postpartum. Infants separated at the earlier ages were maintained in the laboratory nursery by human attendants; those separated later were housed with other animals of approximately the same age. The second comparison group was composed of 38 animals who were reared in conditions that closely parallel some human institutions. They were laboratory-born and separated from their mothers within 2 days postpartum. They were diapered, fed the standard formulae, and received medical attention as needed. These animals had sporadic contact with investigators, daily handling by caretakers, and were occasionally housed with peers.

Overall, mortality rates were low and apparently unrelated to rearing conditions. Three mother-reared infants, five nursery-reared infants, and one restricted infant died before the age of 2 years (five from pneumonia, one from complications resulting directly from a nutritional deficit sustained while with the mother, one of a congenital disorder, and two from undetermined causes).

Weight gain appears to be a reasonable index of health in both human and chimpanzee infants. There were large and statistically significant group differences in weight in a direction contrary to that reported for human infants. The heaviest infants were those who had never experienced natural maternal care but were reared from the first in an institution-type nursery; next were those reared in the restricted and impoverished environments; and lightest were those reared by their own mothers. Almost without exception, the weights of mother-reared infants were below the mean weight of all artificially reared animals.

Also, contrary to what would have been predicted from human reports, infants separated from the mother after varying periods of maternal care did not lose weight; instead, there was an immediate and progressive weight gain and an apparent improvement in general health. In addition, there was no evidence of depression. Although behavioral effects of maternal separation and restricted rearing may be profound, physical health is not similarly affected. When maternal deprivation occurs immediately at birth or after a firm relationship has been established with the mother, it has no adverse effects on weight gain and mortality. The adverse physical effects, if any, resulting from deprivation are more than compensated for by the superior nutrition and hygiene that can be provided by proper artificial conditions.

STEREOTYPED BEHAVIOR

Perhaps the most salient and distressing behavioral abnormality noted in the great ape laboratory is stereotyped behavior. *Stereotyped behaviors* are mechanical-like repetitions of postures or movements that occur with great frequency and vary only slightly in form from time to time. The practice of the stereotypies usually begins early in life and continues into adulthood. The defining characteristics of stereotyped behaviors are: (a) occurrence at a high frequency within a relatively brief period of time, e.g., within a few minutes; (b) persistence over a relatively long time span, e.g., at least several months; (c) they serve no adaptive function from the human observer's point of view; and (d) they involve little interaction with the features of the environment. These stereotyped behaviors are sufficiently peculiar in frequency and form to be noted as such by people who are only moderately familiar with the behavior of apes. The stereotypies of chimpanzees are topographically very similar to repetitive behaviors of some normal children (Kravitz et al. 1960), e.g., head-banging, rocking, etc., and to those exhibited in exaggerated and persistent fashion by some profoundly retarded human beings (Berkson & Davenport 1962; Davenport & Berkson 1963; White 1948). (They are also similar to repetitive activities of some zoo animals, although the etiology is probably different.)

Stereotyped behavior occurs in almost all chimpanzees who are separated from the mother within the first several months of life, and are reared subsequently in artificial conditions (Nissen 1956).

Although all the mother-deprived, restriction-reared animals developed stereotyped behaviors, a great deal of individual variability in the patterns was exhibited. Subsequently, although the amount of time spent in stereotyped behaviors decreased, the actual number of patterns increased with increasing age up until about mid-childhood. During the second year of life, the restricted-reared chimpanzees engaged in stereotyped behaviors in about three-quarters of their waking day.

Infants separated from their mothers very early and raised together in peer pairs or peer groups, where there is much stimulation and interaction after they become mobile, also stereotype, but less frequently than animals who have undergone more severe deprivation.

As far as can be determined from laboratory records, even animals separated from the mothers very early and given a huge amount of substitute human maternal care also engage in some stereotyped behaviors, but these were practiced less frequently and lacked the persistent compulsive quality of the behaviors performed by restriction-reared animals. Obviously human care is qualitatively and quantitatively different from full-time natural maternal care.

Wild-born chimpanzees and those laboratory-born infants who have remained with the mother for a few months postpartum are markedly different from the others. There is no record of stereotypy in laboratory-born animals who were mother-reared for most of the first year, even after separation from the mother. (Only one wild-born subject has developed any repetitive rhythmical activity of the whole body. She was imported into the laboratory at a very early age and is likely to have been separated from the mother a considerable time before that.) These stereotypies, therefore, are clearly typical of artificially reared chimpanzees and are not unique to those raised in conditions of extreme restriction. The difference is that in the latter animals, the frequency is many times greater and the repertoire is much broader.

To determine whether environmental deprivation, per se, resulted in stereotyped behavior or whether the deprivation must occur within a relatively short period after birth, three wild-born, mother-reared animals were put into a condition of extreme environmental restriction for a 6-month period at the age of approximately 12-18 months (Davenport et al. 1966). During this time these animals practiced some of the behaviors of the infants isolated from birth; however, they were very infrequent, brief, lacked persistence, and did not become habitual. As in the case with older mother-reared infants, they simply occurred for occasional brief moments in the course of general activity and could by no means be classed as stereotyped behaviors.

Most restricted animals at about the age of 3 began to live in social groups and received a good bit of social and other stimulation in the course of behavioral testing and general living with wild-born animals. Gradually over a period of years, the frequency of stereotyped behavior has dramatically decreased and, in some cases, these animals appear to be normal insofar as this behavior is concerned, except under conditions of stress. Two restricted animals who did not have the benefit of this range of postrearing experiences with age peers continued to stereotype at a high frequency until well past adolescence. It appears that social experience at an early age with normal animals of approximately the same age has an ameliorative effect.

We have postulated that the development of stereotyped behavior in the infant chimpanzee is related to the absence of a certain level of stimulation ordinarily provided by the mother in her normal day-to-day care. Deprived of this stimulation, the infant engages in activities that are self-stimulatory. Repeated over a sufficient period of time, these behaviors become rather fixed and habitual. We cannot specify exactly the kind of stimulation that is missing; however, the nature of the behaviors that become repetitive in all animals (rocking, swaying, twirling) suggest that kinesthetic and vestibular modalities are involved. This thesis has received some support from Mason's (1968) work with rhesus monkeys.

Stereotypy can be related to stimulus condition and other modes of behavior. In the 2-year-old restricted animal, stereotyping serves no single function; that is, it appears when the subject is not actively engaged in (externally directed) behaviors (locomotion and object contact) and in conditions of high arousal. At later ages, however, the practice of stereotypy tends to become more situation-specific. There is greater probability of its occurrence in high arousal states, for example, just prior to feeding or after an attack from another animal (Mason et al. 1968).

A clue to at least one function of stereotyped behavior (and perhaps to its maintenance) at a later age is suggested by the behavior, in the same situations, of socially reared chimpanzees and artificially reared animals that have been allowed to become attached to a blanket or towel. On occasions of high arousal, especially if some uncertainty or anxiety is present, each will seek out its accustomed object; the socially reared animal will hug a friend, while the other will grasp the cloth or drape it around himself. Both appear to reduce distress. The isolation-reared chimp, having no attachment to an object, engages in accustomed stereotyped patterns and appears to become less distressed. Stereotyped behavior thus may be a mechanism for moderation of anxiety or fearfulness at a later age.

Firm conclusions are limited by small numbers of subjects, isolation periods, and ages, but at this point it appears that maternal care during the first year of life is sufficient to prevent the development of idiosyncratic motor patterns at a later age, even under conditions of extreme environmental restriction.

None of the orangutans has engaged in stereotyped behavior. This fact may be related to the greater mobility of the young orangutan, who can climb at an earlier age than the chimpanzee or gorilla. The one infant gorilla, even though often housed with a young chimpanzee and orangutan, has developed a stereotyped repertoire and frequency similar to that of peer-reared chimpanzees.

LEARNING AND COGNITIVE DEVELOPMENT

Although less spectacular, of particular interest in part because of its relevance to learning ability and intelligence in man, has been the study of long-term intellectual effects of differential rearing in the chimpanzee. It is generally accepted that the characteristics of the early environment have an important influence on the learning ability and intellectual level of human children. Children from deprived and unstimulating environments are said to be poorer learners and score lower on intelligence tests than children from more enriched environments. This notion has been challenged by some nonhuman primate studies which have shown that, although severe social isolation produces profound and practically irreversible effects on social behavior, it has no measurable effects on intellectual factors (Harlow et al. 1965; Harlow & Harlow 1968). Because of the nature of intelligence tests (being highly verbal and school-related), and the poor control of the experience of comparison groups, nutritional differences, etc., research with this problem in human beings is difficult.

In our research, a central focus has been on *long-term* effects of early environments on *intelligence,* i.e., a complex of more or less specifiable factors that are reflected in the ability of animals to perform adaptively on problems devised by human experimenters. By precise specification of the requirements in test situations, we have attempted to identify in a variety of tasks some specific characteristics of performance that are affected by rearing conditions.

We compared wild-born, laboratory-reared, and restricted chimpanzees on a number of tests of learning and cognitive abilities. The groups were comparable in sexual makeup and were tested between the ages of 7 and approximately 13. (During any one test, the animals were at most 2 years different in age.)

The tests used were conventional laboratory learning procedures, including delayed response, discrimination-learning sets, oddity learning, discrimination-reversal learning, and a test of tool-using. The first formal test used was object quality discrimination learning sets (Davenport et al. 1969). This technique permits an examination of improvement in performance with repeated trials on the same problem and an improvement in performance from one problem to another over a large number of problems of the same general type. With sufficient experience, a new problem may be solved in one trial, i.e., if selection of one object of a pair is rewarded on trial 1, selection of that object on all other trials within that problem will be rewarded. If selection of an object on trial 1 is not rewarded, on trial 2 and all subsequent trials within that problem, selection of the other object will be rewarded.

Pretraining is necessary to insure motivation, attention to the task, and acquisition of the necessary component skills, e.g., looking at the problem tray and objects, displacing only one stimulus object, retrieving and eating the candy reward. Since our interests were in studying differences in multiple dimensions of learning between two groups, the experiment and the comparison commenced with pretraining.

In the pretraining phase, the wild-born animals were clearly superior to the restricted animals. Subsequently, 438 problems were given. Both groups clearly improved with experience; however, the wild-borns achieved a higher level of performance much earlier than did the restricted animals. The test was arranged so that, on half of the problems, the animals' trial 1 choice was correct and the other half incorrect. A more detailed breakdown revealed another important difference between the two groups. Throughout, the most striking inferiority of restricted subjects was in performance on problems on which the first choice was not rewarded; that is, their greater tendency to choose an object again after response to that object had gone unrewarded on the preceding trial—a stimulus perseveration error (Harlow 1950; Miles 1965). They appeared to profit less from a nonrewarded trial than from a trial on which they were rewarded. There was no such difference in the wild-born subjects.

Next, the groups were compared on a delayed response-discrimination task (Davenport & Rogers 1968). In essence, it involves indicating to the subject which, among several objects, is correct, imposing a delay of 0, 5, or 10 seconds, then having the subject make a response. Correct performance requires remembering *(bridging the delay)* which response to make.

Before testing, the subjects were pretrained by steps to a criterion of 22 correct out of 25 trials. Although initially the wild-born group was clearly superior to the restricted group, both groups improved significantly with practice. The wild-born subjects' latency to respond after the delay was short to begin with and did not change thereafter. The restricted subjects' latency, however, was extremely long initially and, although it decreased markedly as testing progressed, they still responded more slowly than did the wild-born group at the end of 1200 trials, especially at the 10-second delay. The tendency to break visual contact with the test apparatus followed a similar course.

A reasonable hypothesis seems that the improvement (especially in the restricted animals) was due to learning *techniques of attending,* e.g., remaining in front of the experimental array, maintaining visual regard of the correct position, etc., instead of exploring the cage, responding to the experimenter or engaging in stereotyped behaviors that interfered with test performance.

These two studies suggest that restricted early experience results in *intellectual* limitation in chimpanzees. This finding is consistent with the reports on the influence of early deprivation on learning ability in human children. *Distractability,* the tendency to engage in non-task-oriented behaviors, is said to be characteristic of human children from deprived and intellectually unstimulating environments.

A third test was concerned with oddity learning (Rogers & Davenport 1971). Learning to respond to oddity seems to require a form of concept formation or abstraction since it necessitates the subject to respond, not to specific properties of a stimulus object, but rather to the property of oddity or uniqueness alone, a property dependent on its relationship to other objects. In each trial, the subject was presented with an array of four stimuli, three bearing identical patterns and the fourth bearing a different pattern. To obtain the reward, the subject had to displace the panel displaying the odd pattern.

The differences between the two groups in overall accuracy and rate of improvement were clear. The superiority of wild-born subjects on the oddity tasks appeared initially and persisted throughout testing.

In the first 100 trials, all restricted subjects performed at about chance level (25%), but most wild-borns did considerably better. By the end of 500 trials, five of the seven wild-borns were performing at above 75% accuracy, and then continued to improve thereafter; whereas it was not until 1700 trials that five of the restricted subjects were able to attain approximately this level. Clearly the wild-born subjects learned the task much more rapidly than the restricted subjects. In addition, the wild-born subjects seemed to be more adaptive; that is, low-yield solutions, such as choice by position (which assured 25% reward; chance alone) were abandoned more readily. Results of the present study suggest that restricted subjects were deficient in problem-solving because of a relative slowness in changing strategies and a strong tendency, exhibited previously in discrimination-learning set testing, to perseverate, i.e., to make successive responses, in this case, to position, following nonreward (Davenport et al. 1969).

The final and most definitive assessment of learning ability was given on a test involving discrimination-reversal learning (Davenport et al.

1973). In comparing the two groups of chimpanzees in previous tests, consideration must be given to the possibility that group differences in motivation, emotion, attentiveness, perception, etc., might confound measurements of *intelligence* with the effects of variables that are primarily of a performance nature. This test was designed to control for these factors. The procedure was a variant of the transfer index developed by Rumbaugh (1970). The testing entailed giving a series of object-quality-discrimination problems but, in contrast to learning-set training in which traditionally a small and fixed number of trials are given per problem, criterional training was given on each problem. Only if the criterion is satisfied is the problem acceptable for the administration of reversal test trials. The requisite of prereversal criterional achievement helps to insure that all subjects are equivalently task-oriented, that the possible confounding effects of performance variables have been mitigated, and that all subjects have learned an operationally defined amount regarding the task prior to test. Differences between groups were taken to reflect differences in higher-order cognitive function.

The subjects received a series of two-choice visual discrimination problems, each of which was to be learned to a criterion of 84% correct. When criterion was reached for a given problem, the values of discrimination objects were reversed for the next consecutive ten trials. The testing for that problem was then complete, and the next problem of the series was given in the same manner. The experiment was continued until each subject had achieved satisfactorily the prereversal criterion on 40 problems, each having ten reversal test trials.

The wild-born group's reversal-test performance was substantially and consistently superior to that of the restricted-reared group.

The results of this experiment afford additional strong support for the conclusion that impoverished early rearing results in long-term cognitive deficits. As the method equated all subjects on a performance criterion prior to the reversal-test trials and as both groups of chimpanzees were equivalent in terms of average trials to criterion prior to those reversal-test trials, it is concluded that the restricted-reared chimpanzees were less able than the wild-born subjects in transferring learning from the criterional training to the ensuing test trials. It is maintained, furthermore, that the obtained differences between groups on the reversal test cannot be attributed to group differences of an emotional-motivational-attentional nature.

Although our restricted chimpanzees were generally inferior to controls, it should be noted that their deficit was relative rather than absolute, and it was most marked at the early stages of radically new problems.

The persistence of cognitive deficits in the restricted-reared chimpanzees, even after 12 years of environmental enrichment, prolonged testing, and group maintenance, is interpreted to mean that deficits so acquired are not readily corrected.

MATERNAL BEHAVIOR

The laboratory setting has a number of advantages for the study of maternal behavior. The animals can be kept under intensive observation for prolonged periods, and the interference with ongoing maternal patterns from other animals can be minimized. A precise health and reproductive history can be kept of each subject, including occurrences of miscarriage, abortions, and the ordinal position of any particular infant. Of special concern in the study of maternal behavior are the experiential factors that contribute to the development of proficiency in the care of the infant. One question of particular interest is whether it is necessary for the behavior patterns for good maternal care to be developed through actual practice. A second question is whether the mother's early experience with her own mother has an influence on her maternal ability.

In one study (Rogers & Davenport 1970), we looked at the records of 33 females (103 births) with a range of from one to 13 viable parturitions. All these females were born in captivity or were acquired prior to full sexual maturity, ruling out the possibility of unrecorded prior maternal experience. Twenty-two mothers were separated from their own mothers prior to 18 months of age. These females had 63 offsprings, with a range of one to 13 births per mother.

The other group of 11 mothers was separated from their own mothers between 18 months and 9 years. Most of these animals were wild-born and little is known of their history prior to acquisition by the laboratory. These females account for 40 births, with a range of from one to 12 per mother. It has been standard procedure in the laboratory to record the behavior and physical condition of mother and infant following parturition. Sometimes the actual parturitions were observed. At other times they were begun as soon as possible.

Most of the mothers were isolated a few days before delivery and were kept with their babies for at least a few hours following parturition. (Due to the laboratory requirements for infants as experimental subjects, a number of the infants were separated from the mothers within a few days after birth.)

Analyses of descriptions from the diary records and those observed by ourselves (usually within the first 12 hours postpartum) were scored on a 5-point scale according to criteria of increasing maternal proficiency ranging from avoidance to good maternal care (0-4). Analyses of the data revealed that the rearing condition of the mother had a strong influence on her maternal behavior. Females reared by their own mothers for more than 18 months were clearly superior as mothers to females who received less maternal care. Thirty-five of the 40 maternal experiences occurring in the former group were rated 3 or 4 on the proficiency scale; whereas, only 27 of 63 occurrences were so rated for the less mothered group. The difference between the groups was apparent, even on the first parturitions. Six of the ten initial interactions in the first group were rated 3 or 4; but only two of 19 were rated 3 or 4 for the second group.

Overall, maternal performance improved with repeated births. Although the advance at each ordinal position beyond the third birth was small, there was a clear trend of improvement to this time. By the third birth, most of the motherings were rated 3 or 4. In both groups, even brief exposure to the first baby had a strong influence on improvement on subsequent maternal care.

There is a positive relation between the mother's condition of rearing and amount of experience as an infant on maternal ability. Thus, the longer mothered group were better mothers overall, required much less experience to become good mothers, and were much more stable in their performance once they did perform well than were mothers having less mothering.

Immediate cleaning of the infant was characteristic of good mothers. They cleaned the infant by sucking, licking, and grooming fluids and membranes, particularly from the body orifices. Although (up to a point) the time and method of separation of the baby from the afterbirth are not critical to its survival, the solicitous mother was careful that no tension was put on the umbilicus. Until the cord broke or was severed, she carried either the placenta or a length of the cord as she moved about. The time of detachment from the placenta varied from within minutes to 3 days following birth, but usually occurred within 36 hours. On three occasions the mother was observed to nick the cord with her teeth a few inches from its attachment to the infant. After the cord dried, it broke apart at this point. Occasionally females were observed to bite through the cord (Yerkes 1943), but more frequently in the course of normal movement and manipulation, it parted after drying.

The nursing interaction is not automatic in the chimpanzee and some learning is required for the development of skillful, efficient performance. Chimpanzees do not typically hold and carry the neonate at the breast area and primiparous mothers do not reflexly place the infant on the nipple. The newborn lacks the strength and coordination necessary to reach the breast unaided from the lower ventral surface, where it is usually kept. Before adequate learning, the behaviors of mother and infant lack intention, coordination, and smoothness.

We suggest the following analysis to account for the development of initial nursing and for its subsequent refinement and maintenance. Both mother and infant have a set of more or less automatic responses to signals and states, the coincidence of which results in the initial nursings. From birth the infant reflexly sucks when stimulated near the lips, sporadically bobs and turns its head from side to side, whimpers and squirms when uncomfortable, grasps anything put in its hands, and hugs objects to its chest. Discomfort in the newborn is caused by not having something to grasp, cling to, and suck, and by being squeezed, and presumably by hunger. It should be noted here that the nursing posture, in which the infant grasps the mother's hair or skin, has chest contact with a firm, warm surface, and has the nipples to suck, is a most potent tension-reducer for the newborn chimpanzee.

The mother's main contribution to the process leading to initial nursing is her tendency to be distressed by signs of the infant's discomfort. Even mothers who did not care for the infant at all became agitated by the cries and flailings of the unattended infant. The mother typically responds by inspection, grooming, and by a variety of manipulations of which the most important, in this context, is repositioning of the infant on her body. In the course of the mother's more or less random repositioning or readjustment, the baby eventually reaches a point at which its head movements bring it in contact with the nipple and sucking commences. With repetition, the mother shows rapid learning of component behaviors that get the baby to the breast area. The initial reinforcement for the mother is a reduction in her tension brought about by a reduction in the distress of the infant. With repeated occurrences of the nursing sequence, the physiological drives may play an increasingly important role in nursing. Reduction of these drive states is presumably the reinforcement responsible for the development of coordinated, well-patterned, behavioral sequence of nursing.

Although it is apparent that the infant is born with certain mechanisms that are essential for nursing, of at least equal importance, although less obvious, are the mechanisms by which it influences the mother's behavior. Distress in the infant produces agitation and other indications of distress in the mother. Even in the primiparous female chimpanzee, this heightened drive state results in the expression of a variety of behaviors. Because they are innately calming to it, the infant selectively responds to some of these behaviors by a reduction in the distress signals. This response produces a decrease in the mother's drive state and is thus reinforcing; it also increases the likelihood of appropriate maternal behavior occurring again in that situation. The infant thus, in effect, *shapes* the female's pattern of maternal responsiveness. Herein lies one possibility for overcoming inadequacies of deficient early experience. This process may not insure the survival of a particular animal, but will increase the likelihood of the survival of future births.

We then examined the maternal behavior of five restricted females. They had between one and five offspring and, although there was some slight indication of improvement in maternal care, no animal scored higher than 2 on the maternal care rating scale. A number of these infants were allowed to remain with the mother for up to 4 or 5 hours, but little care was shown. At best, the infant was carried about in a hand or very low on the abdomen, and there was little indication of the mother's responsiveness to the distress cries and flailings of the off-spring. The restricted-environment animals did not show appropriate maternal behavior. As adults, some of the restricted environment animals have been observed playing with or, in one case, carrying 1-2-year-old infants on the back. To what extent this experience with more infants will improve their handling of future offspring is unknown.

SEX BEHAVIOR

Sexual behavior in the captive Yerkes population, especially with chimpanzees, has been difficult to study systematically. The problem is complicated by many factors. Certain animals have been essentially re-moved from sexual studies by their commitments to other investiga-tions. Housing of animals is variable and ranges from a compound that accommodates approximately 16 young adult chimpanzees to cages that comfortably accommodate two adults. The majority of adult apes at the Yerkes Center are housed in two animal cages. Accordingly, the availability of sex partners is limited and is often arbitrarily determined by the people in charge of colony maintenance. All animals born at the Center and reared in the laboratory nursery are diapered to the age of about 2 (the extent to which this has an influence on sex behavior is, of course, unknown).

To begin with, a few generalizations may be in order. Chimpanzee males are considerably more subject to long-term sexual deviation than are females, regardless of the condition of rearing; however, the major-ity of wild-born apes do engage in sexual behavior at maturity. In the laboratory-born and reared chimpanzees (again, particularly the males), the interest and facility in copulation is greatest in those who were mother-reared for a significant part of their childhood. The most sexu-ally damaged (at least in early adulthood) of all laboratory-reared animals are those who were separated early from the mother and reared in some condition of environmental restriction for the first several years of life. Indeed, quite contrary to expectations, there is some evidence during early adulthood that animals separated from the mother and reared with considerable social interaction with human beings were less sexually active than those reared in conditions of extreme restriction.

Almost all females, regardless of rearing conditions, engage in sex behavior as adults. In the wild-born female apes, they are at least willing partners and, for the most part, will actually seek out and incite the male to sex behavior. The latter is less true for restricted females.

Nissen (1954) has reported some interesting results on the sex be-havior of a group of young chimpanzees who were deprived of normal maternal care but received a great deal of human handling during infancy and early childhood. They were allowed some social experience

355

with each other from approximately age 2 to 8 (6 months to 1 year before puberty) when the sexes were segregated. After puberty these animals were observed in male-female pairings. Females were paired with males during a period of their cycle when they are usually receptive. No copulations occurred. After a prolonged series of observed pairings among these animals, it was decided to employ sexually proficient tutors. Experienced males were able, after several attempts, to copulate with two of four naive nursery-reared females, but experienced females were successful in encouraging only one of six naive nursery-reared males to copulate. With age and further experience, the majority of these animals, especially the females, increased significantly in sexual proficiency.

We have intensively studied sex behavior in restricted environment animals (Rogers & Davenport 1969). Of the five males who reached sexual maturity and were given sufficient opportunity to engage in copulatory behavior, all but one have done so. For these animals, considerable learning appeared to be involved. Initial attempts at copulation were very poorly coordinated. For example, males with erections might mount the side or head end of the female and thrust against her, but with experience, particularly with the helpful tutelage of sexually proficient females who assisted with positioning and penetration, these animals have improved in frequency and style to the point that they are approaching normal species-typical copulation, except that they lack the usual signalling systems exhibited by wild-born males. One (now fully adult) restricted male has neither attempted nor solicited copulation, and females rarely approach him. He masturbates frequently, sometimes to ejaculation, and occasionally uses a 55-gallon drum for thrusting.

In summary, distortion of the environment in infancy and early childhood, including social isolation and substitute maternal care, has a significant effect on the sexual behavior of the chimpanzee. It appears that the critical variable is early social experience, particularly with the mother.

Second, some, if not most, chimpanzees reared in total isolation, in contrast to similarly reared rhesus monkeys (Mason 1960; Harlow 1965) can recover from behavioral injuries to the extent that they can copulate. This fact may reflect differences in behavioral complexities. The chimpanzee, in general, has a broader behavioral repertoire and, specifically, a less rigid and stereotyped pattern of sexual behavior; his or her behavior is more modifiable by experience (Mason et al. 1968).

Finally, the sexual behavior of the chimpanzee reared in total isolation is somewhat less drastically affected in early adulthoood than that of the chimpanzee for which human care has been substituted. Although we cannot specify the critical factors, we can point to the adverse effects of this apparently inappropriate interspecies relationship during rearing.

These findings and interpretations underline the complexity of the influence of early experience on later behavior. The important dimension in chimpanzees is not merely of the amount of social experience, but is also the qualitative appropriateness of the social companion during early life to the development of later species-specific behavioral organization, especially during early adulthood.

All the sexually mature (and wild-born) orangutans and gorillas have engaged routinely in copulation.

GENERAL SOCIAL BEHAVIOR

Since adult gorillas and orangutans are maintained for the most part in *compatible* pairs and in rather sterile cages, the opportunity for the occurrence of many behaviors is limited. Most of their time is spent in observing caretakers, solitary manipulation of their cages, sleep, and eating. No gross abnormalities in behavior have been observed.

For the first several years after isolation, restricted chimpanzees were either socially unresponsive or inappropriately responsive; for example, reacting to friendly approach with fear and avoidance. In those chimpanzees housed in groups with mother-reared wild-born animals for a number of years, the fearfulness decreased, and social behaviors have gradually changed toward increased sociability and appropriateness (Davenport & Rogers 1970). This change has apparently been brought about by contacts with the wild-born animals. Gradually they participated in most kinds of social interactions. Although their behavior was not entirely normal, it approached species-typical behavior. Grooming is a good example. Adolescent-restricted animals possessed the elements of grooming, close visual inspection, manipulation of a spot on the groomed animal, etc., but these lacked coordination and patterning of normal grooming. Most of the restricted subjects have recovered to the extent that they engage in some semblance of the grooming pattern. (There are great individual differences here.) This recovery was not true of restricted animals who had extensive experience with old normal animals only in early adulthood. It appears, thus, that the age of experience is very important in normalizing social behavior.

In summary, abnormal behavior in the laboratory is rare in mother-reared animals, and of special importance, it can be ameliorated by experience with behaviorally normal animals. The possible significance of this to human psychopathology should be noted.

E. W. Menzel, Jr.

Communication of Object-Locations in a Group of Young Chimpanzees

The communicative capacities of nonhuman primates have been approached in at least three different ways: (a) *Language oriented approach.* A majority of psychologists seem most interested in whether individuals are capable of learning a human language (Fouts 1972; Gardner & Gardner 1971; Hayes & Hayes 1954) or some other arbitrary task with presumably similar logical properties (Mason 1970; Premack 1971b; Rumbaugh, et al. 1973). (b) *Signal-oriented approach.* Ethologists more often ask what species-specific responses or signals an untrained feral individual emits, and, given a particular class or subclass of signals, how do others respond to it (van Hooff 1967; van Lawick-Goodall 1968b; Marler 1965). (c) *Group-oriented approach.*

The author, following von Frisch (1955) and Carpenter (1969), asks the converse of the last question, namely, what does the "society as a whole" do, and given a particular class of coordinated group action (for example, the ability of primate groups to travel simultaneously to distant goals, such as food, and not become separated or lost), what is the minimum amount of information required, and where does it come from? As Ashby (1970) puts it, any well-defined coordinated activity requires a certain measurable amount of information flow between the parts of the system. The problem is not whether natural groups of primates are capable of communication in the biological sense, but rather *what* they are communicating and *how* they do it.

Group travel and foraging behavior have been described in many general primate field studies. As in signal-oriented studies of communication, the general conclusion seems to be that wild primates are not capable of communicating the nature or location of distant goals. In other words, following the philosophers (e.g., Cassirer 1944), communication is said to be *emotional* rather than *objective,* and animals are said to travel together because they follow each other for the rewards intrinsic to social behavior per se. No field worker, however, has used experimental procedures that would allow one to say for sure what the animals' goals are, or what objects or signals are determining the group responses, or what information about the environment followers gain from their leaders. The field evidence to date must, in short, be judged inconclusive rather than negative.

Previous experiments by the author studied various aspects of group travel and foraging in social units of four to eight captive juvenile chimpanzees, using a large outdoor enclosure. It was shown that the direction of travel of the *group as a whole* varied systematically with the direction of experimentally introduced goal objects; that any member of the group was capable of leading the group to objects that he alone had seen; that this leader was dependent upon his followers for getting to the object and thus foraging was a cooperative enterprise; that whether the animals traveled close together or widely dispersed varied both with the incentive conditions and with the particular leader, or pair of leaders, used on a given trial; and that leaders traveling to a preferred class of goal object attracted a larger number of followers than leaders going to a nonpreferred class of object, regardless of their status in the general leadership hierarchy (Menzel 1971a, 1971b, 1973a, 1974). From these and other closely related findings, it was concluded that the animals traveled together, not only because they liked each other, but also because of the environmental consequences of following, and that chimpanzees (if not many group-living species of mammals) often know from each other's behavior when some goal object is *out there,* approximately where it is, how desirable or undesirable it is, whether it is better or worse than some other goal object (including one that is directly visible at the time of response), and which member of the group best knows the foregoing details. Since asymptotic performance was seen on Day 1 of most of the tests, it would seem that the animals had trained each other before the experiments began; and there is no reason to expect any worse performance in feral chimpanzee groups.

There is a definitely weak point in these studies, and it stems from the psychological organization of young chimpanzees themselves. Probably because they shared food and were quite dependent on each other motivationally, the animals seldom ran far ahead of the one group member who knew where a hidden goal lay.[1] Thus, one might still maintain therefore, that what the followers learned from a leader's behavior was his direction of orientation as such, and possibly the strength of his emotional state; the animals could not predict from this the location or nature of the *object* until the leader walked to the object and uncovered it (Kummer 1971, pp. 31 and 62). In other words, the followers might have perceived the leader's emotional state, from this predicted what their own emotional state would be if they were to accompany the leader, and on this basis decided whether or not to follow (Premack 1975).

It is not clear to me, nor would it be clear to theorists such as Heider (1958), Michotte (1959), Tolman (1932), Gibson (1966) or Sommerhoff (1950), why such a feat of emotional communication should be considered psychologically simpler or more primitive than directly perceiving the goal-directed aspect of another animal's behavior (and hence, by definition, the presence if not the geographical direction of some goal). In any case, the question as to which explanation of performance is more adequate is in principle empirically testable, as purely *emotional* communication provides no mechanism whereby a follower could orient his response geographically in the leader's absence or in the event the leader has not yet actually arrived at the goal.

The present experiments were designed specifically to examine directional, locational communication in foraging behavior. They were concerned less with arguing for or against *language,* as a linguist might define it, than with translating the basic issues outlined above into terms in which they may be evaluated quantitatively. The questions are:

1. How contiguous in space or time must a leader be to a hidden goal object before the followers can get to that object by themselves? That is, what degree of *displacement* (Hockett 1960; Altmann 1967) is possible?

2. Is it necessary that the leader be present in the situation at the time of the followers' response?

3. Can information regarding the direction of a goal be transmitted along a chain of individuals, none of whom has seen the object himself?

4. How closely must a leader's behavior resemble "a chimpanzee walking all the way over to the hidden goal and uncovering it"? Or, more generally, what are the effective cues of direction?

GENERAL METHOD

Subjects

Two male (Shadow, Bandit) and four female (Belle, Bido, Polly, Gigi) juvenile wild-born chimpanzees were used as subjects. All had lived together in a 30.5 x 122m outdoor enclosure for more than a year, and three (Shadow, Bandit, Belle) had lived together since their arrival at the laboratory at the estimated age of 1-1½ years. At the time of the present tests the animals were estimated to be 6-8 years old and they had been observed and tested on up to 700 test days in the enclosure. They formed a very compatible social group, albeit they were much less dependent upon each other and much less cohesive as a group than had been the case at earlier ages (Menzel 1971a; 1971b; 1974). Three of the four animals who were tested as leaders (Belle, Bandit, Gigi) had had extensive testing in similar experiments; the fourth (Bido) had had only a few pretraining trials as a leader.

Apparatus

Testing was conducted in the living enclosure. This enclosure had a heavy carpet of grass on the ground, stumps that were 3-4m high of about 25 pine trees (which had been killed by the chimpanzees), and numerous logs scattered through the area. An elevated runway constructed of 2½ m high posts and 5 x 15 cm lumber and plywood extended for about 110 m down the center of the enclosure. Just outside the field was a set of small cages with guillotine doors opening into the field. These doors could be opened from an observation tower by means of overhead pulleys.

The goal objects consisted of fruit and vegetables, cut into pieces about the size of one-quarter of an apple.

An electric timer that produced a clearly audible click every 30 seconds furnished the experimenters a time cue in their recordings of behavior. Most observations were made with naked eye and recorded with pencil and paper, but 7 x 35 mm binoculars and a tape recorder were always on hand. Sample sessions were filmed with a 16 mm Bolex camera and 35 mm Nikon still cameras, using lenses that varied from 8 mm extreme wide angle to 300 mm telephoto.

Procedure

Provided it was not raining, testing was conducted 5 days a week. Each session consisted of several trials in which the experimenters gave one animal a cue of hidden food (as in a delayed response test), and the other animals had to get their cues from him (as in a discrimination learning test).

On a given trial, all animals were first locked into a release cage, from which they could not see out into the field. (On a few occasions they managed to poke peepholes through the cage wall with a wire, but such an event was immediately detectable from their behavior and all the data of that session were discarded. Periodic control tests assured that the animals got no cues from us when inside the release cage.)

Next, food was placed in one or more locations in the field, 10-75 m from the release cage. These locations had been determined to the first approximation before the session by three adjacent columns of a table of random permutations; one column specified position on the Y axis of the enclosure to about the nearest 3 m and the other two columns specified position on the X axis to about the nearest 3 m. Within this predetermined sector the exact hiding place was chosen on the spur of the moment on the basis of available natural cover; and, if possible, we avoided using the same general type of cover (e.g., grass clumps, trees, stumps, holes in the ground) more than a few times in succession or the same exact place more than once in an experiment. The chimpanzees were extremely acute in detecting any change in the enclosure or any hiding place that would look to a human being as if it had been *rigged* (e.g., a neat pile of tree bark or a large handful of freshly plucked grass on otherwise open ground was invariably investigated); therefore, great care was taken to avoid giving them such cues. To eliminate auditory cues of direction that the animals might get from an experimenter's footsteps, a second experimenter sometimes walked simultaneously in a false direction; also, the experimenter who was hiding the food did not always take a direct route to it, and in subsequently approaching the release cage, he always came in from the same angle.

Next, a single animal—hereby operationally designated as the *leader* for that trial—was taken from the group and either carried to the pile of food and permitted to see but not to touch it (direct method of delay) or given some other cue to the direction of food, such as a manual indicating or pointing by the experimenter (indirect method of delay). After this, the leader was returned to the release cage and the experimenters left the field, ascended an observation tower, and opened the release cage door via the pulley system. The delay between the cue and the release of the animals was not less than 2 minutes. Recording commenced when the release cage door was opened and continued for at least 2½ minutes. If the animals were still traveling or searching the field after 2½ minutes, the trial continued (sometimes up to 10 minutes); otherwise the animals were rounded up and locked into the release cage. Then the procedures for the next trial started. If the food had not been found on the previous trial, however, we first removed it and (in most experiments) got the leader for that trial out of the release cage and showed him the empty pile.

Two (and, in some experiments, three) experimenters rotated the task of giving the chimpanzee leader the cue. In no experiment was there any difference in performance according to who served as cue-giver.

Recording was by a time-sampling procedure. As the electrical timer sounded its click, one experimenter located each of the animals as quickly as possible (usually this required no more than 2-3 seconds) and recorded their positions on a 1:500 scale map. In addition he traced the group's general travel routes over the rest of the 30-second time interval. A second experimenter wrote down or verbally recorded by tape recorder parallel qualitative notes on the same time schedule; the most important response categories were: who found the food at each location; any obvious signaling behaviors (e.g., tapping a partner on the arm; pulling him to his feet, orienting quadrupedally toward the food while glancing back toward others); and various other forms of social interaction (displays, fighting). For the types of data to be reported here, inter-observer reliability is typically better than 90% exact agreement.

PURPOSE AND METHOD OF THE SPECIFIC EXPERIMENTS

Experiment 1: Leadership Toward Fixed Destinations

In this experiment food was located in a limited number of locations and the same locations were used day after day. Thus the followers were given a good chance to learn that when a leader set out in a given direction, the group's destination was the food in a given location. The situation seems quite analogous to foraging situations such as those encountered by Hamadryas baboons (Kummer 1971) or by provisionized Japanese monkey groups.

Experiment 2: Reactions to Changes in Leaders and Destinations

In wild chimpanzee groups, the same animal does not invariably lead each move; any individual other than very young infants might initiate travel (van Lawick-Goodall 1968b; Itani & Suzuki 1967). Even in gorillas and other species in which certain individuals do seem to be stable leaders, these individuals might die off, be deposed, etc. Similarly, food supplies change constantly and the group must adapt to these changes. Finally, food is rarely clumped in one spot; sometimes it is thinly dispersed over a larger area and instead of taking exactly the same direction as the leader, the group might fan out to cover the larger area. How far back in the travel sequence does this fanning out occur? The present experiment studied these problems, using procedures similar to those used in Experiment 1.

Experiment 3: Are the Animals Communicating Within the Release Cage?

Here the question is: Given that followers seem to know where a leader is going almost from the start of a travel sequence (Experiments 1 and 2), is it possible that they in fact know it *before* the leader actually sets off, and can they proceed on his *instructions* even without him (i.e., when he himself is restrained)?

Experiment 4: Is Pointing a Sufficient Cue and How Far Ahead Can the Animals Extrapolate?

Another way to consider the problem posed in the last experiment is to ask what the rest of the group would do if a leader initially gave his usual cues but then, for some reason, did not continue toward the goal himself. The best procedure would be one in which the leader gave a signal and then suddenly vanished without disturbing the rest of the group in the process—but we could think of no way to accomplish this with chimpanzees without introducing confounding variables of one sort or another. We finally decided to enter their communication system and serve as *leader surrogates* or crude ethological models of a chimpanzee. Since our prime interest was *natural* communication between chimpanzees, the signals we used in these experiments (and all subsequent ones) were modeled more after signals chimpanzees themselves might have used in their own interaction with each other, than after human language. Manual pointing was included largely so that we might compare its effectiveness with cues that are more natural to the chimpanzees.

Experiment 5: Walking vs. Manual Pointing as Cues of Object Direction

This experiment was based on the last one; it tried to identify more precisely what social cues (given by a human *leader surrogate*) the chimpanzees were using to infer the direction of distant hidden food. The questions were: Can the chimpanzee leader get the message solely by watching another being at a distance, and without actually coming into tactile contact with him? Is it necessary for the chimpanzee himself to locomote or be carried over a portion of the indicated route before he will remember it? Can the chimpanzee use either manual pointing (more strictly a human mode of signaling, and essentially never observed in chimpanzees) or postural and locomotor cues (common to both chimpanzee and man) as a sufficient indicator of direction? Is a composite of these cues (walking, then manual pointing) more effective as a directional signal than either cue taken separately?

Experiment 6: Variations in Orientation as a Basis for Discriminating Between Two Goals

The present experiment builds on the results of Experiment 5 and asks whether variations in orientation alone are sufficient to convey to the chimpanzee leader whether food is present or absent in a given direction. More specifically, we wished to know: Given two different cues on the same trial, one signifying food and the other signifying the absence of food in a given direction, will the animals choose between them systematically? Is visually attentive, purposive locomotion a more effective cue than simply walking in a given direction (and then orienting at no place in particular)? Will the chimpanzees go in the *opposite* direction from which we walk and then orient at no place in particular —thus using such behavior as a cue of negation?

Experiment 7: Variations in Walking as a Basis for Discriminating Between Two Goals

Several of the previous experiments, especially the last one, had already shown that the chimpanzees could discriminate between two or more cues given by a human leader surrogate. In the present experiment, we offered them a choice of two cues that varied in intensity and (in accordance with their presumed natural system of communication) made the more intense cue positive. That is, a strong cue was given in one direction and it signified a large pile of food; and a weak cue was given in another direction, and it signified a small pile of food.

CAPSULE SUMMARY OF THE RESULTS

The only experiment yielding negative or inconclusive results was Experiment 3 (Is there communication within the release cage?). In that experiment when the leader was restrained and could not accompany the others he whimpered loudly; and the followers spent their time in trying to open the door of his cage, rather than striking off without him. This result was interpreted as proving nothing about communication, but merely demonstrating again the emotional dependence of the group members upon each other.

The chimpanzees used walking, pointing (bodily or manual), and other molar behaviors—of human beings and, presumably, each other— as equally effective cues of the direction and relative incentive value of distant, hidden food. Some social cues were almost as effective as the direct sight of food. They could be remembered for at least several minutes, and cue and goal could be separated by at least 75 m (the limits of our test procedures). Once he had given his cue, it was not necessary that the cue-giver remain in the situation at the time of group response. Information regarding the direction of the goal could be transmitted along a chain of several individuals, none of whom had seen the food for himself. In most tests, effective performance was seen on trial 1, which suggests that the animals already knew the appropriate communication system before the tests began. Species-specific gestures, vocalizations, etc., seemed to supplement or reinforce the information provided by more molar cues—i.e., to reduce the uncertainty of the situation still further.

CONCLUSIONS

The only way to determine whether a signal has primarily an *emotional* meaning or an *objective* meaning for animals is to determine what selective function the signal serves on the range of the animals' possible behavioral or internal states. This selective function may be quite different for different parties in the communicative interaction. That is, it may differ according to whether one takes the point of view of the leader, a follower, the group as a whole (or some *standard observer* from the same community), or the scientist who looks in on the scene from the outside (MacKay 1969). The job of an animal psychologist is, of course, to guess—insofar as he can—the meaning of the chimpanzees' signals for the chimpanzees themselves.

What then—first of all—was the leader trying to do? The most parsimonious way to describe his behavior is to say that he did almost anything he could and performed almost any signal in his repertoire that would achieve the molar objective end state of getting himself to food and getting others to accompany him at least part of the way. Anthropocentrically speaking, the meaning of his signals for him was something on the order of "Come with me!" or, more accurately, "Let's go this way rather than that way"; and this message varied from a mild "request" to a strong "imperative" according to the circumstances of the test. If other group members followed without prompting, no specialized signals were necessary. If others did not follow, the leader might stop and wait, glance back and forth at them and the goal, "present" his back for tandem walking, whimper, "beg" with his hand extended palm up, tap a follower on the shoulder, hold a follower's thumb lightly in his teeth and start walking him thus, scream and bite the follower on the back of the neck, drag him along by a leg, or throw himself on the ground in a "temper tantrum" —according to the precise circumstances and the degree of obtuseness shown by the followers (Menzel 1971, 1974). Obviously many of these behaviors (primarily the latter ones, which were seen only very rarely in the last year of our research, and were common only in the youngest leader) could be called "emotional" from the standpoint of an outsider. But it hardly follows that the leader himself was principally trying to convey his

367

emotional state. This would be just as farfetched as saying that he was literally telling his followers where to go, and what was out there—for differential information for such judgments was just as patently available from his behavior as was differential information about his personal emotional state. In other words, the criterion of the meaning of the leader's messages *for him* lies in how the leader varied his behavior as a consequence of the followers' reactions to him. If his signals had no environmental reference (for him), why did he become more upset when the followers tried to groom or comfort him than when they simply got up and started walking in the indicated direction without even touching him? I do not imply that the leader's message was always conveyed intentionally. However, we have clear evidence that leaders in some cases vary their behavior in such a way as to withhold information from a dominant or nonpreferred follower, or even apparently throw them off the track (van Lawick-Goodall 1971; Menzel 1974; see also Bandit in Experiment 1), and this certainly suggests intent, and shows that normal behavior with preferred companions is not invariable. I do not imply, either, that the leader's behavior was unselfish. If the leaders had been bold enough and independent enough to go it alone and get all the food for themselves, they would most likely have done so.

We come now to the central question, which is how the rest of the group interpreted the leader's behavior, and from which aspects of his behavior they derived their information. Were the followers attending principally to the leader's (or their own) emotional state; or did they know something more objective, namely where the leader was headed, and for what end? I shall answer these questions in the terms in which they were stated in the Introduction, and by summarizing the basic results of the experiments.

(a) As far as spatial displacement is concerned, followers can extrapolate at least 75 m ahead of a leader—this being the limits of our test situation and not necessarily the limits of capacity even for juvenile chimpanzees. It seems reasonable to suppose that with an adult leader, chimpanzees could do far better than this, for there is an enormous increase in the typical length of aimed travel runs made by adults vs. juveniles, adults being more apt to attend to distant events and travel straight to them without being distracted by other events or by the lack of an immediate following (Menzel 1969b; J. Goodall, personal communication). In other words the behavior of adult leaders would provide all the information necessary for a follower to predict long ahead of time where the group is going. Whether followers can actually utilize all of this information is, however, still an unsolved problem which awaits experimental analysis in the field. No laboratory cage is big enough for the test.

(b)   At least when a human being is used as a leader surrogate it is not necessary (for the followers' performance) that he remain in the situation after giving his cue. Chimpanzees who have seen a social signal of direction can remember the signal for a period of at least several minutes, and I would not be surprised if the limits of their capacity for even such indirect method delayed response under field conditions would extend for hours. The fact that we could not regularly get the same results using chimpanzee leaders rather than a human leader surrogate seems clearly attributable to the social and motivational complications that arise from experimentally forcing a separation between group members, and not to a lack of cognitive or communicative ability. Here too the greatly increased social independence of adult vs. juvenile chimpanzees would lead one to predict that adults would perform better than the present animals. The only question is whether adults, who no longer actually need each other's close company in foraging, and who are not overly altruistic about sharing their goals, would have any great incentive to recruit each other or to communicate (i.e. share information) about their goals.

(c)   Information regarding the direction and nature of a hidden goal can be passed along a chain of several animals, *none* of whom has seen the goal for himself. To be anthropomorphic, chimpanzees can bear rumors. Usually the message becomes attenuated or distorted as it is passed along.

(d)   At least in the present context, the aspects of the leader's behavior that followers rely on most are not any simple or limited set of specialized signals or "fixed action patterns", each of which functions as a sort of "word" in a protolanguage and specifies this or that specific parameter of the situation. Instead, followers can use a wide range of mutually-intersubstitutable cues for inferring the presence, direction and incentive value of a goal, and who knows best these details. Some of these cues (e.g., postural "pointing," the direction of visual orientations, direction of walking, rate of walking) are common to most species of animals, including man. The fact that chimpanzees themselves rarely, if ever, point out directions manually is not due to their inability to perceive the cue value of such signals should they be given; it is due to their having other directional signals that are at least as effective as manual ones. In other words, this "deficit" is not in the perceptual ability of the receiver, but in the overt performance of the sender. Although these experiments do not rule out the possibility that gestural, vocal, or other molecular signals, might be used as a sufficient basis for communication in other, less routine contexts, especially by adult wild chimpanzees, I suspect that molecular cues serve largely as a close-range "attention getter" and a supplement to the information that is available at more molar "levels of analysis", and that the leader or sender gives such signals mainly when other cues do not suffice. The better the animals know each other and a given situation (or more generally, the more information they have in common), the less their uncertainty as to what the group is likely to do next; and, consequently, the less their need for specialized communication, i.e. *additional* information. It is very clear from studies on home raised chimpanzees (Fouts 1973; Gardner & Gardner 1971; Hayes & Hayes 1954) that these animals can learn to use other than naturally-occurring signals if this is the only way to get their point across.

In sum, coordinated group travel with respect to a hidden goal about which only a single individual knows may be achieved on the basis of so many different alternative cues, and the same cue may serve so many different selective functions according to its exact context, that the Lorenzian ethological hope of understanding primate group communication systems principally via an identification and analysis of individual classes of signals seems unwarranted (see also Beer 1976). Even if it could be assumed that it is possible to "completely" enumerate all classes of signals that individuals can make, this would still not be equivalent to an exhaustive catalogue of all possible *messages* that can be sent or received—witness the ability of two telegraphers to transmit millions of different messages with only three signals (dot, dash, silence), and the literal impossibility of defining all of the functions and meanings of any one of these signals if it is taken separately and out of context.

Rather than continue to start out most investigations of communication on the structuralistic assumption that a system is described once we know all possible "elements" of individual behavior and the function of each "element", a more effective strategy in the long run might be to first examine coordination at the sociological or ecological level of analysis, and then gradually dissect this gestalt (here, a whole group of animals going together to goals only one of them has seen), into its constituent processes and subprocesses. I do not imply that this strategy has not already been used, or that it necessarily leads to conclusions that could be reached in no other way.

Acknowledgment

This research was conducted at the Delta Primate Center, Tulane University, with support of NIH Grant FR-00164 and written up at the State University of New York at Stony Brook with support of NSF Grants BO-38791 and GU-8350. Because of space limitations the article prepared for this volume could not be published in its entirety; this is a synopsis only. The complete article will be published elsewhere.

Notes

1. Some leaders screamed and attacked a follower who beat them to their goal—so, socially speaking, getting too far ahead of others can be negatively reinforcing in more ways than one.

PART FIVE

Special Topics

Emil Menzel's experiments demonstrate communication in apes under the condition of a large cage and with strictly limited subject matter. Roger Fouts and Richard Budd describe much more complicated communication based on close human association and the use of a computerized system. The quantity communicated, the cognitive system implied, and the social use of signs all suggest a much higher level of intelligence than was anticipated. After the failures to teach apes to speak, the degree of communication by other means is a surprising achievement. Again, just as is the case with Menzel's experiments, the laboratory study of communication suggests that far more detailed study is needed in the field.

Many papers in this section deal with the results of ape communication, although the communicative acts are not stressed. For example, predation and tool use (McGrew) pose important problems of communication. Grooming behavior, dominance, and social structure (Bauer; Bygott) involve communication. The intercommunity transfer of chimpanzees (Pusey) is related to both agonistic (territorial) and accepting behaviors.

Differences in sexual behaviors are the subject of five papers, and such behavioral differences are related to grooming, dominance, consortships, predation, tool use, dispersion, and transfer to new groups.

For the first time in the history of primate studies, a variety of scientists, after careful preparation, have joined in the study of one species in one locality. The richness of these papers shows the importance of such an approach. It was fortunate that the first such major combined study was on one of our closest relatives, the chimpanzee. Hopefully this cooperative model will soon be repeated in the field study of many other kinds of primates.

Washoe signing to Dr. Fouts, "You, me ride in car."

Roger S. Fouts and Richard L. Budd

Artificial and Human Language Acquisition in the Chimpanzee

Studying the behavior of the great apes allows one to examine non-human beings who are very similar to and, at the same time, very different from human beings. A scientist can examine the similarities or the differences and obtain a wealth of data concerning the mental and behavioral capacities of the great apes, as well as comparative data that assists in understanding human behavior. This paper will deal extensively with the similarities between the communication capacities of chimpanzees and human beings. We must emphasize that, although the similarities are stressed, there are differences, that reflect the respective individuality of the two species. It would be naive and incorrect to assume that differences do not exist between two species that have as many millions of years of separate evolution; likewise, we cannot assume that two species, as close as these two, do not have similarities, and especially similar bases for a behavior such as language. Language is a complex behavior subject to the same laws as other behaviors, and to assume that language developed solely in our species is to push the mechanisms we know of genetics and evolution to extreme, and most probably incorrect, assumptions.

Myers (1976, p. 755) states that, "From a neurologic view, the evolution of speech must represent the evolution of those mechanisms of the cerebrum located posteriorly in zones of cortex that function to analyze the information of the senses, to establish memories thereof, and to organize voluntary responses which proceed from these analyses or memories." By claiming that speech developed in humans after the evolutionary split from the lower primates, Myers is denying nonhuman primates the ability to perform the mental functions he describes. To accept this conclusion one must accept his premise that the evolution of speech *must* represent the evolution of those specified brain functions. As an alternative, we propose that those specified functions, i.e., analyzing the information of the senses, establishing memories thereof, and organizing voluntary responses that proceed from those analyses or memories, were present before our evolutionary split with the nonhuman primates and the evolution of speech represents *only* the evolution of those mechanisms of the cerebrum located posteriorly in zones of cortex that function to control the complex motor movements necessary to produce speech. This hypothesis is supported by the research of Kimura and Archibald (1974) and Kimura, Battison, and Lubert (1976) who propose that the specialized functions of the left hemisphere may be related primarily to the control of complex motor behavior and not to language per se.

Speech is only one possible mode for the expression of language. A system of manual gestures, e.g., Ameslan (American Sign Language—to be discussed in detail later) is another mode. The capacity of the brain described by Myers may have found expression through the gestural mode long before the evolution of speech. The idea of gestures preceding speech has been advocated by Hewes (1973), among others.

Considering the remarkable physiological similarities between humans and chimpanzees in such areas as blood protein and type, chromosomal characteristics, and structural similarities in the brain, it is probable that language capabilities would have developed along similar lines. The behavioral characteristics and their functions further support this conclusion; however, the function of behavior, by itself, is not sufficient to support this conclusion. For example, Schnierla (1972) points out that ant caste systems bear remarkable similarity to human caste systems but have an entirely different basis. He warns that when comparing phenomena one should look for the bases for the similarities and differences and emphasize these according to their respective importances. If Schnierla is correct, it seems likely that when considering two organisms that are extremely different in physiology (as humans and ants) but show a similar behavior (such as caste systems), drawing a conclusion of a *similar* basis for this behavior may be a gross error. On the other hand, it may be an error in the opposite direction to assume *dissimilar* bases when considering two organisms that are physiologically very similar (e.g., humans and chimpanzees) and show a similar behavior (e.g., language). In this vein, Lashley (1951) states that

> I am coming more and more to the conviction that the rudiments of every human behavioral mechanism will be found far down in the evolutionary scale and also represented even in primitive activities of the nervous system. If there exists, in human cerebral action, processes which seem fundamentally different or inexplicable in terms of our present construct of elementary physiology of integration, then it is probable that the construct is incomplete or mistaken, even for the levels of behavior to which it is applied.

This statement leads to the opinion, in regard to human language as compared to chimpanzee language acquisition, that the differences are one of degree rather than kind, and we are still in the process of discovering the mental capacities of the chimpanzee that will hopefully indicate to what degree they are similar and different.

EARLY ATTEMPTS AND SPECULATION CONCERNING LANGUAGE CAPACITIES IN THE GREAT APES

It appears that almost as long as man has known about chimpanzees he has speculated about the language capacity of this species. Hewes (1973) states that on August 24, 1661, Samuel Pepys stated in his diary that he viewed a great baboon much like a man in most things (to the point that he mentions that it probably is a monster born of a man and a she-baboon). It was most probably a chimpanzee. Pepys went on to state that he believed the creature already understood much English and might even be taught to speak or make signs.

An early attempt by Furness (1916) to teach an orangutan vocal speech had limited success in teaching the orangutan two vocal words: *papa* and *cup*. The orangutan used these in a correct manner, which indicated the potential capacities for language acquisition in great apes. For the next 50 years, however, the speculation was concerned with the chimpanzee's ability to use the vocal mode of communication. In 1921, Köhler (1971) makes the following statement concerning chimpanzee vocalization: "During his crying and howling or other acoustical expressions of feeling, the chimpanzee produces, on the whole, so many phonetic elements which resemble those of our speech that it is certainly not for peripheral-phonetic reasons that he is without speech." Köhler was most probably impressed by the similarities in the vocalizations of chimpanzees and humans, but did not appear to understand the crucial structural differences between the vocal tract of the chimpanzee and that of man. He seems to imply tacitly that the reason for the lack of speech is most probably in the central nervous system rather than in the peripheral mechanisms for speech.

Yerkes (1925, p. 53) seemed to come to a different conclusion. He states that "If the imitative tendency of the parrot could be coupled with the quality of intelligence of the chimpanzee, the latter undoubtedly could speak." Later Yerkes (1927, p. 180) begins to point toward the future with the following statement: "I am inclined to conclude from the various evidences that the great apes have plenty to talk about, but no gift for the use of sounds to represent individual, as contrasted to racial, feelings or ideas. Perhaps they can be taught to use their fingers, somewhat as does the deaf and dumb person, and helped to acquire a simple, nonvocal sign language."

It was not until the late 1940s that another attempt was made to establish communication with a nonhuman primate. Keith and Cathy Hayes (1951; 1952, and C. Hayes 1951) attempted to teach vocal English to a female chimpanzee they raised in their home for over 6 years. Their chimpanzee, Viki, learned to produce four spoken English words—mama, papa, cup, and up. Viki's pronunciation of these words was largely voiceless. Up to that time it had been the most successful attempt. Like the Hayeses, however, other people assumed that the chimpanzee's vocal apparatus had the structural capacity to produce human sounds. It was generally agreed, based on this premise, that the chimpanzee did not have the capacity for language. We now know that there were other reasons for the limited success of attempts to teach vocal language to chimpanzees; namely, their vocal apparatus is quite different. Lieberman, Crelin, and Klatt (1972) point out that chimpanzees lack the supralaryngeal pharyngeal region comparable to that of the human adult. Because of this, chimpanzees cannot produce the human vowels of [a], [i] and [u]. The chimpanzee also appears to lack the tongue mobility, which enables humans to change the shape of their vocal tract allowing the production of several human sounds.

Myers (1976) and Robinson (1976) present data to show a difference in function of certain brain areas in human and nonhuman primates. They found that humans possessed two separate brain areas for the control of facial muscles and vocal apparatus, one for emotive (involuntary) control and one for volitive (voluntary) control. Their research with nonhuman primates (mainly rhesus monkeys) has found emotive control similar to that of humans but failed to find any volitive control. To quote Myers (1976, p. 750), "The overall experience with vocal conditioning in rhesus monkeys supports the view that the vocal apparatus (and, in all likelihood, the facial musculature as well) of non-human primates is poorly accessible to those mechanisms of the brain that organize and control voluntary movements."

Fouts (1975, p. 138) concurs with this view, stating that ". . . their vocalizations appear to be elicited by their environment. For the most part chimpanzees are quiet. They make vocalizations in specific situations (food, greeting, etc.) and these vocalizations are highly stereotyped in their sound across chimpanzees. Whatever the physiological explanations may be; the chimpanzees' vocalizations by themselves do not appear to be sufficient for language." Some degree of control, however, does seem to be present. Fouts observed a chimpanzee (Booee) incorporate a voiceless pant (generally used by chimpanzees when they are in comparable situations that would produce laughter in humans, e.g., tickling) into a signing phrase by using the pant to take the place of the *tickle* sign. For example, instead of signing *tickle Booee*, the chimpanzee will instead pant, and then sign *Booee*. He also uses this to attract people to his cage to play with or tickle him through the wire. Booee has shown that chimpanzee vocalizations and sounds may have more plasticity than what had previously been assumed. This finding suggests a possible experiment in which chimpanzee vocalizations and sounds are used as the mode for two-way communication, possibly resulting in a larger degree of success than the earlier attempts to teach chimpanzees to produce vocal speech.

More recently the attempts to establish two-way communication with chimpanzees have taken two quite distinct and separately conceived approaches. One method involves devising an artificial language in a visual mode, either with plastic pieces of various shapes or lexigrams represented on keys. The other approach involves establishing communication with the chimpanzees by teaching them to use an established human language made up of gestures. Recent findings in these approaches will be examined and compared.

ARTIFICIAL LANGUAGE STUDIES IN THE CHIMPANZEE

Sarah and the Plastic Language

The Premacks (1970; 1971a; 1971b, and Premack & Premack 1972) were the first to devise an artificial system for two-way communication between two species by using pieces of plastic to represent words and relationships. Sarah is a wild-born female chimpanzee who was estimated to be 6 years of age when the Premacks began this particular project. Since their artificial language was visual and written, they used the chimpanzee because of the similarities of the chimpanzee's visual system to that of man.

The artificial system was made up of plastic pieces that varied in size, shape, texture, and color and these were used to represent words. The pieces of plastic were metal-backed and could be arranged in a linear fashion on a magnetized board. In this manner the pieces of plastic were displaced in space, but not in time. This method avoided the use of memory, which is an integral part of human languages, gestural or vocal. Also, this system enables the human experimenter to control the availability of the vocabulary, i.e., which pieces of plastic were available to the subject at any time, and therefore controlled for the individual difficulty of any problem, at the expense of removing the spontaneity of usage typically found in human language. It should also be pointed out that, when an experiment is strictly controlled, it limits the possible findings to those preconceived by the human experimenter rather than exploring the mental capacities of the subject. That factor will be discussed later when comparing the various approaches to this behavior in chimpanzees.

Their approach emphasized the functional aspects of language by breaking language into behavioral constituents and then by providing environmental contingencies for the constituents of language selected by the Premacks to train Sarah. Briefly, Premack and Premack (1972, p. 92) summarize their work with the following statement: "We have been teaching Sarah to read and write with various shaped and colored pieces of plastic, each representing a word. Sarah has a vocabulary of about 130 terms that she uses with a reliability of between 75 and 80 percent."

The Premacks appeared to work from the premise that the relational and logical functions of language are derived from operant procedures, and therefore they used training procedures based on operant methodology. The procedure is of the same type used by psychologists to train pigeons or rats to peck keys or press bars. They reduced the constituents to very simple steps and then used standard operant techniques to teach them. For example, training was initiated by placing some fruit on a board and then allowing Sarah to eat it. Next, Sarah was required to place a piece of plastic representing the particular fruit on a magnetized board before she was allowed to eat the fruit. Training continued in this manner, new pieces of plastic were added simultaneously with new aspects in the situation; e.g., a different kind of fruit or a new person. For example, a new piece of plastic was added when Sarah was changed from bananas to apples; then she was required to place a piece of plastic representing the trainer's name (e.g., Mary or Jim) on the board. In the next step a piece of plastic representing the word *give* was introduced, and required to be placed between the piece of plastic representing the trainer's name and the piece of plastic representing the particular fruit. Finally a piece of plastic representing Sarah's name was introduced and this had to be placed at the end of the sequence of plastic pieces. Sarah appeared to have little difficulty making these conditional discriminations. Sarah was also induced to respond to a sequence of pieces of plastic representing *Sarah give apple Mary* by offering her a piece of chocolate if she gave up her apple. By using a more preferred item (chocolate), they were able to induce Sarah to place the pieces of plastic representing *Mary give apple Jim.*

Using these procedures, the Premacks claim they were able to train Sarah to use and understand the negative article, the interrogative, *wh* questions, the concept of *name of,* dimensional classes, prepositions, hierarchically organized sentences, and the conditional.

Premack and Premack (1972, p. 99) sum up their research with the following statement: "Sarah has managed to learn a code, a simple language that nevertheless included some of the characteristic features of natural language. Each step of the training program was made as simple as possible. The objective was to reduce complex notions to a series of simple and highly learnable steps. . . compared with a two-year-old child Sarah holds her own in language ability."

Lana and the Computer

Rumbaugh, Gill, and Von Glaserfeld (1973) have devised a computer controlled training situation, which objectively examined some of the language capacities in a chimpanzee. They used Lana as their subject, a 2½-year-old female chimpanzee. After 6 months of training they have found that Lana is able to read projected word characters and is able to complete incomplete sentences based on their meaning and serial order or reject the incomplete sentences if they were grammatically incorrect.

Rumbaugh et al. are using a PDP-8 computer which has two consoles containing 25 keys each, on each key is a lexigram in *Yerkish,* the artificial language developed by Rumbaugh et al. Symbols in Yerkish are made up of white geometric figures. Each symbol is composed of nine stimulus elements, which singly or in combination are used to make up the white symbol. The symbol is displayed on a key that has a colored background made up of three colors used singly or in combination. When the key is available for use by Lana, it is softly backlit. When she presses a key, it becomes brightly lit. When a key is not available for use, it has no backlighting. When Lana depresses a key, a facsimile of the lexigram appears in serial order on a projector above the console. There are seven projectors above the console available for the lexigrams to be shown in a visual display of the communication. The computer also dispenses appropriate incentives to Lana when she depresses the keys in the correct serial order in accordance with the grammar of

Yerkish. Lana may ask for such things as food, liquids, music, movies, toys, to have windows opened, to have a trainer come in, and so on, when they are available (defined by when the appropriate keys are softly backlit). There is also a console available only to the human experimenters so that the computer can mediate conversations between the experimenters and Lana.

Lana's training was begun by requiring her to press a single key in order to receive an incentive. Next she was required to begin each request with a *please* and end it with a *period*. The depression of the period key instructed the computer to evaluate the phrase for correctness of serial order. If it was correct, a tone would sound and Lana would receive what she had asked for; if not, the computer would erase the projector display and reset the keys on the console. The next step was to require Lana to depress holophrases (e.g., machine give M&M) in between a please and a period. The next step was to teach Lana to depress each key represented in the original holophrase (e.g., Please/Machine/give/M&M/period). Then the keys were randomized on the console and finally she had to select and press the keys in the correct serial order from the randomized keys on the computer.

One of the more interesting findings in this study was that Lana began to attend to the lexigrams on the projectors without training. The experimenters reported that Lana, by pressing the period key, would erase sentences in which she had made an error, rather than finishing them. Lana's spontaneously learning to attend to the projected lexigrams and their order suggested further experiments. The experimenters could examine Lana's ability to read sentence beginnings, discriminate between the valid and invalid beginnings, and write the completion of the incomplete sentences. In the first experiment, Lana was presented with one valid sentence beginning (please machine give) and six invalid beginnings. Lana could either erase them or complete them. If she completed them, she had to choose from correct lexigrams (e.g., juice, M&M, or piece of banana) and incorrect lexigrams (make, machine, music, Tim, movie, Lana, etc.). Music and movies were incorrect since the computer was programmed to accept these with *make* rather than *give*. Lana's performance on the various aspects of this test ranged from 88% correct (please X give, an invalid beginning) to 100% correct (please machine give, the valid beginning). The second experiment was the same except that make was substituted for give in the sentences with the valid beginning. Lana performed at 86% correct or above in this experiment. The third experiment used only valid beginnings varying in the number of words in them, e.g., please; please machine; please machine give; please machine make; please machine give piece of, and so on. Lana ranged from 70% correct to 100% correct on the various beginnings.

The fourth experiment was the same as the third except that additional options were added (apple, juice, water, music, tickle, and Lana). Lana's correct responses ranged from 63% to 100% correct.

Rumbaugh et al. conclude that Lana accurately reads and perceives the serial order in Yerkish and is able to discriminate between valid and invalid beginnings of incomplete sentences in order to receive an incentive.

HUMAN LANGUAGE ACQUISITION IN THE CHIMPANZEE

Washoe and the Language of the Deaf

Project Washoe was the first successful attempt to teach a nonhuman primate a human language. Gardner and Gardner (1969; 1971) began this project with a 10-14-month-old female chimpanzee, Washoe, in June 1966 at the University of Nevada in Reno. By the time the project ended in Reno (October 1970), Washoe had acquired a vocabulary of over 130 signs. Even though this young chimpanzee had a comparatively small vocabulary, she used it well. She quite readily produced the vocabulary in spontaneous combinations that reflected her syntactic ability and her combinations were contextually correct. She was able to transfer her signs and combinations to novel situations with ease and in a reliable manner.

Washoe was raised in the Gardner's backyard (5000 sq. ft.) in an 8-x-24-foot house trailer that was completely self-contained (toilet facilities, kitchen, bedroom, etc.). The researchers who worked with her used only American Sign Language for the Deaf (Ameslan) to communicate with her and with each other when in her presence. Washoe was immersed in a linguistic environment of Ameslan throughout her daily activities.

Ameslan is a gestural language used by the deaf in the United States and Canada. It is made up of gestures that have specific movements and places where they begin and end in relation to the signer's body. The signs are analogous to words in a spoken language.

Several methods of acquiring the signs were examined. The Gardners looked for *manual babbling* (e.g., playing with one's hands) but were unable to observe very much of it in Washoe. This was most probably because Washoe was almost a year old when the project started, and as she began to acquire signs and use them the babbling stopped. Of course, the same is true of human children; when vocal speech appears, babbling stops. The Gardners also used *shaping*, which involved administering a reward when Washoe made an approximation of a sign and then shaping the response into a closer approximation of the sign. The *open* sign is one example of a sign acquired in this fashion. Washoe would bang on doors with her fists when she wanted them to be opened. This is similar to the open sign in Ameslan, which involves two flat hands placed parallel to each other and then moved up and apart while rotating the wrists to the outside. By shaping her original banging, the Gardners were able to train Washoe to make the open sign. Washoe also acquired signs by observational learning. She apparently acquired signs in this fashion by observing the human signers and then imitating their signs without any intentional training on their part. Guidance proved to be an efficient method of teaching Washoe new signs. It involved forming Washoe's hands and putting them through the required movements for the sign. Fouts (1972) compared guidance or molding to imitation. In the experimental situation, Washoe acquired signs much quicker when molding, rather than imitation, was used. Washoe also made natural gestures that are typical in chimpanzees. When a natural gesture was similar in form to a sign in Ameslan, Washoe was encouraged to use it.

Daily records were kept of Washoe's signing. She was reviewed on her vocabulary each day and a record of her responses was kept. In addition, a record of her combinations and the context in which they occurred was kept. Later a tape recorder with a whisper-microphone was used as a recording device.

The Gardners tested Washoe's vocabulary in several tests with blind conditions to control for cueing. The most efficient test was the double-blind slide test. This test involved the presentation of transparencies of objects representing signs in Washoe's vocabulary. The transparencies were several different representations of the objects. Washoe got 53 items correct out of 99 presentations, which is above a chance expectancy of 3 correct out of 99. Washoe's errors on this test fit into meaningful categories. For example, a picture of an animal had a high probability of being responded to with a sign for another animal. Pictures of objects in the food category generally received incorrect responses that were within the food category; the same was true for pictures of objects in the grooming category. These errors indicate that Washoe understood the concepts of the various categories in which objects belonged.

Some of Washoe's combinations were examined for their rule following behavior or syntactic qualities. In examining the combinations in which Washoe used the pronouns *you* and *me* with an action verb, it was found that you preceded both the action verb and the pronoun me over 90% of the time. Initially in the data collection she used her previously preferred sign order of you-me-action verb, but she then changed to a preferred order of you-action verb-me; therefore, me preceded the action verb 60% of the time and followed it 40% of the time. The reason she changed from her preferred order in the middle of this data collection to the sign order preferred by her human companions is open to speculation.

Washoe and the Oklahoma Chimpanzees

Project Washoe ended in 1970 and Washoe was taken to the Institute of Primate Studies in Norman, Oklahoma, directed by Dr. W. B. Lemmon. This is a primate laboratory with numerous chimpanzees in its main colony and several chimpanzees being reared in private homes around the area. Chimpanzees in both situations have been taught to use Ameslan in several experiments.

One of the first experiments done in Oklahoma (Fouts 1973) examined the acquisition of signs in four young chimpanzees. The experiment compared the ease and difficulty of acquiring signs among the four chimpanzees, and it also found individual differences between the chimpanzees in their rates of acquiring the signs.

Two female and two male chimpanzees were used as the subjects. They were taught ten signs during 30-minute training sessions. The acquisition rate for each sign was compared on the basis of the number of minutes in training to reach a criterion of five consecutive unprompted responses. Molding (Fouts 1972) was used as the method of teaching the signs.

After the chimpanzees acquired the ten signs they were tested on nine of the signs (all nouns) excluding the *more* sign (an adjective). They were tested in a double-blind box test, similar to the test described by Gardner and Gardner (1971).

It was found that some signs were consistently easy or difficult for the four chimpanzees to acquire. The mean times for acquiring individual signs ranged from 9.75 minutes to 316 minutes. Individual differences were found between the chimpanzees. The mean times to reach criteria for each chimpanzee were 54.3, 79.7, 136.4, and 159.1. These differences were interpreted as possibly being partly due to the chimpanzee's individual behavior in the training sessions.

All the chimpanzees performed above the chance level during testing. The percentage of correct responses in the double-blind box test were 26.4%, 58.3%, 59.7%, and 90.3% correct. The low score of 26.4% in one chimpanzee may have been a result of the difference between acquisition and testing. This particular chimpanzee seemed to require much praise and feedback for her correct responses in acquisition. Since testing was a double-blind condition, the observers were unable to give any positive feedback to her responses; as a result, her performance would begin to deteriorate noticeably after the initial trials in the test were completed.

One other important finding of this study was that chimpanzees other than Washoe had the capacity to acquire signs in Ameslan.

In a study (Mellgren et al. 1973) examining the relationship between generic and specific signs, we were also able to examine the conceptual ability of a chimpanzee in regard to the category of items. This was done with a 7-year-old female chimpanzee (Lucy). Lucy had been raised in species-isolation in a human home since she was 2 days old. She had been receiving training in Ameslan for over 2 years and had a vocabulary of 75 signs.

We wished to determine whether a new sign would become generic or specific. Lucy had five food-related signs in her vocabulary: *food, fruit,* and *drink,* which she used in a generic manner; *candy* and *banana,* which she used in a specific manner. The object of the study was to teach her a new sign and then determine if it became specific or generic relative to a category of items. The sign we chose to teach her was *berry,* and the category of items were 24 different fruits and vegetables ranging from ¼ of a piece of watermelon and a grapefruit to small berries and berry-like items such as blueberries, cherry tomatoes, and radishes. The exemplar for the berry sign was a cherry.

The procedure was to record Lucy's responses to these 24 different fruits and vegetables in a vocabulary drill. Each one was presented to her and she was asked *what that* in Ameslan. She could pick it up or do anything she wished with it. The items were presented to her in a different random order for each day of data collection. These items were interspersed with at least two other items that were not in the fruit or vegetable category but for which she had a sign in her vocabulary. For example, the experimenter would ask her what a shoe was and then a string and then he would place a piece of fruit or vegetable in front of her and ask her what that in Ameslan. Following her response he would then question her on two other items such as a book and a doll and so on. Before she was taught the berry sign, she was given four days of baseline responding to the 24 items to determine her usual responses to these items. After day four, she was taught the berry sign using a cherry as the exemplar. For the next four days the berry sign remained entirely specific to cherries. After day eight, she was taught the berry sign again, but this time blueberries were used as the exemplar. During the next four days, she called the blueberries berry for the first two days and then went back to what she previously called them; but she still continued to label the cherries with the berry sign. Her response indicated that she preferred to use the berry sign in a specific sense.

In addition to those findings, we were able to examine Lucy's conceptualization of fruit and vegetables. For example, she appeared to dichotomize the two categories in her responding. When comparing the *fruit* responses to the *food* responses, she showed a preference to label the fruit items with the fruit sign (85%) as opposed to the food sign (15%). The reverse was true for the vegetables since she preferred to use the food sign (65%) instead of the fruit sign (35%).

She also demonstrated that she was able to combine various signs in her vocabulary into novel combinations in order to describe various items in a striking manner. For example, when presented with a radish, she labeled it a *fruit food* or *drink* for 3 days. On the fourth day she took a bite of it, then spit it out, and called it a *cry hurt food.* She continued to use cry or hurt to describe it for the next eight days. She predominantly preferred *candy, drink,* and *fruit* signs to describe a watermelon by calling it a *candy drink,* or a *drink fruit;* whereas the experimenter referred to it with entirely different signs that Lucy did not have in her vocabulary *(water* and *melon).* Lucy used 65% of the smell signs in her vocabulary to describe the four citrus fruits, which she labeled *smell fruit.* Probably she was referring to the strong odor of the skin of citrus fruits.

Lucy's novel combinations demonstrated her ability not only to form new combinations using signs in her vocabulary, but also to use her vocabulary to map various concepts she had concerning the category of fruits and vegetables she was presented with.

Since many of the institute's home-reared chimpanzees had a great deal of exposure to vocal English before they were taught Ameslan, they also appeared to have a good understanding of vocal English. The next study was conducted to determine the relationship between their English understanding vocabularies and their Ameslan vocabularies. This study (Fouts et al. 1973) tested a young male chimpanzee, Ally, who was being home-reared. In pretraining, we tested his understanding of ten vocal English words by giving him vocal English commands, such as give me the spoon, pick up the spoon, find the spoon, and so on. Ally had to obey the command by choosing the object from a group of several other objects. After he met a criterion of obeying the command for five consecutive times, he was considered to have an understanding of the vocal English word. Training was begun by dividing the list of ten signs into two lists of five. One experimenter would attempt to teach a sign to Ally using only the vocal English word as the exemplar. Then a second experimenter would test all five of the objects corresponding to the vocal English words without knowing which one had been taught or if any had been acquired. The reverse was done for the second list of five words. It was found that Ally was able to transfer the sign he was taught to use for the vocal English word to the object representing that word. This transfer is similar to second language acquisition in humans, in addition to having cross-modal implications.

We know now that chimpanzees can produce novel combinations of signs in their vocabularies. Humans have this capacity, as well as the ability to understand novel combinations of words produced by someone else. The study by Fouts et al. (1976) examines the understanding of novel combinations in the chimpanzee. This study also used Ally as the subject. In the initial training, Ally was taught to play a game of picking one of five objects from a box and putting it in one of three

FIGURE 1
Bruno and Ally simultaneously sign "hat" to the appropriate exemplar.

places. For example, the command might be *put baby in purse*. After Ally was sufficiently adept at this game, new items were placed in the box that had not been used in training, and a novel place for him to put them was added to the other places. The items were placed in a box so that they were occluded from the experimenter who gave Ally the command and the place was also hidden from his view by a screen placed between the experimenter and the specified places. This was done so that the experimenter would not be able to cue Ally in regard to choosing the correct item or putting it in the correct place. With five items to choose from and three places to put them, chance responding would be one correct in 15 trials. Ally was able to obey the novel commands far above the chance level. For example, on the last 4 days of testing, Ally got 7 correct out of 16 trials, 11 correct out of 17 trials, 6 correct out of 16 trials, and 10 correct out of 18 trials. The number of trials on any day ranged from 16 to 18 depending on Ally's cooperativeness.

We have also begun to examine intraspecific communication using Ameslan in chimpanzees. In a preliminary report (Fouts et al 1973), various conditions for exploring chimp-to-chimp Ameslan communication were examined. We used Booee and Bruno, two young male chimpanzees who had vocabularies of 36 signs apiece. We have found the following situations are conducive to this type of communication: tickling and play, mutual comforting, and food sharing. The two chimpanzees often seemed to prefer their natural form of communication to Ameslan; however, they have had much exposure to other chimpanzees. It is almost comparable to teaching two Frenchmen a few words of English and then expecting them to use English to communicate with each other. This condition is quite different from that of Washoe and another home-reared chimpanzee (Ally), who has just been introduced into the main colony; Ameslan appears to be their preferred mode of communication in dealing with humans or chimpanzees. In April 1974, we introduced Ally to Bruno and were able to record and observe a good deal of chimp-to-chimp communication from Ally to Bruno. These communications were mainly about food or play. We hope in the near future to house Ally and Washoe together, since they are both excellent signers, and observe their interactions. Should Washoe have a baby, we will be able to study the mother-infant relationship and the use of signs in that relationship.

Other characteristics of the chimpanzee's ability to use Ameslan have been observed in situations other than experiments. For example, both Washoe and Lucy have been observed to invent signs for objects for which they did not have signs. Washoe invented a sign for bib by making an outline of one on her chest (Gardner & Gardner 1971). Lucy invented a sign for leash by making a hooking action with her index finger on her neck. The experimenter had no sign for leash, so he referred to it with the *string* sign.

FIGURE 2
Ally signs "ball" in answer to experimenter's question, "What this?"

FIGURE 3
(a) Ally signs to Bruno, "You feed Ally;"
(b) Bruno responds by giving Ally a slice of orange.

Recently Washoe was observed to invent a new combination to describe some Brazil nuts by referring to them as *rock berry*. Her meaning was explained when the box she was referring to was found to be full of Brazil nuts.

On another occasion Washoe referred to a rhesus monkey she had been fighting with earlier (through the bars of his cage) as a *dirty monkey*. Until this time she used the dirty sign as a noun to refer to feces or soiled items. Now she uses it quite readily as an adjective to describe people who refuse her requests. For example, when she signed to me from an island to *Roger out me* and I refused by replying *sorry, you must stay there*, she responded by signing *dirty Roger* repeatedly as she walked away.

A COMPARISON OF THE ARTIFICIAL LANGUAGE APPROACH TO THE HUMAN LANGUAGE APPROACH

The studies that use artificial languages are similar to the studies using Ameslan as the mode of communication in that both approaches avoid the problems surrounding the vocal mode by using visual modes of communication. The artificial language approach, in one case (Rumbaugh et al. 1973), has removed the human element by using a computer as an intermediary; because a computer is used, an exact record can be made of all the chimpanzee's communication. This refinement is expensive, however, and structures the chimpanzee into a strict and rigid paradigm, which allows only those behaviors to appear that will fit into the experimental situation. For example, the computer is not programmed to accept novel or innovative uses of the particular language by the chimpanzees. This approach, nonetheless, has succeeded conclusively at finding and confirming such behavior as a responsiveness to word order (syntax). In terms of truly exploring the mental capacities of the chimpanzee, however, the artificial language approaches are limited to examining those behaviors that are conceived of by the experimenters rather than the chimpanzee. In 1921 Wolfgang Köhler (1971, p. 215) made the following points:

> . . . Lack of ambiguity in the experimental setup in the sense of an either-or has, to be sure, unfavorable as well as favorable consequences. The decisive explanations for the understanding of apes frequently arise from quite unforeseen kinds of behavior, for example, use of tools by the animals in ways very different from human beings. If we arrange all conditions in such a way that, so far as possible, the ape can only show the kinds of behavior in which we are interested in advance, or else nothing essential at all, then it will become less likely that the animal does the unexpected and thus teaches the observer something.

The human language approach incorporates the best of both methods. We are able to do highly controlled experiments; at the same time, we can allow the chimpanzee to show us many of his capabilities since the chimpanzee has control over his language—he carries it with him in his head at all times. He is not bound to the program of the computer or to the limits of the experimenter in making only certain words in his vocabulary available to him.

An example is the study we did with Lucy on the 24 different fruits and vegetables (Mellgren et al. 1973). If we had limited her possible responses to only the five food-related signs in her vocabulary, we would not have made the discoveries concerning her conceptualizations of the items. Washoe would not be able to insult people by calling them *dirty* if she had not been allowed to change a noun into an adjective, or to refer to Brazil nuts as *rock berry*.

Both approaches have their individual advantages. It is up to the scientist examining the behavior to decide if he wants to examine only those things of which he can conceive, or if he is willing to accept some help from the chimpanzee when it comes to examining the mental capacities of that chimpanzee.

Harold R. Bauer

Agonistic and Grooming Behavior in the Reunion Context of Gombe Stream Chimpanzees

Results from extensive studies of chimpanzee populations in Tanzania indicate that chimpanzee social organization is characterized by the spatial flow of community members into and out of temporary associations here called *parties*. At any given moment several parties may exist and be scattered over several kilometers, although they all belong to the same community. Just after a reunion during the formation of a new party, members who traveled to the location eat less and socialize more than those who did not travel (Bauer 1975). Prominent forms of social behavior in adult males after reunion include agonistic charging displays and social grooming. This study concerns the relationship between agonistic charging displays and social grooming in the reunion context of chimpanzee parties.

The relationship between agonistic charging displays and social grooming has been suggested by three different studies of chimpanzees. Captive adult chimpanzees who display more frequently are reported to receive some sort of *associative action,* such as grooming, play, and sexual interactions (Reynolds & Luscombe 1969). In another case, a study of adult male chimpanzee grooming in the Gombe Stream provisioning area reported that higher-status adult males, who did more charging displays and supplanting of peers, also were groomed relatively more frequently (Simpson 1973a). An older study of captive adult female chimpanzees also indicated that, in general, food-dominant chimpanzees received more grooming from the subordinate than vice versa (Crawford 1942).

The positive relationship between the frequency of performance of chimpanzee charging displays and involvement in social grooming has been explained by a theory of *attention structure* (Reynolds & Luscombe 1969). The theory proposed that behaviors in primate groups were directed toward central and outstanding members and that the pattern of this direction characterized the structure of the group. A general account of chimpanzee social structure has used this theory in noting that chimpanzees have a *hedonic mode* of attention structure (Jolly 1972). Hinde (1974) has criticized the attention structure theory as unexceptional in just translating social dominance theory and over-stating the *positively magnetic* and binding nature of agonistic displays. With few clear exceptions (Kummer 1970), the immediate consequences of agonistic displays in wild primates have not been examined.

STUDY 1

The first study considers only reunion contexts in the woodland-forest habitat of the Gombe National Park, Tanzania and has two objectives. One is to evaluate the relative amount of postreunion social grooming that involves those performing agonistic charging displays. The second is to evaluate the relative amount of postreunion social grooming between those party members who have just met and those party members who were together before the reunion.

Methods

The subjects in the study were five male chimpanzees (*Pan troglodytes,* named Faben, Figan, Jomeo, Sherry, and Evered) of the habituated Kasakela community. These chimpanzees were habituated to human observers, which made it possible to follow them through a variety of temporary associations of the chimpanzee community (Goodall 1975).

Each subject was followed as a target animal for 60 hours over a period of a year; behavior patterns and party composition were recorded on checksheets or into a tape recorder for later transcription.

A party was defined as a temporary association of independently traveling chimpanzees including the male target chimpanzee and was recorded on 5-minute intervals. A *reunion* was the point in time when one or more new chimpanzees joined to form a party after at least one half-hour separation. A postreunion sequence was the temporal structure of behavior beginning at the time of reunion up to one-half hour postreunion. Postreunion sequences were put into one of the four categories: only social grooming; postcharging display social grooming; postcharging display social grooming involving the displayer; and no social grooming following the charging display. A *charging display* of chimpanzees is used here as a general term including at least one sequence of rigorous and exaggerated locomotion with hair erect, occurring with a number of display elements, as branch-waving, dragging, flailing, throwing, and stamping (Goodall 1968b, plate 8a; Bygott 1974). Social grooming is the repeated combing of another chimpanzee's pelage.

Circumstances in the field sometimes terminated postreunion sequences before an hour had elapsed. To check for the possibility of differences between social grooming and charging display postreunion sequences being artifacts of the duration of sequences, a comparison was made between the mean length of display and groom only sequences. Postreunion sequences with only charging displays had a mean length of 40 minutes and those with only social grooming had a mean length of 41.9 minutes (range 15 to 60 minutes). This close comparability of sequence length allowed differences in sequence samples not to be considered artifacts.

To check for the possibility that differences in the relative occurrence of charging displays and social grooming were attributable to differences in party size, postreunion sequences in which only displays or social grooming occurred were compared. The results showed that median party sizes of charging display and social grooming sequences were very similar (Mann Whitney $U = 93$, $N = 12$, $p > 0.10$).

To compare target and party measures of charging displays followed by social grooming, postreunion sequences were divided into two cases. In the first case, the target chimpanzee did the postreunion charging display and subsequent social grooming was noted. In the second case, target animal social grooming was noted after prior charging displays. In the first case, 46.1% of the displayers were involved in social grooming and, in the second case, 40% of the social grooming involved prior displayers (mean party size = 4.04).

If a simply distinguished or outstanding behavior pattern could not be recorded accurately, a notation was made and reunions including such sequences were rejected from the sample. This situation rarely occurred. Sequences analyzed included at least one new party member and were terminated if a displayer left.

Results

The results showed that a male doing a charging display was involved in social grooming in more postreunion sequences than other nondisplaying males in the same parties. (These results are compared with additional data in Study 2.) When displaying males were involved in social grooming subsequent to the display, the social grooming lasted for significantly shorter periods of time in each postreunion sequence than for nondisplayers (Mann Whitney $U = 1$, $n_1 = 3$, $n_2 = 6$, $p < 0.05$, 2-tailed, Siegel 1956).

Milder threats than a charging display, such as hunched shoulders, might explain why most of the postreunion social grooming occurred without a prior charging display. Only 17% of the postreunion sequences without prior charging displays were preceded by mild threats such as *hair erect* (Table 1).

The second question was to see if social grooming of new party members was preferred to grooming of old party members. This question is of special importance, since 83.1% of the social grooming observed did not involve a displayer. The results are shown in Table 1 and indicate that social grooming between new and old party members was chosen more often than grooming between old members or between new party members. Almost all *old party member* grooming was between sibling pairs. A test was done to compare new and old party member social grooming excluding cases in which the only choices were either new party members or siblings. The results showed that new party members were chosen by old members as a grooming partner in significantly more postreunion sequences than another old party member (Mann Whitney $U = 0$, $n_1 = 4$, $n_2 = 5$, $p < .016$, 2 tailed; Siegel 1956).

STUDY 2

This study's first objective is to make a general comparison between reunions in two situations; namely, the charging display and social grooming interrelationships of woodland-forest parties; and parties in the Gombe Stream Research Center provisioning area. A second objective is to describe a telephotographic sequence taken during a post-reunion period, which involved a prior charging display before the social grooming, to provide a perspective on the complex set of behaviors sometimes associated with such interactions.

TABLE 1

Postreunion sequences in which social grooming was observed without prior charging displays

CHIMPANZEE	OLD PARTY KIN	OLD PARTY NON-KIN	NEW PARTY	PRIOR THREATS	TOTAL REUNIONS
Sherry	1	0	2	0	3
Jomeo	4	0	5	1	7
Faben	0	0	3	0	3
Evered	0	0	3	1	3
Figan	1	1	3	1	3
Totals	6	1	15	3	19

Methods

For the comparison between the forest and provisioning area postreunion sequences, data were collected and studied from two periods of time. The first period was between July 1970 and June 1971, and the second period was between January and September 1972.

For the first period, daily provisioning area data indicated the time of arrival, the direction of arrival and departure, the periods in camp attendance, and the charging displays and social grooming of chimpanzees present (Goodall 1968b; 1975). Observers, who were trained in recording on these behavioral checksheets, included David Bygott, Anne Pusey, Richard Wrangham, Margaretha Thorndahl, and Harold Bauer.

A party consisted of independently traveling chimpanzees in the provisioning area with at least one male. A reunion refers to the time when one or more chimpanzees from the forest joined chimpanzees in the provisioning area—after at least a half hour of separation when they arrive from different directions or a one-hour separation when they arrive from the same direction. No successive meetings of the same individuals were included. A postreunion sequence began at the time of reunion and continued up to an hour, as in Study 1. Postreunion sequences were divided into four categories, as was done in Study 1. No observations were included that were made when provisioned food was present or had been available in the previous hour (Wrangham 1974). The woodland-forest data from Study 1 was matched to the provisioning area data by the definition of party and initially for party size. The data from the second period was collected on voice tape recorder and later transcribed and scored separately for woodland-forest and provisioning area conditions, in a manner comparable with Study 1, except that social grooming of any party member was noted.

In these studies, chimpanzees in the provisioning area and followed in the field are habituated to the presence of trained human observers. They do not treat human observers as social peers and, for the most part, ignore them. Observers try to maintain a distance of at least 5 meters from the animals, avoiding contact with them.

Results

The comparison of forest and provisioning area data indicates that simple social grooming, without prior charging displays, predominates postreunion sequences in which social grooming was observed. This result was consistent in the two time periods and in the forest and provisioning area conditions. In both forest and provisioning area conditions, when prior displays occurred, social grooming with the displayer after the display was more common than social grooming between non-displayers after the display (Table 2).

In the second period of observation during 1972, it was more characteristic of the adult male Humphrey (HM) to do a charging display when entering a party of Kasakela community members than the other adult males. A graded response to these charging displays—ranging from occasional ignoring to intense vocalizations, avoidance, crouching, and other submissive behavior—was observed (See article opening photograph M; Goodall 1968b). For these reasons, Humphrey was known as the alpha male during the second period of study in the Kasakela community. The article-opening photographs of Gombe Stream chimpanzees is an example of a postreunion sequence that is representative of the social relationship of Humphrey with the adult male Figan (FG). What was interesting was that Figan's submissive response to Humphrey's arrival was often delayed, relative to other chimpanzees present, and Humphrey would often persist displaying until Figan was submissive or left the party.

In the postreunion photo sequence, Humphrey (HM) has just reunited with Figan (FG) and performed three charging displays before the first photo. This sequence was taken in August 1972 when no food was available in the provisioning area. During the actual reunion the sequence was as follows:

a. *First minute:* HM has his hair erect, shoulders hunched, and thoracic-lumbar vertebrae convex in a quadrupedal walking gait.

b. HM stands, has hair erect, looks away from FG, as FG looks towards HM and sits with hair sleeked holding himself.

c. HM stands, hair mildly erect, looks away from FG, as they both avoid looking toward each other simultaneously (HM foreground).

d. HM has hair erect and begins to walk upslope, as he glances towards FG, who remains sitting and holding himself.

e. HM has hair erect and walks upslope behind FG, who glances over his shoulder toward HM and holds himself (left).

    f. *Second minute:* FG turns, extends hand (not shown), grunts, while looking toward HM, who sits down, hair erect and scratches.

    g. *Third minute:* HM begins self-grooming and FG remains sitting holding himself.

    h. *Fourth minute:* FG begins self-grooming as HM continues self-grooming.

    i. *Sixth minute:* HM remains sitting, pauses in self-grooming, and FG stands and walks toward HM.

    j. As Figan approaches grunting, HM looks away from FG and has hair erect.

    k. FG has a pout face, stands and extends hand to HM, who sits looking away from FG with hair erect.

    l. FG continues to extend hand and touches HM's face, while HM remains sitting, looking away from FG with hair erect.

    m. FG crouches and pant-grunts in front of HM, who continues to sit (right) and look away from FG with his hair erect.

    n. HM goes to lean on his elbow, maintains hair erect, while FG gives a full open grin at this movement and begins grooming HM.

    o. FG grooms HM, as Flint (FT) approaches and puts his face close to HM's face. Later FT sits and watches his brother FG groom alpha male HM.

The complexity of a postreunion sequence is difficult to appreciate from verbal descriptions alone. Visual interactions are an important part of the initiation and regulation of many sequences of social behavior in chimpanzees. In the article-opening photographs, Humphrey has just arrived in the provisioning area and completed several charging displays in Figan's presence. Here we see Humphrey walking at an oblique angle to Figan with hair erect and rigid posture, while both animals appeared to avoid looking at one another simultaneously (See b-f). Figan holds himself and does not self-groom until Humphrey has sat down and begun self-grooming. Pauses are often important in the structuring of such contexts. For example, Figan gets up to begin ap-

TABLE 2

Distribution of postreunion charging displays and social grooming in reunions of (1) social grooming with no prior display, (2) social grooming postdisplay not involving the displayer, (3) social grooming postdisplay involving the displayer, and (4) charging displays and no social grooming afterward

| | | | SOCIAL GROOMING | POST-DISPLAY GROOMING | DISPLAYER GROOMING | DISPLAY ONLY |
PERIOD	PLACE	NUMBER REUNIONS	(1)	(2)	(3)	(4)
1970-1971	Forest	37	51%	12%	16%	24%
1972	Forest	17	53%	6%	18%	23%
1970-1971	Provision area	30	67%	13%	17%	3%
1972	Provision area	49	51%	14%	28%	7%

proaching Humphrey, who has just broken his bout of self-grooming (See f-n). The slightest movement by one animal moving toward another at an oblique angle and sitting nearby in such a context can lead to an extreme response in the animal being approached. Such is the case of Figan's full-open-grin face at Humphrey's move to lean on his elbow (See n). These relationships were so complex and varied in context that they could not be substantiated quantitatively.

DISCUSSION

A number of the findings were of interest in relationship to previous reports. First, adult males doing charging displays were in relatively more frequent postreunion sequences with social grooming than those males in the same context but not displaying. Nondisplaying grooming pairs groomed for longer periods than displaying males (Table 2). This is comparable to the finding that higher-ranking males that do more charging displays had more frequent but briefer grooming encounters in the Gombe provisioning area (Simpson 1973a). This finding could be interpreted to support either the social dominance or attention structure viewpoints. Neither of these viewpoints would have predicted that new party members are preferred for grooming encounters and that most social grooming encounters after a reunion occur without prior threat (Table 1). These results are better understood from the viewpoint that long-term social relationships are renewed after reunions (Hinde, this volume).

A second finding of interest is that the provisioning area at Gombe is like a *waiting area* where social grooming occurs as an alternative activity to feeding in the absence of natural foods (Wrangham 1975). This is suggested by the finding that charging displayers in the provisioning area are more often groomed than are forest displayers, and that relatively more postreunion sequences occur with social grooming in the provisioning area. The differential response elicited under forest and camp conditions to charging displays followed by social grooming (Table 2) qualify the positive correlations between social rank and social grooming as possibly exaggerated, since these observations were made in the Gombe provisioning area (Simpson 1973a). The relatively lower social grooming response to charging displays in the forest condition also suggests that social dominance is overemphasized or distorted from its naturalistic importance as has been suggested (Rowell 1974). Similar qualification would apply to the theory of attention structure, which may be considered to be a reinterpretation of social dominance theory.

In Gombe chimpanzees, the labile, transient, and possibly deceptive nature of communicative behavior is seen. The lability of behavior was apparent when one motivational set of behavior elements was immediately transformed into a reciprocal set. For example, strong submissive responses were sometimes immediately followed by aggressive ones in adult interactions. The transient nature of communicative behavior was apparent, for example, when a resting chimpanzee briefly gave loud, submissive vocalizations and returned to a prone resting position. The deceptive nature of some behavior patterns was clear in cases of mothers distracting infants from desired objects and adults giving *misinformation* to manipulate rivals. These three qualities suggest that chimpanzee communication under field conditions involves cognition and intention rather than just simple emotion.

A third observation of interest is that there seems to be a contextual basis to chimpanzee communication that is part of a long-term social relationship. This contextual basis may be called *communicative structure*. In any one context, the temporal structure of soft vocalizations, gestures, postures, and visual orientations of the actors appears important in regulating the course of charging display and social grooming sequences observed, as for example, article-opening photographs b-n. Single response contingencies out of the social context or social relationship of the individuals involved are not helpful in estimating the outcome of an interaction. The social effect of being in the provisioning area, compared to the forest, is suggested by the differences in the consequences of charging displays (See article-opening photograph n). Such contextual relationships in the communication of chimpanzees must be studied experimentally before the apparently paradoxical results of chimpanzee communication about the environment without stereotyped patterns can be understood (Menzel 1974). Descriptive techniques must be developed to "place" the significance of a behavior that may not be "obviously social" in its original social context (Hallett 1969). Quantitative techniques are being developed that can help with this problem (Reynolds 1972). Such developments will eventually lead to a more comparative understanding of chimpanzee social communication.

Acknowledgment

I would like to thank observers mentioned in Study 2 for their help in collecting G.S.R.C. records, Dr. Jane Goodall, G.S.R.C. Director, and Dr. David A. Hamburg for making this study possible, the Wenner-Gren Foundation for original support, and the Grant Foundation for support since 1971.

Females displaying aggressively (left); two others keep low profile (right).

J. David Bygott

Agonistic Behavior, Dominance, and Social Structure in Wild Chimpanzees of the Gombe National Park

The need to understand the roots of human violence has stimulated much research into the cause of aggression in nonhuman primates. Laboratory studies have demonstrated amply that many proximal environmental factors increase the tendency to attack; e.g., overcrowding, aversive stimuli, presence of strangers, or competing for rewards. There is also evidence that ontogenetic factors such as early learning and early social deprivation may influence aggressiveness. The profound disturbances caused by rearing in isolation prevent any direct testing of the innateness of a primate's behavior; consequently, the genetic basis of aggression has been the subject of much controversy.

Although field studies are inevitably too incomplete and uncontrolled to elucidate the cause of aggression, they can provide data about its natural frequency and contexts. Such information may permit us to speculate about the possible adaptiveness of such behavior, i.e., whether its consequences are such that an aggressive animal may survive longer or produce more offspring than a less aggressive one.

Significantly, the chimpanzee received little mention in discussions of the origins of aggression during the 1960s (e.g., Lorenz 1966; Russell & Russell 1968; Ardrey 1966). Although detailed field studies were beginning, they were hampered by the difficulty of observing and following the shy apes in their forest habitat. The early reports from these studies suggested that adult chimpanzees formed small temporary associations with no clearly defined social units beyond mother-offspring families and that aggression was extremely rare (Goodall 1965; Reynolds & Reynolds 1965).

Prolonged field studies in Tanzania (in the Gombe National Park and the Mahale Mountains) have yielded more detailed data on the social structure and behavior of wild chimpanzees (Goodall 1968a; 1968b; Nishida, this volume). In both these study areas, chimpanzees were initially habituated to the presence of observers by means of artificial provisioning; although this may have distorted their natural behavior and social structure, the groups in both areas show close similarity and are unlikely to differ greatly from pristine populations in these respects.

This paper is based on a study undertaken at Gombe in 1970 and 1971 (Bygott 1974). It summarizes the form and contexts of the most common types of agonistic behavior patterns, shows how social relationships between individuals (particularly adult males) can be described in terms of such behavior, and discusses the functional value of such relationships.

METHODS AND STUDY POPULATION

Methods have been described more fully elsewhere (Bygott 1974). The data come from two sources:

1. A daily record of the attendance and interactions of all chimpanzees who visited the camp (provisioning) area from mid-1966 through 1971. About 20 different observers participated; prior to 1970, data were recorded as a typed narrative, and subsequently standard checksheets were used.

2. About 800 hours of observation of adult male target animals, in which the target was followed wherever he went and his associations and interactions recorded in detail. Adult males were selected as target animals because they were the class most frequently involved in agonistic interactions.

Definition of the study population requires some explanation of chimpanzee social structure, a subject that has been reviewed by Sugiyama (1973) and elaborated by Wrangham (this volume). Wrangham's model of chimpanzee dispersion proposes that the ranges of females are considerably smaller than those of males, and that the females' core areas are dispersed more or less evenly throughout the available habitat, without discontinuity. Adult males form stable, mutually exclusive *communities;* the members of such a community associate freely only with one another and with as many females as their large, shared range encompasses. Females may commonly transfer from the range of one male community to another, but males do not.

By this criterion, we can regard the study population as a single community of 14 adult males (in 1971), within whose range live at least 14 adult females and 24 adolescent, juvenile, and infant males and females, all more or less habituated to observers, plus a few definitely unhabituated females. Completely unhabituated communities occupy areas to the north and south. Within the habituated community, a split was apparent during 1971, and by the end of 1972 the northern and southern subcommunities had formed the effectively separate communities described by Wrangham.

RESULTS

Patterns of Agonistic Behavior

Aspects of the behavioral repertoire of chimpanzees have been described and classified by Goodall (1968b), Nishida (1970), and van Hooff (1971). For the purposes of this study it seemed desirable to record not only attacks and signals which indicated the imminence of attack but also the responses evoked by aggression. The term *agonistic behavior* (Scott & Fredericson 1951) covers these forms of behavior, and also was used by Nishida (1970); it corresponds with van Hooff's categories of *aggression* and *submission* and with Goodall's categories of *aggression, flight and avoidance, submission, reassurance,* and also some elements of *frustration.*

The choice of behavior units recorded in this study was guided by two considerations: first, that they should be easy to detect in natural surroundings, where visibility was often poor; and second, that they could be recorded reliably by the various observers who contributed to the long-term data pool. Such behaviors had to be either common or conspicuous, and preferably both; they are outlined as follows.

Attack consisted of any form of sudden, deliberate contact with another individual in a manner potentially capable of causing injury, e.g., hitting, biting, stamping on, or rolling the victim. The attacker normally had *hair erect* and was silent, but occasionally *screamed.* The victim of an attack almost invariably screamed with fully bared teeth and made some attempt to break free or (more rarely) to fight back. The great majority of attacks were mild—merely a few blows on the back.

407

Attackers (especially males) often preceded an attack with a *directed charge*—rushing with hair erect toward the victim, who usually attempted to *avoid*. Such a charge was frequently accompanied by display elements, i.e., exaggerated motions such as slapping and stamping on the ground, throwing loose objects, swaying or dragging branches, flailing handfuls of foliage, and drumming on resonant objects, all of which appeared irrelevant to normal locomotion. Often many of these display elements were performed in a sequence that, although usually shorter than 1 minute, sometimes continued for more than 10 minutes. Which display elements were performed depended in part on the surroundings and in part on the individual; thus one male typically slap-stamped quadrupedally, another ran bipedally and sometimes beat his chest, a third slowly stamped around throwing many sticks and stones into the air at random. As all these patterns occurred in similar social contexts and evoked the same responses, it seemed convenient to describe as a *charging display* any behavior sequence involving exaggerated locomotion with hair erect. Displays were classed as *vocal* and *nonvocal* according to whether or not they were accompanied by a *pant-hoot* call. Although both were equally common, 14% of nonvocal displays but less than 1% of vocal displays were accompanied by an attack; the two kinds of displays were otherwise similar in form. Vocal displays included a higher proportion of audible elements, such as drumming, slapping, and stamping, than nonvocal displays, which incorporated more visual effects such as bipedalism, branch-swaying, and throwing; these facts suggest that vocal displays may be primarily concerned with long-distance communication between parties, whereas nonvocal displays are for short-range communication of aggressive intent to members of the same party.

Common responses to directed charges and other nonvocal displays included screaming, avoidance (ranging from stepping aside to fleeing at full speed), and pant-grunting. The latter consisted of a rapid rhythmic series of grunts or barks directed toward the other individual, sometimes accompanied by crouching or by a vertical bobbing of the head or whole body.

Two fairly uncommon behaviors that seemed aggressive, because they sometimes elicited avoidance, squeaking, and screaming, were *head-tipping* (an upward jerk of the head, sometimes accompanied by a soft bark) and *arm-waving* toward another individual. Neither was usually accompanied by hair erection, nor appeared to lead to attack.

The characteristic shared by the more common agonistic behaviors among the chimpanzees was that none necessarily involved body contact. In contrast, the next most frequent gestures can be regarded as *contact-seeking* (e.g., *presenting* and *holding a hand toward* another individual). We can say that these gestures and the less frequent *excitement contact* behaviors *(touching, kissing, embracing, mounting)* are only partially agonistic because they normally seem to terminate the agonistic interaction sequence in which they occur. The contact-seeking gestures were seldom directed at an individual who was in the process of attacking or displaying because to approach him would be to risk being hit. They sometimes occurred between individuals who had just met, often when one individual had hair erect and looked as though he might soon display. They also occurred between *spectators* at a display or attack and sometimes between the victim of an attack and the attacker himself or a third individual. In none of these situations was a touching, embracing, or mounting interaction normally followed by a display or attack; the excitement-contact behaviors were usually followed by the maintenance of peaceful proximity between the participants, e.g., grooming, feeding, or just sitting together.

A data study showed the relative frequencies with which all these behaviors were observed (in a random sample of interactions among all age-sex classes) and it demonstrated striking sex differences, particularly in the frequency of aggressive behaviors.

Although this study was concerned only with intraspecific behavior, almost all the behaviors were also directed at animals of other species—particularly baboons *(Papio anubis),* with which chimpanzees had a broad spectrum of social interactions ranging from play to predation. The interesting exception was *pant-grunting,* which appeared to be completely species-specific and, as we shall see, individual-specific as well.

Contexts of Agonistic Behavior

A preliminary survey indicated that 90% of all agonistic interactions involved at least one adult male. Consequently, data on the natural contexts of agonistic interactions were collected by recording the interactions of adult male target individuals away from camp.

Attacks. The most frequent context of attack was at a *reunion*—a term used by Bauer (1974) to mean a meeting between two individuals who had been separated for at least 30 minutes. In this present sample, 32 out of 83 attacks by adult males (39%) occurred within 5 minutes after a reunion. It must be emphasized, however, that in only 1% of reunions did attacks occur.

Twenty-five percent of the attacks accompanied excitement during the consumption of certain popular foods, mainly meat. Often, these attacks did not seem to be aimed at taking food by force—indeed, in 12 out of 21 cases the victim had no food, and in only one case did the attacker obtain food as a consequence of his aggression. Teleki (1973), in a detailed study of chimpanzee predation, also commented on the futility of attack as a means of obtaining meat.

Attacks tended to occur in clusters; the same attacker might attack the same or different victims in quick succession, or different attackers might simultaneously or successively attack the same or different victims. Goodall (1968b) also mentioned this, and suggested that the screams of an individual who had just been attacked might stimulate another to attack him. Altogether, about 30% of attacks occurred within 5 minutes of a previous attack. The most extensive attack series was seen during a predatory episode, when two old males with large portions of meat were attacked eight times in 2 minutes by other males who had no meat. (They made no attempt to fight back, and their attackers made no attempt to take their meat.) Simultaneous attacks by two or more males on the same victim were only seen in the context of reunions; in the most severe case witnessed in this study, five adult males attacked an adult female (perhaps unfamiliar to them), capturing and killing her small infant (Bygott 1972).[1]

These three contexts covered a total of 76% of the observed attacks, and the remainder occurred in a variety of less common situations; some were directed at play partners, others at individuals who interfered with copulations, others at apparently innocent bystanders during a charging display.

Wrangham (1974) noted that the frequency of attacks per individual per hour increased with group size, on days when chimpanzees were provisioned in camp. My data confirm that this was also true of groups observed away from camp; the overall median group size was 3, and the median size of groups in which attacks were seen was 9.

Displays. Bauer (1975) found that 42% of displays occurred within 15 minutes after a reunion. He did not distinguish between vocal and nonvocal displays; data from this study indicate that 64% of nonvocal displays and only 26% of vocal displays occurred during the first 15 minutes after a reunion.

Sometimes displays appeared to be a response to the distant calls of other chimpanzees; here, however, 34% of vocal displays and only 16% of nonvocal displays began within 5 minutes after distant calls had been heard. This data further confirms the suggestion that vocal displays are mainly a nonaggressive form of long-range communication. Often, a lone male or a party of males would perform several vocal displays in succession, pausing to listen carefully after each display, apparently trying to evoke a response from other chimpanzees within hearing range.

As one context of attack, displays often occurred in the feeding situations described earlier. Typically, they were nonvocal displays, performed by individuals prevented from access to a limited food source such as meat or bananas; perhaps it is for this reason that Goodall (1968b) described displays as *frustration* behavior. In the camp area, such displays were sometimes instrumental in obtaining bananas; for example, if an adult male reached camp to find several females eating bananas in low trees, he might display from tree to tree, shaking each tree until its occupants dropped some bananas, and then gather up the bananas. This behavior was not observed in natural surroundings.

Curiously, prolonged displays were sometimes performed in heavy rain or near running water. Goodall (1968b; 1971) has described *rain displays* and several were observed during this study. When rain began, chimpanzees usually sat huddled and waited; but if rain became very heavy, an adult male might slowly work up to a display, rocking and branch-swaying rhythmically and continuing at a slow tempo for up to 17 minutes. These displays were mainly nonvocal, with sporadic hoots. Heavy rain usually lasted only for a few minutes, and when it eased off, displaying would stop. Not all males in a party necessarily displayed, but if one started others often joined in. The motivation of these displays remains a mystery; one is tempted to suspect that some kind of *superstition* is involved, e.g., if rain showers often stopped after a bout of displaying (which they do), this might reinforce rain-displaying in the future.

Males were also seen to display as they approached a stream, throwing or rolling rocks in the streambed at the climax of the display, and on several occasions males (sometimes completely alone) performed prolonged displays in front of a waterfall, swinging from side to side in vines growing across a cliff.

Some displays were directed at other members of a party who were creating a disturbance; a few fights were broken up in this manner. In many situations, when one male started to display, it seemed to stimulate others to display as well; males who frequently associated together often displayed in parallel when they joined another party. Some spectacular instances of this were seen during this study. During 1971 the northern and southern males mainly stayed in separate core areas, but at intervals of a few days or weeks, a party from one subcommunity made an expedition into the other's area and for a while associated with them fairly peacefully. These reunions were often preceded by one party stealthily stalking the other through dense cover and then suddenly displaying all together into their midst. The initial response of the party taken by surprise was usually to flee.

Finally, the composition of the group seemed to be an important context of displays; the presence of a specific individual might increase or decrease a male's display frequency, as Simpson (1973a) has already shown. In 1970 Mike and Humphrey performed nonvocal displays with about the same frequency, but Humphrey virtually never displayed if Mike was present. On one occasion when he did, Mike instantly chased him and attacked him.

Screaming, Avoidance, Pant-Grunting, and Presenting. As explained earlier, these behaviors were commonly performed as responses to some aggressive act by another individual. We can easily illustrate the response by keeping the *stimulus* animal constant. Data were recorded in 1971 during 105 hours of observation of Humphrey, the aggressive adult male; his agonistic interactions with other individuals were recorded in detail. The different kinds of response were related to the imminence of attack: thus by doing nothing, Humphrey elicited mainly pant-grunting and presenting, and by directed displays, he elicited mainly screaming and avoidance. Despite their wide variation in absolute frequency, a very similar proportion of each of these behaviors (40-50%) was performed by individuals within 5 minutes after a reunion with Humphrey—about the same as the proportion of Humphrey's displays and attacks occurring in the reunion context.

Head-Tipping and Arm-Waving. These behaviors were usually directed by a stationary individual (resting, feeding, or a female nursing an infant) at an approaching one. The latter typically stopped approaching, or retreated, screamed, or presented. The gestures could thus be interpreted as a mild form of defense of personal space. They were almost never followed by attack.

Arm-waving was sometimes directed at an aggressor by an individual who had just been chased or attacked; at the same time, the victim might give scream or *waa* calls.

Excitement-Contact Behaviors. Touching, kissing, embracing, and mounting all occurred in similar contexts. These have been described by Goodall (1968b). Most often they occurred at reunions or during excitement over food, but sometimes they were elicited by an aggressive interaction involving one, both, or neither participant, or by distant calls.

Intercommunity Interactions. Mainly, everyday agonistic interactions were observed between members of the same community. Members of different communities seldom came close enough together for face-to-face interactions of this kind; they more often avoided one another at a distance. For example, on one occasion a large party of over 20 habituated chimps was feeding in a valley when they heard a distant chorus of pant-hoots further up the valley. Immediately, they burst into a deafening chorus of waa-calls and hoots, and two males displayed for several hundred meters in the direction of the distant calls before turning back to rejoin their party. A few minutes later, about seven unknown adult males were seen traveling rapidly out of the valley. In another observation, the southern males traveled toward the southern limit of their range, and paused on a ridge, listening. Hearing calls from a large party ahead, they crept quietly back into the valley from which they had just emerged; when safely out of hearing range, all displayed noisily down the slope and then traveled back to the northern part of their range.

The *gang attacks* mentioned earlier were all directed at isolated individuals discovered by *patrolling* parties of adult males, and would probably serve to reinforce intercommunity avoidance. Nishida (this volume) describes very similar forms of intercommunity interactions in the Mahale mountains.

Agonistic Relationships

By recording the frequency, direction, and types of agonistic behavior exchanged by two individuals in their interactions over a period of time, we were able to form some idea of their *agonistic relationship*. This recording of data collected in camp was done for the study population throughout 1970. The results were described and discussed in more detail elsewhere (Bygott 1974). For simplicity, only the more frequent behavior patterns were considered—pant-grunting, nonvocal displays, avoidance, presenting, and screaming; even though it occurred infrequently, attack was also included because it was central to the definition of agonistic behavior.

A separate matrix was compiled for each behavior, and it was found that each tended to be directional; i.e., in the interactions of two chimps, if a behavior occurred at all, it was largely or exclusively performed by one partner. In the case of displays, it was usually impossible to decide which chimp (if any) was the target of the display; but if we considered all the occasions when each chimp displayed in the presence of the other, there was usually a tendency for one partner to display more when both were together.

The most strongly directional behavior was pant-grunting, which was also the most common. In 354 out of 357 different pairs of individuals (99%), only one member of each pair was seen to pant-grunt at the other during the year. Attack was the next most directional behavior (98%) and then avoidance (96%). As a general rule, where chimp A consistently attacked chimp B, chimp B consistently avoided, pant-grunted, screamed, or presented to chimp A, but A seldom or never directed any of those behaviors at B. It is convenient to follow the tradition of describing this as a *dominance-subordinance relationship* (e.g., Scott 1956); in the following account, the term *dominance* will only be used in this strict sense.

Some pairs of individuals were seen to have very few agonistic interactions, or none at all, and it seemed less valid to describe such relationships in terms of dominance because of insufficient data, even though in reality such individuals may have had occasional but unobserved interactions in which one partner consistently dominated the other.

Female and immature chimps had more frequent agonistic interactions with adult males than with one another, and were almost invariably subordinate to males. Some exceptions should be noted; the youngest adult males received very little pant-grunting and avoidance from the older females, and were occasionally chased or attacked if they attempted to display at such females. Two adult males (Figan and Faben) were not seen to have any agonistic interactions with the old female Flo, who was their mother; although they were both larger and stronger and more aggressive than she was, their relationship with her seemed to be one of mutual respect. They were clearly dominant to all other adult females, and Flo was clearly subordinate to all other adult males.

Very few agonistic interactions were seen between adult males and infants because mothers were especially protective of their infants in situations involving even a slight possibility of aggression. It was not unusual for adult males to attack females with infants, and in such cases the infant was passively involved in the fight, although the mother generally protected it by crouching over it. Juveniles, being too large to be protected in this manner, had more frequent interactions with adult males and were attacked no less often than most adults.

The relationships between immature males and adult females could not be described satisfactorily in terms of dominance and subordinance, inasmuch as they were in a state of change. Adult females were dominant to juveniles in the sense of winning fights, but did not attack juveniles unless provoked. (If a juvenile initiated a quarrel with an adult female, however, its mother sometimes intervened in support of the juvenile, in which case the latter effectively won.) Juvenile males were never seen to pant-grunt to females of any age. Adolescent males were closer in size to adult females and could generally intimidate young adult and adolescent females, but could still be attacked by older adult females. Although young adult females sometimes pant-grunted to adolescent males, they pant-grunted more to adult males.

Among immature males, fights often occurred in the context of play, and were usually won by the oldest (and, therefore, the largest) participant. Although immature males pant-grunted to adult males, they were not seen to pant-grunt to one another.

Agonistic interactions among females were rare. Where clear dominance-subordinance relationships did exist, the older of two females was usually dominant. In 1970, the female involved in most agonistic interactions with other females was a young adult who had just given birth to her first infant and was particularly cautious in her social interactions with any individuals. She pant-grunted frequently to all older females. For the most part, the direction of dominance among females could only be inferred from the direction of pant-grunting, inasmuch as they so rarely attacked or displayed at one another. Analysis of attack data recorded in camp over a 4-year period showed that the proportion of two-sided fights (in which there was no clear winner or loser) was much higher in fights between females (23%) than in male-female fights (2%) or male-male fights (3%); during that period, female-female fights constituted only 4.5% of all fights seen. These data suggest that dominance-subordinance relationships among females are either not as clearly defined or not as important as those between males and females.

Agonistic Relationships among Adult Males

The pattern of agonistic relationships among the 15 males can best be described by dividing them into four groups according to the number of individuals to whom they were dominant or subordinate. Within each group, agonistic interactions were seldom seen and it was difficult to rank individuals, but between members of different groups, interactions were more frequent and dominance-subordinance relationships clearly defined. The groups were as follows:

1. *Alpha Male.* In 1970, one elderly male (Mike) was clearly dominant to all others. Although he was not seen to attack all the males, all pant-grunted to him and he pant-grunted to none.

2. *High-Ranking Males.* Three middle-aged males (Hugh, Charlie, and Humphrey) were subordinate only to Mike and dominant to all other older and younger males. Hugh and Charlie were thought to be siblings, on the basis of physical similarity and a long history of close association. In 1970 it was impossible to say which of the three was dominant.

3. *Middle-Ranking Males.* This group consisted of two old males (Hugo, Leakey), a middle-aged male with a paralyzed arm (Faben), and three young males (Evered, De, and Figan). Figan, by the end of 1970, appeared to be dominant to all the other middle-ranking males; although he was the youngest of them, he might thus be regarded as transitional between middle and high rank. Figan and Faben were known siblings, and on one occasion during the year they combined to attack Evered, who had previously been dominant to both and was subsequently subordinate (at least to Figan).

4. *Low-Ranking Males.* These included one old male about the same age as Mike (Goliath), one middle-aged male with a paralyzed leg (Willy Wally), and the three youngest males (Godi, Jomeo, and Satan). These appeared to be subordinate to the middle-ranking males, except that during the year Godi began to display at, and attack, Leakey. They were all dominant to adolescent males.

This distribution of age and rank suggests that a male's ability to dominate others has a U-shaped relationship with age. As he grows through adolescence, he becomes dominant to adolescent and then adult females, as a young adult he begins to dominate the oldest males, and by middle age, he reaches his highest rank and can dominate most—if not all—younger and older males. When he enters old age and his physique deteriorates, younger individuals may begin to dominate him for the first time.

Analysis of agonistic interactions among these males over the period 1966-1971 supported this hypothesis, but suggested that there was considerable variation in the age at which certain ranks were attained. These differences might be due to a variety of factors, such as health, coalitions, kinship, aggressiveness, and other personality traits.

Physical fitness seemed important; the two partially paralyzed males were lower ranking than their age would predict, and were clearly at a disadvantage in fights with intact males of the same age. Goliath, one of the lowest ranking and most senile-looking males, had been alpha male until 1964, and remained high ranking for some years after being defeated by Mike. In 1968, following a period of illness, his rank declined swiftly.

Coalitions have been described in other primate species (Kummer 1968; Hall & DeVore 1965; Bernstein 1969) and seem to be important in chimpanzee societies. Goodall (1968b; 1971) describes persistent alliances between the males David and Goliath (who was then alpha male), J. B. and Mike (who became alpha male), McGregor and Humphrey (who later became alpha male) and the siblings Faben and Figan (Figan is currently alpha male). In each instance, it was the younger member of the partnership who seemed to benefit most from the association. During this study, the coalition between Hugh and Charlie, the high-ranking southern males, was important in their relationship with Humphrey, the high-ranking northern male. Although Humphrey was larger and more aggressive than either of them, he behaved in a very subdued manner when both were together and avoided their parallel displays. Alone, they were more vulnerable; Figan was seen to attack Hugh in 1971 when Charlie was absent, but later in the year Figan himself was severely beaten up by Hugh and Charlie in combination. A similar alliance may ultimately develop between Sherry (at the time of writing, a young adult) and his elder brother Jomeo. A similar kind of relationship between the males KA and KS is mentioned by Nishida (this volume).

The phenomenon of *dependent rank* (Kawai 1958) has been described in some primate species; the sons of high-ranking mothers become high ranking themselves because their mothers supported them in agonistic interactions with peers when young. There is certainly evidence that chimpanzee mothers support their offspring in fights, but none that this influenced the rank of any adult male; the agonistic relationships between the mothers of the adult males in this study were largely unknown.

Aggressiveness—as measured by the frequency of attack—did not seem to be obviously correlated with dominance rank, nor with age, although age to some extent determined the choice of possible victims. In 1970, middle-aged Humphrey and old Hugo attacked more often than any other males, but Hugo mainly attacked females, whereas Humphrey's attacks were more equally divided among males and females. Mike, who was then alpha male, attacked about half as often as Hugo or Humphrey did. Within any age group, there was some evidence that the large heavy males attacked more often than the smaller males, who displayed more than attacked. The correlation between the dominance rank of adult males and their display-attack ratio was quite significant (rS = + 0.64, p < 0.01) which would not be expected if this ratio was more or less constant across individuals. The ratio was highest for Mike (8.6), Figan (6.8), and Godi (6.2), who were the most dominant high-, middle-, and low-ranking males, respectively. Goodall (1968b; 1971), describes the way in which Mike's frequent and novel displays led to his attainment of alpha rank in 1964. During this study, Figan and Godi both displayed repeatedly at older males, thereby successfully harassing individuals who could probably have defeated them in a fight. Displaying, because it avoids physical involvement with other males, is probably an optimal strategy for a male to use to test the responses of others and ultimately to dominate them.

An exception to this rule was Humphrey, whose attacks were very frequent (display-attack ratio = 3.2) despite his high rank. This behavior may have occurred because his large size enabled him to attack others with impunity, and attacks may have involved less effort for him than displays. Although he did perform strenuous and impressive displays, they often left him panting exhaustedly. It may be significant that he only remained alpha male of the northern community for 20 months before being defeated by Figan.

Personality is difficult to discuss in quantitative terms. Some males, however, behaved in a way that implied they were not only interested in maintaining and increasing their dominance rank but also were using opportune moments skillfully for display or to avoid harmful encounters. Mike's opportunism in incorporating kerosene cans into his displays, at a time when other males were wary of human beings and their artifacts, has been described by Goodall (1971). When Mike was finally defeated by Humphrey in 1971, he seemed to abandon any attempt at retaining high rank, and quickly became a mild and subordinate male, using his intelligence to avoid situations in which he was likely to be attacked (such as reunions with Humphrey). Figan, a high-strung young male, showed a similar degree of social awareness; his *tactics* were to use display to harass persistently any senior male who showed signs of ill health or was unaccompanied by high-ranking allies and to sit inertly through the displays of high-ranking males without pant-grunting. In contrast, Jomeo seemed largely uninterested in dominating others. Slightly younger than Figan, he was the largest member of the entire community, but was easily intimidated by the displays of Faben and of Willy Wally (who was three-quarters his size and partly crippled) and even by some adult females. A male's confidence in social situations may in part stem from his upbringing; for example, Figan's mother Flo was known to be relatively gregarious and often associated with other males and females, whereas Jomeo's mother was apparently much less sociable. An infant who had frequent opportunity to play with, or even to watch other individuals, would presumably develop social skills more readily than one who had had a more sheltered infancy.

Relevance of Dominance to Other Activities

Feeding. The feeding behavior and diet of Gombe chimpanzees is discussed in detail by Wrangham (1975). Most food sources are dispersed in such a way that competition between individuals for food items would be unnecessary and inefficient. Some sources are localized, however, and bring individuals into close proximity (e.g., palm-nut clusters, *Dorylus* ant nests, termite mounds) and some are portable as well (mammalian prey); in the latter case it was particularly difficult for all members to obtain an equal share. As we have seen, meat-eating accounted for most agonistic interactions that were observed over food. In such a context, we might expect a dominant individual to gain prior access to the food, as so many captive studies have demonstrated. Teleki (1973) found, however, that although dominant individuals were given consistent priority of access to meat, that action "does not conform rigidly to a predictable social pattern, nor is access determined by overt competition." Wrangham (1975) has shown that the eight northern males could be ranked linearly on the relative amount of meat that each obtained in predatory episodes during a period of 3¼ years. This ranking was significantly correlated with age *(rS* = + 0.833, *p* < 0.01); however, it could not be correlated with dominance rank because the direction of dominance changed in 12 out of 28 relationships during this period. The oldest male, Hugo (who had the highest success rate), was middle ranking throughout the period, and Humphrey (who ranked fifth on meat-success) was high ranking throughout in terms of dominance. When females were included in the ranking, it was noteworthy that some adult females were more successful in obtaining meat than were some of the young adult males.

Wrangham's data also indicate that the priority ranking observed during meat-eating was adhered to on the few occasions when competitive interactions over other types of naturally occurring food were seen.

Thus overt competition for food seems too rare to justify dominance-subordinance relationships; also it must be remembered that one source of food (meat) can only be exploited, for the most part, through cooperative hunting.

Sex. The sexual behavior of Gombe chimpanzees has been described in detail by McGinnis (1973; and this volume; Tutin, in preparation). Estrous females may either copulate more or less promiscuously in multi-male parties, or associate with a single male for a consortship lasting for several days or weeks.

McGinnis found that the frequency of consortships initiated by adult males was not significantly correlated with their dominance rank. He did not examine relative copulation frequencies of adult males in promiscuous groups; my analysis of all copulations recorded in camp during 1970 showed no significant correlation between copulation frequencies and dominance, although the lowest-ranking males tended to have low copulation frequencies.

Both Tutin and myself, nevertheless, observed occasional instances of the highest-ranking male in a party monopolizing an estrous female, by displaying toward, or merely approaching with hair erect, any other male who tried to copulate. Such action requires constant vigilance, especially in a large party, and a less stressful means of ensuring exclusive access to a female—an option also open to middle- and low-ranking males—is to consort with her away from all other males.

Behavioral data alone cannot answer the question of whether dominance is correlated with reproductive success, since there is no way of proving paternity among wild chimpanzees, nor of assessing the relative viability of different males' sperm, so for the time being this must remain an open question.

Allogrooming. Goodall (1968b) showed that adult males were the class that spent the most time grooming others and being groomed. Simpson (1973) studied in more detail the grooming relationships between 11 of the 15 adult males at Gombe during 1969 and 1970. He ranked the males on five different measures of grooming and three agonistic measures. From the results of his correlations he concluded that "the males most frequently involved in grooming usually had the highest (agonistic) statuses"; "the higher an individual's status, the less long he usually groomed his fellows"; and "whatever his status, a male tended to groom others for longer, the higher their statuses."

Simpson did not consider the possible relationship between age and the various grooming measures; moreover, Simpson had to omit one individual (Goliath) from all his calculations in order to obtain any significant rank correlations between *agonistic* and grooming measures. When correlations between his grooming rankings and a conservative age-ranking were performed (Bygott 1974) the results showed that "age" could validly be substituted for "status" in the quotations just given on the basis of equally (or more) significant correlations between age and grooming ranks. In my study there was no significant correlation between age and any of the agonistic status measures; therefore, the inclusion of Goliath in correlations between *age* and grooming measures did not affect the significance of the results.

Data from my 1974 study, although not of the same quantity as those collected by Simpson, supported the finding that old males are groomed at least as much as dominant males, so the possibility of being groomed more would not seem to be a strong incentive to increase dominance rank.

DISCUSSION

Forms of Agonistic Behavior

Comparison of chimpanzee agonistic patterns with those described for gorillas (Schaller 1963) and orangutans (MacKinnon 1974) shows that chimpanzees have the richest repertoire of the three species, but the aggressive signals of each species are somewhat similar.

Hair erection is a sign of aggressive arousal in many primate species and is particularly noticeable in chimpanzees. It has not been observed in gorillas, but may accompany the apparently aggressive *long call* and branch-shaking displays of orangutans.

Branch-shaking displays are widespread among the primates, and are performed (with variations) by all the great apes. Orangutans, being mainly arboreal, typically display in trees, swaying rhythmically and sometimes dropping or throwing broken branches. Displays may be accompanied by the long call. Chimpanzees, although semi-arboreal, usually display on the ground, but when bushes or low branches are available these will be swayed, broken, or dragged. A common element in chimpanzee displays is some form of percussion, such as stamping on the ground or drumming on trees. Loose objects may be brandished or thrown. On the ground, a display generally involves a forward charge, but in the trees it may closely resemble the orangutan pattern. Gorillas, like chimpanzees, may begin a display with a hoot or hoot series; the displayer may place a leaf in his mouth, then rise bipedally, sometimes throwing vegetation in the air, beat his chest, slap or drag plants as he charges, and finally slap the ground. All these components except the chest-beating and symbolic feeding resemble chimpanzee display elements. Chest-beating has only been observed in two adult male chimpanzees at Gombe (Goodall 1968b) but is commonly performed by gorillas of any age and sex.

Threats include head-jerking, arm-waving, and staring. A jerking of the head, together with a sudden exhalation, is a common mild threat in many primates (Andrew 1963); gorillas have a pattern very similar to the head-tip of the chimpanzee. All three apes perform arm-waving threats. Goodall (1968a) lists a fixed stare as a threat in chimpanzees, and Schaller describes it in gorillas. Staring also accompanied some *submissive* patterns of chimpanzees, however, such as pant-grunting, and usually only *elicited* submission if combined with hair erection. Gorillas turn the face away or shake the head to indicate submission or nonaggression; adult male chimpanzees, when approached by a pant-grunting subordinate, were also seen to turn their face away, with the effect that pant-grunting ceased. On the other hand, prolonged eye-to-eye contact often occurred during mutual grooming and food-begging interactions, with no indication that this was perceived as a threat. Staring, therefore, may be more important as a threat in gorillas than in chimpanzees.

MacKinnon and Schaller describe no submissive gestures and calls other than screaming with bared teeth (widespread among primates), avoidance, and crouching (in gorillas). The frequent pant-grunting and presenting of chimpanzees seem to have no parallel among other apes.

Proximity among orangutans is unusual, but MacKinnon reports a few instances of tactile gestures such as embracing and kissing, noting that these were much rarer than in chimpanzees. Schaller did not observe embracing, kissing, or mounting between gorillas, and Fossey (personal communication) confirms the virtual absence of such behavior.

Chimpanzees thus differ from the other apes in their frequent use of submissive and contact gestures. They also spend relatively more time grooming one another (especially adult males). Schaller and MacKinnon both comment that gorillas and orangutans (respectively) are less excitable than chimpanzees. The frequency and variety of contact and submissive behavior unique to the chimpanzee may, therefore, be related to its increased need for de-arousal (Mason 1967).

Comparing the chimpanzee's agonistic signals with one another, we find a good example of Darwin's principle of antithesis (Darwin 1872). A chimpanzee *likely* to attack is silent, and increases his apparent size by hair erection, hunching of the shoulders, approach, bipedalism, and other conspicuous display patterns. A chimpanzee *unlikely* to attack vocalizes loudly (pant-grunting or screaming), keeps hair flat, tends to crouch, and may orient away from the other individual (by presenting or avoiding), thus keeping his apparent size to a minimum. This behavior implies that it is important for individuals to indicate unambiguously their aggressive or nonaggressive intentions before they ever come within contact range of each other.

Frequency of Aggression

The chimpanzee has been portrayed in the primate literature as a relatively unaggressive species, and so it is interesting to compare the frequency of attack by individual adult males in this study with data

421

from studies of other primate species. This study showed that chimpanzees fall within a range of variation shown by baboons and macaques, even allowing for the fact that values obtained in this study may be slightly exaggerated.

It may not be valid to speak of baboons and macaques as *more aggressive* than chimpanzees, however, members of large cohesive groups will inevitably *see* more fights going on around them than will members of small temporary parties, even if individual attack frequencies are similar in both cases. Individuals in the former situation may indeed suffer more psychological stress as a result of intragroup aggression than in the latter case.

Relevance of Social Structure to Agonistic Behavior and Dominance

By considering the unique social structure of chimpanzees, we may begin to understand why it is important for adult male chimpanzees to be aggressive and to form clear dominance-subordinance relationships, and why so many agonistic interactions accompany reunions between familiar individuals.

Wrangham (1975; this volume) has discussed the adaptive significance of chimpanzee social structure from an ecological viewpoint. Chimpanzees live in a habitat (typically deciduous woodland) where food resources are widely scattered and seasonally abundant, necessitating that all individuals should have large ranges and low locomotion costs, a factor that would limit body size. The relatively large size of chimpanzees is puzzling, but their morphology suggests that they originally evolved in a habitat where size was less limiting (e.g., rain forest) and later expanded into drier, more open habitats, or were forced to do so by the Miocene reduction of the African rain forest (Kortlandt 1972; 1974).

In any case, large size imposes a relatively long period of pregnancy and infant dependency, requiring females to maximize their feeding efficiency; thus individual females occupy relatively small ranges with which they are presumably very familiar, and their core areas appear to be evenly dispersed throughout the available habitat (Wrangham 1975).

In this respect, chimpanzees resemble orangutans (Rodman 1973a) but differ in that competition for females is between communities of males rather than between individual males; within a community, mating is more or less promiscuous, and the larger an area that a group of males can control, the more females will be available to them. Why should there be this difference? Wrangham proposes that males will combine to defend an area if it is larger than the sum of the areas they could independently defend, because each male will then (on average) have access to more females.

In orangutan habitat, the distribution of food permits a relatively sedentary life and adult males have evolved a strategy of territorial defense by long-distance calls, which appear to repel males but attract females (MacKinnon 1974). Body size has escalated, presumably as a result of inter-male competition, so male locomotion costs are high, and indeed they feed more and move less than other age-sex classes (Rodman, this volume). Fights between males, though rare, may be severe (Galdikas, this volume) and presumably the costs of invading another male's territory (locomotion costs + risk of being attacked by resident male) may outweigh the potential benefits, since the probability of finding an estrous female would be relatively low.

In chimpanzee habitat, the need to be able to find food in small trees and to travel long distances would presumably favor more agile males. Even if individual territories were similar in size to those of orangutans (and they might need to be larger, if food were scarcer) they would then be much more difficult to defend against highly mobile intruders, to the detriment of feeding efficiency and reproductive success.

The chief advantage of collective territorial defense to an individual is that he need be involved in very few potentially harmful confrontations with competitors (males from other communities). A group of males can be a more powerful deterrent to intruders than can an independent male, since the group can inflict a severe or lethal attack with minimal risk to its members. When parties from different communities hear each other's calls, therefore, it will be in the smaller party's interests to retreat, as has been observed. By merely accompanying other males on border patrols (which can be combined with foraging), an individual male can help to maintain his continued access to a large number of females.

This model implies that there would be strong selection for males to be rapidly aroused to attack strangers, particularly males, on sight. Wilson and Wilson (1968) noted that in a captive chimpanzee colony, the most severe attacks were directed at new individuals who had just been introduced into the enclosure, and aggression against strangers has been noted in many other primate studies (e.g., Hall 1962; Southwick 1967). This paper mentions two severe attacks on males who, although once familiar to the attackers, had not associated regularly with them for over 2 years; Nishida (this volume) describes a similar incident.

The transference of receptive females from one community range to another, temporarily or permanently, seems to be common among chimpanzees (Nishida, this volume) and is presumably important in preventing inbreeding, as males do not emigrate. It is therefore to a male's advantage not to attack a strange estrous female, and to the female's advantage not to be attacked. The conspicuous estrous swelling, found in chimpanzees but in no other apes, may serve to indicate the sex and condition of a female from afar; although estrous females are certainly not immune from attack by males, no gang attacks by males on strange estrous females have (to my knowledge) been recorded.

It might, however, benefit males to attack strange females who were pregnant or had small infants, since by doing so they might destroy the offspring of males of another community, and increase their own chances of conceiving a child by the same female when she subsequently came into estrus. This possibility might account for the infant-killings observed by Suzuki (1971) and Bygott (1972), although of course in neither case was the paternity of the dead infant known to the observer.

From the female's viewpoint, therefore, as soon as she becomes pregnant it is imperative that she settle down either in her paternal community's range (where all the adult males know her) or in the range of her infant's father, in which case she must ensure that all males become familiar with her (and with her infant when it is born). One might expect some aggression by established resident females toward new transfers, since the latter will be competing with them for food if they stay permanently.

During this study, at least two incidents were seen in which the old female Flo and her adult daughter Fifi repeatedly chased and attacked adolescent estrous females who presumably had come from other areas. In one of these cases, an adult male (Figan, Flo's son) broke up a fight between his sister and the new estrous female, but showed no aggression toward the latter, who subsequently followed him around for a while. Unfortunately, the process of female transfer has not yet been studied in depth, and we still do not fully understand what a female gains from leaving familiar surroundings and companions and establishing a new range among less familiar and probably hostile females.

Given that the gross social structure of chimpanzees is as has been outlined, there may be advantages for males not to spend all their time together, provided that they can occasionally combine to patrol the community range (and their well-developed system of individually recognizable long-distance calls must enable them to find each other easily). By foraging alone or in small parties they may increase their feeding efficiency; data from this study and from Wrangham (1975) show that males in groups spend less time feeding than when alone, and it is also possible that different individuals may have specialized knowledge of different areas as a result of early familiarization with their mothers' core areas. They may also increase their reproductive success by consorting with females in exclusive pair relationships.

With this kind of fusion-fission association pattern, reunions between familiar individuals will be frequent. If males have a tendency spontaneously to attack or flee from strangers on sight, one would expect even reunions between familiar chimps to be accompanied by tension, as there is always a chance of nonrecognition. This possibility may partly account for the frequency of agonistic interactions at reunions, in which postures, calls, and orientation (especially by subordinate individuals who risk being attacked) maximize the chances of individual recognition. The high incidence of grooming that follows reunions (Bauer 1975) may also be important in reestablishing and maintaining friendly relationships. Since individuals will naturally differ in strength and aggressiveness, it is in the interests of both aggressive and unaggressive individuals to avoid fighting as much as possible in such situations to favor the development of dominant-subordinate relationships. The subordinate, by deference or avoidance, avoids being hurt in a fight, and the dominant chimp avoids hurting someone who may be related to him. Such relationships may facilitate association between individuals (Nowlis 1941), but this does *not* imply that without aggression there can be no friendship, as Lorenz (1966) proposes.

This model implies no selective advantage in being dominant per se. Popp and DeVore (this volume) have discussed the theoretical advantages that individuals may gain from dominance. In this study, however, the advantages were by no means obvious; although dominant males *could* monopolize food or females by force, they very rarely did so. Nishida (this volume) reports a different situation, in which the dominant male of a very small community frequently exercised his prerogatives. Moreover, the age-relatedness of dominance meant that most males had the chance to be high ranking at some time during their life. One would expect overtly *selfish* activities by a dominant male to be penalized through the noncooperation of other males in assisting in the defense of the community range, which would jeopardize his own reproductive success. Hence among chimpanzees one would expect selection for the inhibition or ritualization of intracommunity competitive aggression and for retention of *affiliative* behaviors developed during the mother-infant relationship (such as grooming and various forms of contact) which can be generalized to other members of the community.

Popp and DeVore point out that although all members of a coalition (in this case the male community) must benefit from its existence in order to remain within it, this advantage will not prevent a certain amount of interindividual competition (provided it does not jeopardize the stability of the coalition). Such competition could take subtle forms; although all males might have equal access to estrous females, which would promote cohesion, there could be intra-vaginal selection which could favor the male who could ejaculate most frequently or copiously or who could produce the most viable sperm. It is perhaps relevant that chimpanzees have much larger testes in relation to body weight than do most other primates. If males could in some way increase their own fertility by their own actions—e.g., through becoming dominant to others and undergoing hormonal changes as a result—selection would favor behavioral traits that enhanced a male's dominance rank.

It is quite possible that there is some selective advantage, undetected by this study, in being outstandingly high ranking within a community, i.e., alpha male, but it is interesting to reflect that any such selection might also favor the ability to form close cohesive bonds with another male, since the evidence indicates that the majority of alpha males may attain their position with some assistance from a close associate such as a sibling.

In conclusion, the growing body of information about chimpanzee social structure indicates that cooperation among male chimpanzees may have been forced upon an originally more solitary species by environmental changes, and that much of the agonistic behavior observed within a community may be more important in preventing physical fighting than in gaining individual priority to incentives. Male chimpanzees, perhaps more than males of any other primate species, have developed the ability to spend long periods in peaceful proximity to one another, to take collective action against intruders, and even to cooperate in the hunting of mammalian prey.

Finally, it must be emphasized that this study concerns only one particular community during a limited period of time, and focused mainly on one section of the community. Although readers may wish to draw their own comparisons between chimpanzees and humans, it would not be valid to generalize from this study to the species as a whole, and even less valid to make cross-species generalizations.

Acknowledgments

I am very grateful to the Tanzanian government for permission to work in the Gombe National Park, and to Dr. Jane Goodall for providing facilities. I would also like to thank the many observers and assistants who contributed to the data. I am indebted to the Grant Foundation of New York for financing the work, and to Professor R. A. Hinde (F.R.S.) for his supervision and advice. I am grateful to the other contributors to this symposium and to Dr. Jeannette Hanby for some stimulating discussions.

Notes

1. In 1974, observers witnessed two very severe attacks by gangs of northern males on the young males De and Godi of the now separate southern community. Both victims were injured; Godi was never seen again and De underwent extreme physical deterioration and disappeared about a month later (Riss, personal communication).

Patrick R. McGinnis

Sexual Behavior in Free-Living Chimpanzees: Consort Relationships

A pattern of sexual behavior common to many vertebrate species is the tendency for a female to consort and copulate with only one male at the period of the sexual cycle when conception is most likely to occur. Field investigations have verified the occurrence of this behavior in a number of primate species: ring-tailed lemurs (Jolly 1967), rhesus macaques (Carpenter 1942; Conoway & Koford 1964; Altmann 1962; Kaufmann 1965; Southwick et al. 1965), Japanese macaques (Tokuda 1961-2; Hanby, personal communication), baboons (Hall & DeVore 1965), howler monkeys (Carpenter 1965), langur monkeys (Jay 1965; Yoshiba 1968). Although the time courses of consortship vary from hours to days between species, in only a few species have consortships been reported to be conspicuously infrequent or absent. Mangabeys (Chalmers 1967), bonnet macaques (Simonds 1965), and chimpanzees (Kortlandt 1962; Reynolds & Reynolds 1965; Nishida 1968; van Lawick-Goodall 1968; and Sugiyama 1969) have been observed to exhibit lack of consortship with little apparent possessiveness of partners. In field studies of chimpanzees, however, patterns of association (Reynolds & Reynolds 1965) and forms of behavior (van Lawick-Goodall 1968) that suggest the occurrence of sexual consortships have also been reported. This paper presents a preliminary description and analysis of one form of chimpanzee consortship, and discusses some of the factors involved.

STUDY AREA, SUBJECTS, AND METHODS

The study was conducted at Gombe National Park, Kigoma Region, Tanzania, East Africa. The eastern or long-haired chimpanzees *(Pan troglodytes schweinfurthi)* that range freely within the park were studied for the first time by Jane Goodall in 1960. In 1963, Dr. Goodall established a long-term artificial feeding station where chimpanzees periodically receive a highly preferred food (bananas) in the presence of human observers (van Lawick-Goodall 1968b; 1971). As a result, chimpanzees that frequent the station have become habituated to the unobtrusive movements of humans at distances as near as three meters.

Standardized records were kept of the attendance and behavior of all chimpanzees that visited the feeding station. With rare exception, there was at least one observer at the station throughout the daylight hours of every day. Behavior was recorded either directly onto data sheets or narrated into a tape recorder and subsequently transcribed to data sheets. Chimpanzees were also followed and observed away from the feeding station where their behavior was recorded in a similar manner.

THE FEMALE SEXUAL CYCLE

The sexual cycle of chimpanzees is manifested externally by conspicuous changes in the size and turgidity of the perineum. Since November 1, 1966, the Gombe Stream Research Centre has kept daily records of perineal swellings of every female chimpanzee observed. A female's perineum is recorded to be in one of three states: no swelling, maximal swelling, or partial swelling. No swelling is recorded when the perineum is shriveled, small, and inconspicuous in profile. Maximal swelling is recorded when the perineum is fully turgid and void of wrinkles. In this condition it is noticeably enlarged and protrudes well beyond the contours of the rump. Partial swelling is recorded as an estimated fraction of the largest size attainable for the perineum of the female in question. Day-to-day changes in size and turgidity are normally continuous between stages of no swelling and maximal swelling. Thus, the cycle of perineal swelling can be divided into four phases:[1] no swelling (13.6 ± 5.5 days), tumescence (7.6 ± 3.4 days), maximal swelling (16.3 ± 0.5 days), and detumescence (4.5 ± 1.3 days). Ovulation is thought to occur during the latter half of maximal swelling or possibly during detumescence (Graham 1970). Copulation normally occurs only during the maximal swelling phase.

CONSORT BEHAVIOR

On 40 separate occasions between November 1, 1966, and May 31, 1969, observers at the feeding station witnessed patterns of behavior that suggested that a particular male, upon departure from the station, was coercing a particular female to follow. Typically, the male displayed elements of behavior normally associated with courtship, such as gazing, hair erect, branch-shaking, rocking, and horizontal extension of one or both arms. Contrary to the pattern in courtship, however, when the female began to advance, the male turned and walked away from her. Failure of the female to advance toward the male resulted in increases in the intensity, frequency, and concurrence of elements of the male's display until, ultimately, the male alternated displays with attacks on the female. The most effective consummatory stimulus for the male's displays and attacks was the female following the male without hesitation.

The consequent absences of the male and female from the feeding station were longer than normal and coincided one with the other to a greater degree than would be expected by chance (see next section).

SIMULTANEITY OF ABSENCE AS EVIDENCE FOR CONSORTSHIP

In an attempt to quantify the occasions on which Gombe chimpanzees form exclusive heterosexual pairs that tend to avoid areas frequented by other chimpanzees, records of individuals' daily attendance at the feeding station were examined for concurrent periods of absence. It often happened that individual chimpanzees were not seen at the feeding station for periods of up to 3 days. Less frequently, they were absent for longer periods of time. It is with the longer, less frequent absences that this analysis is concerned.

A Criterion for Absence. The longest durations of absence that accounted for no more than 5% of the total number of absences were selected for analysis. Four days was chosen as the minimal length of periods of absence to be considered. The term *absence* henceforward in this paper refers to a chimpanzee's continuous nonattendance at the feeding station for a period of 4 or more consecutive days.

A Criterion for Coincidence. In all twelve pairs of absences (4 or more days) consequent on departures in which a male was observed to display consort behavior toward a female, the absences of the male and female coincided, one with the other, by at least 75% of the duration of each. That is, the period in which both were absent constituted at least 75% of the male's absence and at least 75% of the female's absence. This mutual level of concurrence was selected as a criterion of coincidence for analysis of all other absences of 5 or more days. A criterion of 100% was demanded of 4-day absences, due to their relatively high frequency of occurrence.

Calculation of a Coincidence Index (CI). For each absence of each chimpanzee, the attendance record of every other chimpanzee in the community was examined for an absence during the same period of time. If the absence of the chimpanzee in question met the criterion of coincidence with the absence of another chimpanzee, a coincidence index (CI) was calculated for the pair of absences:

$$CI = \frac{\text{The no. of consecutive days both chimpanzees were absent}}{\text{The no. of consecutive days either chimpanzee was absent}} \times 100$$

Selection of the Mutually Highest Coincidence Index (MHCI). For each absence of a particular chimpanzee, a number of coincidence indices could be calculated if more than one other chimpanzee was absent concurrently. For each absence for which one or more CI's were calculated, the highest coincidence index (HCI) was examined to determine whether or not it was also the highest coincidence index (HCI) for the absence of the other animal to which it applied, i.e., the mutually highest coincidence index (MHCI) for the pair of absences in question.

Selection of the Singular Mutually Highest Coincidence Index (SMHCI). Since this analysis is designed to uncover exclusive pair-associations, and since more than one MHCI is possible for each absence of a chimpanzee, each MHCI was examined to determine whether or not it was a singular MHCI (SMHCI) for each of the two absences from which it was calculated.

SMHCI's are inferred to be occasions on which the two animals in question were together for long periods of time (4 or more days) while away from other members of the chimpanzee community. Obviously, animals may have been away together for periods shorter than 4 days, but short periods would have been missed by this analysis.

HYPOTHESIS

If simultaneous absences from the feeding station reflect real associations between individuals, and if chimpanzees tend to form exclusive heterosexual consort pairs related to copulatory behavior, then an examination of the frequencies of female SMHCI's would be expected to reveal two basic trends: (a) During periods of maximal perineal swelling, the number of heterosexual SMHCI's should be greater than the number of homosexual SMHCI's; (b) the number of heterosexual SMHCI's should be greater during maximal swelling than during other phases of the sexual cycle.

ANALYSES

Analyses for SMHCI's were conducted on 923 periods of absence among 23 postmenarchal females for the period from January 1, 1965, through October 31, 1971. Nonparametric statistical tests were applied to the data (Siegel 1956).

RESULTS

Simultaneous Absences During Phases of Maximal Perineal Swelling.

1. *The Number of Females with Heterosexual vs. Homosexual SMHCI's.* All the 18 females that exhibited cyclic perineal swellings during the study period had SMHCI's with males during phases of maximal swelling, but only eight of them had SMHCI's with females during maximal swelling. The difference is significant with $p < 0.005$ (one-tailed chi squared test).

2. *The Numbers of Heterosexual vs. Homosexual SMHCI's for Individual Females.* When the number of absences with males and the number with females were compared for each individual during maximal swelling, 15 females had more SMHCI's with males than with females and only one had more with females than with males. Two females had ties. This difference is significant at the 0.001 level of probability (one-tailed sign test). Only eight of the females had samples large enough (i.e., five or more) to be assessed individually. Four of them showed significant differences in the number of heterosexual vs. homosexual SMHCI's. All four had greater numbers for males (one at $p < 0.0001$, one at $p = 0.0005$, and two at $p < 0.05$; one-tailed binomial tests).

3. *Ratio of Females with a Greater Proportion of Heterosexual SMHCI's to Females with a Greater Proportion of Homosexual SMHCI's during Maximal Swelling vs. Other Phases of the Sexual Cycle.*

433

A greater proportion of cycling females had more SMHCI's with males than with females during phases of maximal swelling (15:1) than during other phases of the sexual cycle (9:4). The difference is significant with $p < 0.05$ (one-tailed Fisher Exact Probabilities Test).

From these data, the occurrence of exclusive heterosexual consortships related to the sexual cycle can be inferred.

INFERRED CONSORTSHIPS
AND THE MALE DOMINANCE HIERARCHY

A tendency for sexually attractive females to consort more with high-ranking males than with low-ranking males around the time of ovulation has been reported in wild baboons (Hall & DeVore 1965), and free-ranging rhesus macaques (Kaufmann 1965). In this study, the availability of long-term records of agonistic interactions at the feeding station permitted analysis of a measure of male *dominance* with respect to consortships.

Analysis.

1. *An Agonistic Index: A Measure of Dominance between Males.* The numbers of attacks and chases that occurred between individual males were assessed to determine which member of each pair-combination chased and attacked the other more often. That member of the pair was given a positive (+) point for that pair-combination, and the other member of the pair received a negative (−) point. Some pair-combinations had ties (=). Others did not interact (0). An agonistic index [AI = No. (+) − No. (−)] was calculated for each sexually mature male in the community. Males were then ranked according to their agonistic indices. When two males had equal agonistic indices, the male with the greater number of positive points was assigned the higher rank. When the numbers of positive points were equal at a tied rank, the male that interacted with more other males (i.e., scored fewer 0's) was assigned the higher rank. When the number of +'s, −'s, ='s and 0's were equal for the two males, their ranks were tied.

In order to minimize the effects of dramatic short-term changes in frequencies and direction of chases and attacks for individuals, the study period was divided into four parts: July 1966 through December 1967; January through December 1968; January through December 1969; and January through December 1970. Agonistic indices were calculated separately for each part of the study period, and the ranks of each individual during the four parts were averaged. Twenty-three sexually mature males were considered in the analysis.

2. *Frequencies of Inferred Consortships.* Males were also ranked according to their numbers of SMHCI's (with sexually cycling females) per month, and compared with the mean agonistic ranks. The analysis examined 129 SMHCI's.

3. *Proportion of Inferred Consortships beginning during Tumescence and Maximal Swelling Phases of Female Sexual Cycles.* Males were ranked again according to the proportion of their SMHCI's that began during tumescence and maximal swelling phases of the sexual cycle. These ranks were also compared with their mean agonistic ranks.

Statistical Tests. Spearman rank order correlation coefficients (r_S) (Siegel 1956) were calculated for the rank order of male agonistic indices vs. the rank orders of their frequencies of SMHCI's and their proportions of SMHCI's beginning during tumescence and maximal swelling. For individual males, one-tailed binomial tests were used to test the significance of the proportions of SMHCI's beginning during tumescence and maximal swelling.

Results. No significant correlation was found between male agonistic indices and their frequencies of SMHCI's with cycling females ($r_S = 0.252$; $p = $ n.s.). No correlation was found between agonistic indices and the proportions of SMHCI's beginning in tumescence and maximal swelling phases of the sexual cycle ($r_S = 0.129$; $p = $ n.s.). Eleven males had enough (five or more) SMHCI's to be assessed individually. Of these, only three showed significant differences with the sexual cycle. All three had significantly greater numbers of SMHCI's begin during tumescence and maximal swelling than during detumescence and no swelling ($p < 0.05$). All three males had agonistic ranks above the median.

Conclusions. Although the frequencies of inferred consortships do not show significant correlation with agonistic rank, there is a tendency for certain high-ranking males to begin consortships significantly more often during tumescence and maximal swelling than during detumescence or no swelling phases of female sexual cycles.

MOTIVATIONAL FACTORS IN CONSORTSHIP

Insights into the motivational factors that underlie consortships may be gleaned from an examination of the behavior that occurs during the formation and maintenance of consortships and an examination of the behavior that occurs when groups of males aggregate around a sexually attractive female.

Behavior during the Formation and Maintenance of Consortships. In most of the consort behavior observed, the frequency, intensity, and concurrence of elements of male display were related positively to the audible proximity of other males. Once the male succeeded in getting the female to follow, the constancy and directedness of the male's travel were also determined by the audible proximity of other males.

Females varied in their responsiveness to consort invitations. The two most relevant variables in this regard seemed to be the stage of the sexual cycle and the identities of the individuals involved. The young female, Pooch, was often seen to follow the young male, Figan, with little or no coercion when the female was in maximal swelling. It should be noted, however, that this is apparently the immediate goal of male consort behavior, and the female may learn very quickly to follow the male's lead without hesitation. Thus, without a knowledge of preceding events, it is difficult for an observer to ascertain whether the female behavior of following the male is mainly voluntary or a consequence of behavioral shaping by the male. Continuous observation of a pair, however, sometimes revealed that females that had been shaped occasionally needed reminding. In one pair that was observed for 5 days, the female was continually reluctant to follow the male, stopping frequently in travel, causing the male to display repeatedly, and often incurring attacks. The male in this case was low-ranking, and the female was in the no-swelling phase of the sexual cycle and had just finished a consortship with another male. The responsiveness of the female in some cases was also determined in part by the audible proximity of other males. Females often protested initially with loud vocalizations and great hesitancy to respond to the consort invitation of a male when the pair were within audible range of other males. Once the male succeeded in getting the female out of earshot of the others, however, the female usually followed without much hesitation. It is of note, also, that males in consort with females exhibited total cessation of loud hooting vocalizations and often threatened or attacked if the female vocalized loudly. On occasions when female vocalizations attracted another male of higher rank, the consort attempt was usually terminated. On two occasions when a higher ranking male came upon a would-be consort in the act of intimidating the female, the high-ranking male attacked the suitor. On other occasions when a higher-ranking male came upon a would-be consort between bouts of intimidation of the female, the suitor failed to repeat the displays as long as the higher-ranking male was present.

These observations suggest that the immediate motive of the male in consortship is to be with the female but away from other males, and that consort behavior is inhibited by the presence of higher-ranking males.

The Behavior of Males in Groups around a Sexually Attractive Female. Subjective impressions of the behavior of males in groups aggregated around sexually attractive females suggest that aggression between males over copulatory rights to a female is more likely to occur when the group is small than when it is large, indicating that the number of males present may inhibit overt possessive behavior. The one occasion on which a male in a large group was seen to attack another male over copulatory rights to a female suggested futility in such behavior. The highest-ranking male was staying close to the only sexually attractive female in the group. Eight other males were in the group. When one of them, with hair and penis erect, approached the female, the highest-ranking male charged and attacked the suitor. While the two were engaged in a brief fight, three other males, in rapid succession, copulated successfully with the female. It would appear, then, that fighting over the female in large groups of males is ineffective in preventing other males from copulating. It also expends energy and may unnecessarily jeopardize a high-ranking male's social status. Some males were seen to get around this problem by charging or chasing and sometimes even attacking the female if the latter responded to courtship of other males. As other investigators have pointed out, however, overt possessiveness is the exceptional pattern of behavior in large groups.[2]

In large aggregations of males around an attractive female, as time passes, individuals tend to wander out of sight of the group for variable periods of time. Although factors associated with males leaving the group have not been investigated, it seems reasonable to postulate that low food availability in the vicinity of the female may cause a hungry male to seek food elsewhere. There is also the consideration that males in large groups where an incentive is present may be under degrees of social stress, and may leave the group to remove themselves from the stressful situation. Stress would seem a likely factor if it could be shown that the number of attacks and threats incurred by individuals increased with increases in group size. Wrangham (1974) has found that the number of attacks per individual increases with the number of chimpanzees present at the feeding station on days when bananas are provided. He has also found that such an increase did not occur with an increase in group size on days when bananas were not provided. He has not, however, distinguished between aggressive incidents that involved the possession of bananas and those that did not. If an increase were found in the latter, it would suggest that increases in aggression per individual are likely to occur in the presence of an incentive even though not related directly to its acquisition.

Whatever the factors involved, the tendency of individuals to stray from the group surrounding the female may provide an opportunity for other males to attempt to abduct the female. Not surprisingly, the likelihood of an individual to attempt a consortship was often previewed in the persistence with which it maintained proximity with the female in the group.

THE ONTOGENY OF CONSORTSHIPS

Adult males usually copulate freely in the presence of other adult males in large groups, but juvenile and adolescent males show a good deal of caution in their attempts to copulate with the female. Young males usually await periods when most of the adult males are resting before they attempt to court the female. Their courtship displays are usually performed at the periphery of the group, often at some distance from the female. The displays are typically simple and subdued, conspicuous enough to attract only the female's attention, but not so conspicuous as to arouse the adult males. If able, the female will leave the resting adults long enough to accommodate the young male. As the female advances, the juvenile or adolescent male will typically turn and lead the female to a point farther away from (and, frequently, out of sight of) the adult males, and the pair will copulate in relative privacy. In this behavior can be seen the rudiments of consortship.

SUMMARY

In the study community of free-living chimpanzees at Gombe National Park, exclusive heterosexual consortships are inferred to occur in relation to the female sexual cycle. Although male agonistic rank is not correlated with the frequency of inferred consortships with cycling females, there is a tendency for certain high-ranking males to initiate consortships significantly more often during phases of the sexual cycle that immediately precede or include the period of ovulation than during other phases of the cycle. The male of the consort pair generally initiates and maintains a consortship by forcefully inducing the female to follow him away from the rest of the group. This abductive form of consortship appears to be more effective in maintaining exclusive copulatory rights with an estrous female than does a form that attempts aggressively to repel other suitors while remaining within the group.

Acknowledgment

I wish to thank the Tanzanian Government for allowing me to engage in research at Gombe National Park. I am grateful to Hugo van Lawick and Dr. Jane Goodall for providing living accommodations and research facilities at the Gombe Stream Research Centre, and for access to the Centre's records of chimpanzee behavior. I also thank Dr. J. David Bygott for allowing me to generate agonistic indices from his primary analyses of agonistic interactions between male chimpanzees. I am deeply indebted to Prof. Robert A. Hinde for his invaluable criticisms, and I am grateful to the National Geographic Society of Washington, D.C. for providing financial assistance in the field.

Notes

1. Means and standard deviations of median durations of phases of the sexual cycle for seven postmenarchal nulliparous Gombe chimpanzees.

2. In a more recent study of the same community of chimpanzees, Tutin (1975b) reports that a newly established alpha male has been successful in maintaining exclusive copulatory rights to estrous females while remaining within large groups of males.

Aphrodite uses hand and mouth to reach for a banana peel held by her mother, Athena.

William C. McGrew

Evolutionary Implications of Sex Differences in Chimpanzee Predation and Tool Use

The upsurge in field studies of nonhuman primates over the last 15 years has contributed much valuable information of relevance to the reconstruction of the human evolutionary past (Jolly 1972; Pfeiffer 1972). Previously, anthropologists seeking to draw conclusions about protohominid behavior were limited to inferences from a scanty fossil and artifactual record and to extrapolations from nonhuman primates in captivity. This paper will deal with the animal portion of the diet of wild chimpanzees, *Pan troglodytes*. Predation upon vertebrate, and especially mammalian, prey has been previously dealt with (van Lawick-Goodall 1968b; Teleki 1973; Wrangham 1974) and will only be reviewed in passing here. The paper will focus upon tool use in obtaining social insects for food, especially termites (Goodall 1968b) and driver ants (McGrew 1974b).

The main variable under examination will be gender. Sex differences in various aspects of the behavior of free-ranging nonhuman primates are well known, but in the areas of feeding and tool use, these have been mainly by-products of more general studies. Harding (1973), for example, showed that male olive baboons, *Papio anubis,* in his study area in Kenya made practically all the predations upon small mammals. Similarly Nishida (1973b) presented data suggesting female predominance in fishing for arboreal ants by chimpanzees in western Tanzania.

After presentation of material on chimpanzee sex differences in tool use and feeding on animal matter, the sharing of plant foods between individuals will be introduced. This information consists of quantitative records of the inter-individual distribution of bananas in an artificial observation area and qualitative notes on the similar transfer of natural fruits during daily foraging. Because sharing of food necessarily involves social interaction, the important social variable of kinship will be emphasized in addition to sex.

In discussing the evolutionary implications of these chimpanzee behaviors, another source of information will be utilized: studies of extant Old World tropical foraging peoples (e.g., Lee & DeVore 1968). The customs of living hunter-gatherers can tell us nothing directly about protohominid adaptation processes. It is thought, however, that peoples such as the Bushmen of southwestern Africa and the aborigines of Australia represent the closest surviving approximations to the life style of foraging, savanna-living protohominids.

METHODS

The observations on chimpanzee behavior presented here were made at the Gombe Stream Research Centre in northwestern Tanzania, which is located in the Gombe National Park on the eastern shore of Lake Tanganyika. Descriptions of the park's habitat and the research center's history and activities occur elsewhere (van Lawick-Goodall 1968b; 1971; Wrangham 1974a). The study population is one of eastern or long-haired chimpanzees *(Pan troglodytes schweinfurthi).*

In general, observations of increasing specificity and volume have been made since the research began in 1960. Since the late 1960s, the observation methods have been increasingly standardized and systematized, so that all observers contribute to a common pool of accumulated data. The findings presented here could not otherwise have emerged, as no single observer could amass the amounts of data required.

In almost all cases, pairs of observers followed a target chimpanzee for periods of up to 13 hours of diurnal activity. Typically an expatriate research worker recorded a field check sheet and a Tanzanian field assistant recorded group membership, travel routes, and feeding. Observation conditions were excellent and most records were taken at a distance of 5-10 m. Just over half of the observation periods *(follows)* began in an artificial feeding area *(camp)* and continued into the forest. The other observations began away from camp when chimpanzees were followed as they left their nests of the previous night or after they were found by human searching parties.

FEEDING ON INSECTS

Van Lawick-Goodall (1968b) reported that chimpanzees at Gombe ate insects of one type or another almost daily. (No wild chimpanzee has yet been reported to feed upon any representatives of the other invertebrate classes.) The insects eaten represent five orders: Isoptera, Hymenoptera, Diptera, Lepidoptera, and Hemiptera. Only the first two are eaten sufficiently often to provide systematic quantitative data. In addition Gombe chimpanzees feed on the products of insect labor, e.g., the honey of wild bees, the earth of termite mounds. One species from each order is subjected below to detailed examination as a prey type: the termite, *Macrotermes bellicosus,* and the driver or safari ant, *Dorylus (Anomma) nigricans.* Chimpanzees manufacture tools from vegetation to obtain these species in termite-fishing and ant-dipping, respectively. Sex differences in chimpanzee feeding on social insects have not previously been reported. Human hunting and gathering people regularly feed upon insects, among them termites and ants (Coon 1971; Dart 1964). Humans also parasitize insect labor: The women of the Wailbri aborigines in central Australia raid the grain stores of harvesting ants (Sweeney 1947). Wherever available, honey is a desirable commodity occurring in limited amounts, and this factor combined with the dangers and pain usually involved in obtaining it make it a predominantly male activity (Murdock & Prevost 1973). Finally many of the tool-use techniques used by hunter-gatherers to obtain insects resemble those of chimpanzees in simplicity and materials (e.g., Roth 1901).

Termite-Fishing

The tool-use technique of termite-fishing has been described elsewhere (initially by Goodall 1963), and so will be only briefly reviewed here. The termite-fishing chimpanzee opens a hole on the bare earthen surface of a termite mound. Into this it inserts a long thin object of plant material, e.g., blade of grass, strip of bark, length of vine or twig, etc. In most cases these objects have been modified by the chimpanzee, i.e., shortened, narrowed, stripped of leaves, etc. The termites inside the mound (and unseen to the chimpanzee) attack the intruding object, clamping onto it with the mandibles. The chimpanzee predator carefully withdraws the tool and plucks from it with the lips the insects and eats them in leisurely fashion.

Individual, age, and kinship variability in termite-fishing techniques and volume of intake will be examined in a future paper. Here the interest is in basic questions: How often, for how long, and with what periodicity do chimpanzees fish for termites, and are there sex differences in any of these parameters?

The results are based on almost 7500 observation hours of data collected over 19 months from July 1972 through January 1974. They represent the pooled records of 37 observers using a standardized recording method: the travel-and-group chart. This gives (among other things) the frequency of feeding bouts of a target chimpanzee to the nearest 5 minutes, with time of day and duration. The data comprise 1443 observation periods averaging 5.17 hours in length (range: 0.50 to 13.25 hours). Follows of less than 30-minutes duration are excluded.

Thirty chimpanzees contributed to the observation totals—16 females and 14 males ranging in age from 4.5 to approximately 45 years. The sex ratio of the study population as a whole remains at virtually 1:1. The chimpanzees exhibited 217 hours of termite-fishing in 495 feeding bouts, giving a mean bout length of 26.3 minutes (total range: 5-200 minutes; inter-sextile range: 5-45 minutes).

A study indicated results in terms of percentage of observation time that chimpanzees spent in fishing for termites. For the community as a whole this averages just under 3% of the time, with wide seasonal variation (i.e., in monthly means) from 0.25% to 13%. In all 19 months, female frequency exceeded male frequency, and female termite-fishing occurred three times as often (4.3% vs. 1.4%) as male termite-fishing over the study period. Looked at another way, termite-fishing occurred on 249 (17%) of 1443 follows of chimpanzees. The monthly frequencies ranged from 1% to 57%. Again female rates predominated: in all 19 months of study, females fished for termites on a higher percentage of follows than did males. Overall, females showed termite-fishing on 22% of follows and males on 10% of follows.

Van Lawick-Goodall (1968b) implied a sex difference in chimpanzee fishing for termites by reporting that males were never seen to fish for more than 2 hours at a time while females often exceeded this. This study confirms this point: lengthy termite-fishing bouts (60 minutes or longer) were exhibited almost twice as frequently by females as males. Mean bout lengths over all durations were similar for the sexes: female $\bar{x} = 26.8$ minutes; male $\bar{x} = 24.8$ minutes. In other words, the female predominance in termite-fishing is due to fishing more often (especially in bursts) as well as fishing for longer periods.

The picture is not as simple as it first appears (see Table 1). Although males and females were followed for roughly the same total

TABLE 1

Sex differences in fishing for termites exhibited by chimpanzees in the Gombe National Park, Tanzania

	FEMALES	MALES	TOTAL
Number of Observation Hours	3864(52%)	3597(48%)	7461
Number of Follows (Observation Periods)	835	608	1443
$\bar{x}$ Follow Length (Hr.)	4.63	5.92	5.17
Number of (%) Follows Started in Camp	519(62%)	295(42%)	774(54%)
Number of Hours of Termite Fishing	166.25	50.75	217
Number of Termite Fishing Bouts	372	123	495
$\bar{x}$ Termite Fishing Bout Length (Mins.)	26.8	24.8	26.3
Number (%) of Follows with Termite Fishing	186(22%)	63(10%)	249(17%)
% of Observation Time in Termite Fishing	4.3	1.4	3.0

amounts of time, significant differences existed in observation procedures for the two sexes. Females were followed more often (835 female follows vs. 608 male follows) for shorter periods (female $\bar{x}$ = 4.63 hours vs. male $\bar{x}$ = 5.92 hours). More female follows started in camp (62% of female follows vs. 42% of male follows). Finally, more male follows started earlier in the morning than did female follows: 52% of male follows began before 0900 vs. 40% of female follows. Any of these factors could be responsible for the apparent sex differences, e.g., if termite mounds were nonrandomly distributed vis-à-vis camp or if termite-fishing shows circadian periodicity. (I thank R. Wrangham for calling these points to my attention.)

Further analysis, however, reveals that these factors cannot account for the observed sex differences in termite-fishing. Follows on which termite-fishing occurred were not significantly different in duration between the sexes (female $\bar{x}$ = 5.66 vs. male $\bar{x}$ = 5.88 hours). Females showed significantly more termite-fishing than did males whether or not follows began in or out of camp. Finally, across the 13 one-hour periods of the day (0600-1900) females showed higher rates of termite-fishing than did males in 12 of these.

It is concluded that a genuine sex difference exists in the frequency of termite-fishing by Gombe chimpanzees, with females predominating.

Ant-Dipping

The tool-use technique of ant-dipping was first described in brief by Goodall (1963) and later in detail by McGrew (1974b; see Figure 1). In summary, the chimpanzee predator finds a subterranean driver ants' nest and excavates an opening in it with the hands. It makes one or more long smooth wands of woody vegetation by modifying branches. When inserted into the nest, the tool is attacked by ants which stream up it. The chimpanzee quickly withdraws the tool and while holding it in one hand, sweeps the length of the wand with the other. The ants are momentarily collected on the sweeping hand (the *pull-through)* and directly popped into the mouth, then chewed frantically. In response to the massed active defense of the ants, the chimpanzee predator exhibits various tactics of differential positioning and technique.

The results are based on the same 7461 hours of observation collected by 37 observers over 19 months from July 1972 through January 1974, as referred to in the previous section. Twenty-four chimpanzees (11 females and 13 males) were seen to dip for driver ants; none were younger than 4 years of age. They exhibited 75 bouts of ant-dipping in 44 ant-dipping episodes. These totaled just under 16 hours of ant-

FIGURE 1

An adolescent male chimpanzee (Goblin) withdraws a grass tool during termite-fishing. Photograph by C. Tutin.

dipping activity, and the overall bout length of time actually spent consuming ants averaged 15 minutes (N = 27, range = 3-48 minutes; inter-sextile range: 5-29 minutes).

Given the paucity of data, it is impossible to make as detailed an analysis of possible sex differences in ant-dipping as was done with termite-fishing, but indications of similar sex differences emerge. Chimpanzees of over 4.5 years of age were seen 127 times (chimpanzee-nest-occasions) at driver ant nests at which successful ant-dipping occurred. These were approximately equally divided by sex: 67 males and 60 females. Yet the proportions of these animals that actively consumed ants versus those that only watched were significantly different: 45 of 60 (75%) females dipped but only 30 of 67 (45%) of males dipped. Thus, although the female frequency of dipping for ants did not significantly exceed the male frequency, females seemed much more highly motivated to dip for driver ants when given the opportunity.

Similarly females showed a slightly greater mean bout length of ant-dipping than did males (females: $\bar{x}$ = 15.5 minutes, N = 19, range = 3-48; males: $\bar{x}$ = 13.6 minutes, N = 8, range = 5-32). The sample sizes are small, however, and the difference is not statistically significant.

The same potential biases due to differential observation of the sexes as mentioned for termite-fishing could be operating in the case of ant-dipping. Small numbers of observations prevent systematic examination of the variables of bout length, periodicity, and in/out of camp starts of follows. The latter, however, seems unlikely to be a biasing factor. Although the locations of termite mounds seem to be concentrated in valley bottoms and on lower hillsides (where camp is located), known locations of driver ants nests are randomly scattered about valley bottoms, hillsides, and ridge tops. The distribution of ant-dipping bouts over the daylight hours is approximately the same as termite-fishing, with a peak at midday.

It is concluded that a sex difference exists in Gombe chimpanzees feeding on driver ants, with the females more likely to utilize this insect prey species.

Analysis of Fecal Samples

Between June 1964 and December 1967, Goodall periodically collected and analyzed the contents of fecal samples from Gombe chimpanzees. The results were presented mainly in terms of seasonal variation in diet (van Lawick-Goodall, 1968b, p. 187). Ten months of the data (June 1964 through March 1965) are amenable to reanalysis for sex differences: 194 of the 572 samples are identified by the sex of the defecating individual. The sample represents 30 chimpanzees of juvenile age or greater (11 females and 19 males). Each sample was collected randomly and independently, and a food type was noted as either present or absent. This method provides an independent measure of food intake for sex difference comparison: the dung data were collected not less than 8 years earlier than the behavioral data by different observers of a somewhat different study population of chimpanzees.

The results show a marked sex difference in insect eating: 45 of 81 (56%) female fecal samples contained at least one type of insect remains. Only 31 of 113 (27%) male fecal samples did so. The difference is highly statistically significant ($\chi^2 = 16.84, df = 1, p < 0.001$).

Table 2 presents the results in more detail. The incidence of three insect prey species are presented individually and all remaining insects are lumped into a miscellaneous row. Finally, all insect cases (N = 82) are combined for comparison with all vertebrate cases (N = 14). Termite *(Macrotermes)* remains occurred significantly more often in female fecal samples. Driver ants and miscellaneous insects were too infrequently represented to allow statistical testing. The most common insect species present in the fecal samples was the weaver ant, *Oecophylla longinoda.* (This ant species lives in small groups that construct and inhabit arboreal nests of green leaves stuck together with silk. In feeding on them, chimpanzees pluck the entire nest, crush it in a variety of ways, then peel it apart to ingest the occupants. Chimpanzee feeding on weaver ants infrequently appeared on travel-and-group charts, as bouts were short and difficult to observe.) Weaver ant remains occurred significantly more often in female fecal samples, and when all insect foods are pooled, the sex difference is highly significant. Taken together, the behavioral and fecal data indicate that female chimpanzees are more insectivorous than male chimpanzees at Gombe.

TABLE 2

Sex differences (observed *vs.* expected) in feeding on insect and vertebrate prey by feral chimpanzees in Gombe National Park, Tanzania. N = 194 fecal samples taken over 10 months in 1964-65.

PREY	TOTAL CASES	MALE OBS(EXP)	FEMALE OBS(EXP)	χ^2 k=2, df=1	p
Weaver ants *(Oecophylla)*	42	16(24)	26(18)	6.23	<.02
Termites *(Macrotermes)*	29	9(17)	20(12)	9.09	<.01
Safari ants *(Dorylus)*	5	4(3)	1(2)	—	—
Miscellaneous insects	6	4(3)	2(3)	—	—
All insects	82	33(48)	49(34)	11.31	<.001
Vertebrates	14	13(8)	1(6)	7.29	<.01

FEEDING ON VERTEBRATES

Of the vertebrate classes, only Aves and Mammalia have been seen to be eaten by Gombe chimpanzees. Several species of birds, especially their eggs and nestlings, are taken, but insufficient data are available to me at present to make sex difference comparisons on birds alone. The bottom row of Table 2 combines avian and mammalian prey found in Gombe chimpanzee dung. The remains show a significant sex difference: 13 of 14 cases were eaten by males.

Chimpanzees in the Gombe National Park have been seen to prey upon seven species of mammals: *Papio anubis* (olive baboon), *Cercopithecus ascanius* (red-tailed monkey), *C. mitis* (blue monkey), *Colobus badius* (red colobus monkey), bushback *(Tragelaphus scriptus),* bush pig *(Potamochoerus porcus),* and a bat *(sp?).* Teleki (1973) has presented detailed analysis of Gombe chimpanzee predatory behavior, so only findings relevant to the topic of sex differences are given here. It should be stressed at the outset, however, that Gombe chimpanzee predations are more than casual events: Wrangham (1974b) has estimated that kill rates of the prey of chimpanzees are comparable to those suffered by African ungulates from large carnivores.

From 1960 through 1967, 28 chimpanzee predations were seen by observers at Gombe (van Lawick-Goodall 1968b). Although detailed information is not available in each case, all described instances of hunting and killing involved males. In the 30 predation incidents seen between March 1968 and March 1969, "only adult male chimpanzees were observed to initiate predation and to pursue and capture prey" (Teleki 1973, p. 56). Adult males also divided the carcass, most often attended the feeding sessions after a kill, and consumed most of the resulting meat. Finally, the author scanned the Gombe "predation file" for the period of March 1968 to November 1973. Of the 71 successful predations observed during that period, the sex of the killer was known in 49; 48 by males vs. 1 by females.

The inescapable conclusion would seem to be that predation on mammals by Gombe chimpanzees is largely a male domain.

EVOLUTIONARY IMPLICATIONS: DIVISION OF LABOR

Given that a sex difference in animal prey intake exists in Gombe chimpanzees, with males concentrating on mammalian prey and females on social insect prey, what does this mean? First of all, the behaviors involved in obtaining the prey would seem to be of crucial importance. Male chimpanzees obtain meat by stalking, pursuing, capturing, killing, dividing, and distributing a single mammalian prey. This behavior often involves primarily male groups roaming relatively great distances and acting cooperatively when the appropriate situation fortuitously arises—in short, *hunting*. On the other hand, female chimpanzees (predominantly) obtain ants and termites by prolonged, systematic, and repetitive manipulative sequences. Several chimpanzees may work side by side, but basically the activity consists of an individual accumulating a meal of many small units that are usually concentrated at a few known permanent sources—in short, *gathering*. Viewed in this way, the sex differences might be a preadaptation which could lead to a more advanced form of division of labor by sex. It might have been this type of preadaptation that was amplified by savanna selection pressures to produce eventually the female-gatherer and male-hunter prototype of human social organization (Etkin 1954; Washburn and Lancaster 1968).

Numerous examples of this type of sexual division of labor exist in extant tropical foraging peoples, e.g., Mbuti pygmies of Zaire (Turnbull 1965); Kung San (Bushmen) of Botswana, Hadza of Tanzania, aborigines of Australia (Lee & DeVore 1968). This type of division of labor is strikingly consistent across many human foragers, with the males typically ranging widely to pursue and kill occasional mammalian prey while the females regularly collect plant and infra-mammalian foods from a limited area. Often the unspectacular gathering activities provide the bulk of the group's subsistence (Gould 1967; Lee 1968). Large-scale comparisons across the range of nonforaging human cultures reveal similar tendencies. Even in agricultural and industrial societies, the males tend to do whatever hunting and females whatever gathering that occurs (Murdock & Provost 1973).

Why should chimpanzees exhibit this sex difference in diet? It seems likely that primary sexual dimorphisms selected for at earlier evolutionary stages combined with even earlier general mammalian adaptations to favor an incipient male-hunter/female-gatherer division. Selection pressures involving intraspecific agonistic interaction (especially in competitive hierarchical structures) and/or interspecific protective responses against predators may have produced greater body size, physical strength, and dental armament in males.With the monosexual co-option of the mammary glands as organs of nutrition for the offspring, the female became the indispensable sex in mammalian child-rearing. In the higher primates, the bond originating in lactation necessarily becomes prolonged with increasing selection for neoteny. The combinations of lengthy gestation and suckling ensures that most of the adult life of a female pongid is defined and restricted by reproductive activities.

The picture sketched above is familiar, both in terms of descriptions of human foraging peoples and speculation about protohominids. Males are more suited for hunting because their bigger size and strength enables them to dispatch more efficiently larger prey animals. Gombe chimpanzees concentrate on young animals as prey but may take monkeys of up to 10 kg (the estimated weight of an adult male red colobus monkey). Male canine teeth are larger and more differentiated, enabling them to deal more efficiently with the prey's antipredator responses. For Gombe chimpanzees, this may mean having to resist the threats or counterattacks of adult male baboons, which are themselves formidably equipped with large canines (Wrangham 1974b). Being unencumbered with dependent offspring, males are freer to roam widely in search of prey and perform pursuit activities requiring speed and agility. At Gombe, chimpanzee males' ranges much exceed females (Wrangham 1974b). The exceptions to this rule are sexually cycling females, which by definition have no offspring or have offspring old enough to be left temporarily to locomote independently. The Gombe prey animals are nonrandomly distributed, e.g., the three monkey species are more often found in the upper rather than lower valleys, but at any given time a monkey troop could be anywhere in a large home range (Clutton-Brock 1973). Generally, Gombe chimpanzees capture their primate prey high in the trees; acute balance, sudden direction change, concentrated bursts of exertion, and manual dexterity are very important. The activity is erratic, irregular, and hurried. Transporting an offspring, pre- or postnatally, would hamper all of these.

In contrast, social insects represent a stable, localized animal food source perhaps more suited to female species-specific characteristics. By focusing on an alternative source to meat, females may avoid competition with the more dominant males. Termite mounds may remain productive for many years. Mounds that were being successfully fished by the chimpanzees at the beginning of the Gombe study in 1960 are still productive 14 years later. It seems likely that by the time a chimpanzee begins to travel independently of its mother, it knows the location of many of the main termite mounds in its range, having accompanied its mother to them over a minimum of four termite-fishing *high seasons*. Females continue traveling with their mothers for longer than do males, so they are probably particularly well-informed about termite mound locations. Certain keen termite-fishing chimpanzees at Gombe follow termite mound circuits, going from one to another by the easiest and most efficient routes. For females with dependent offspring, this may be an important economic procedure in terms of energy expenditure.

Once at the mound, the chimpanzee's termite-extracting process is sedentary and interruptible. The fishing technique requires little more than forelimb motion, and young infants may cling ventrally and sleep, suckle, or watch while the mother fishes. Older infants may safely wander, play, or attempt fishing without concern. The passive, limited defense of the prey species presents no danger of pain or injury. In nursery groupings, several mothers may fish simultaneously, while infant play groups clamber on or about the mound, with apparently no ill effects on the fishing. If an infant becomes distressed or otherwise requires attention, the mother may temporarily break off fishing to attend to it, then return to resume feeding. None of these activities is compatible with hunting. Rather, as in human angling, some conditions are more conducive to success than others, but the fish are always there.

In dipping for driver ants, the majority of feeding bouts are re-uses of nests dipped at least once previously, often by other chimpanzees (McGrew 1974b). Since these nests are temporary bivouacs of a few days, however, most chimpanzee contacts with driver ants appear to be fortuitous. The ants are nomadic, moving throughout the range of Gombe habitats, and it would seem impossible for a chimpanzee to predict their locations accurately. Either one returns to a known nest site or stumbles across a colony by accident in daily travels. Both ways of making contact seem to be efficient for pregnant or lactating animals. Once at the nest site, the chimpanzee exploits an animal dietary source that stays focused at one spot for the duration of feeding. The dipping individual can adjust its exposure to the prey by moving in and out of the limited range of the ant defenders. Younger infants usually cling tightly as the mother moves about the nest opening, adopting various positioning strategies (McGrew 1975). Older infants stay outside the ants' defensive perimeter, watching their mothers or playing in the trees. Periodically the mother retires to rejoin the infant, and they may engage in grooming or play before the mother returns to dipping.

Social (as opposed to solitary) and terrestrial (as opposed to arboreal or aquatic) insects are an economical animal protein source for an encumbered individual with near-perpetual child-care duties. Acquisition of the food item depends upon a repertoire of elegant techniques rather than size-strength-endurance factors, and the task is self-paced. The activity is energistically economical and unhurried. Females obtain termites throughout the year, and even in the leanest times, the termite mound locations are such that little energy is wasted in checking a mound, even if it is inactive.

These conclusions resemble closely those previously put forward by various authors, e.g., Etkin (1954), concerning division of labor by sex in a variety of human cultures. Brown (1970) suggested that the limiting factor on the degree of female participation in subsistence activities is the compatibility of the activity with child-care demands. Nerlove (1974) extended this to the key restricting factor, breast-feeding, and found that early supplementary feeding is positively correlated with participation in subsistence activities. (A further limiting factor, intra-uterine pregnancy, seems to be assumed.) Brown posited two solutions to the limitations of child care: reduction of the care-taking load on the mother, or, maternal engagement in compatible activities concurrent with child care.

The chimpanzee solution to the problem presumably resembles that of the protohominids and involves the latter alternative. Unless technology develops extrauterine pregnancy, female gestation is necessary. Until alternative milk sources can be domesticated, female breast-feeding is indispensable. All primate infants require milk at least for the first few months of postnatal life, and only female mammary glands can provide it. Presumably those populations best survive(d) whose females specialize(d) in exploiting suitable sources of essential nutrients (e.g., amino acids, lipids, vitamins, etc., available only from animal sources) that were compatible with safe and efficient reproductive *and* nurturant activities.

FOOD DISTRIBUTION

How would a protohominid population make the evolutionary transition from sex differences in diet to sexual division of labor in food acquisition and distribution? In a general sense the two situations overlap: it seems likely that individuals in those populations in which subsets of its members exploit different food resources will be more successful than individuals in those populations in which all members compete for the same food resources. Sex differences in feeding might be considered incipient division of labor in that energy wasted in inter-sex competition for food is therefore minimized. In anthropological terms, however, societal division of labor means more than this: it means complementary activities that, taken together, consistently and predictably benefit individuals of both sexes and all ages in a group. It seems to exemplify the phenomenon of reciprocal altruism as described for both human and nonhuman animals by Trivers (1971). An important evolutionary step in the development of advanced human social organization, therefore, would seem to be systematic inter-individual food-sharing (Le Gros Clark 1967).

There have been incidental reports of food-sharing in a variety of nonhuman primate species (Dare 1974; Kavanagh 1972), but the chimpanzee is the only one on which detailed knowledge is available. Nissen and Crawford (1936) experimentally investigated sharing of food and food tokens in captive pairs of juvenile chimpanzees. Teleki (1973) described the specific distribution of meat among Gombe chimpanzees in ten episodes of predation upon mammalian prey (see also Suzuki 1971). He found extensive transfer of parts of the carcass to many individuals by recovery, taking, and requesting. Large groups of chimpanzees, including many absent at the kill, assembled around the adult male hunter(s) and obtained meat. Attendance at the kill did not guarantee receipt, neither was distribution egalitarian nor systematic, but recognizable patterns emerged. Eighty percent of sharing interactions involved adults, i.e., males and females receiving from males. Female chimpanzees in estrus were more successful in obtaining meat than nonestrous females. Meat transfers among males did not strictly follow dominance rank-order lines; instead, age was highly positively correlated with frequency of success (Wrangham 1974b). Kinship ties between individuals appear to be important factors in meat distribution, but this has not yet been clearly elucidated. Finally, van Lawick-Goodall (1968b) and Wrangham (1974b) have stressed that meat distribution by Gombe chimpanzees is not always peaceful and that intense bursts of competition may occur.

More frequently, wild chimpanzees share vegetable foods (McGrew 1974a; Nishida 1970). These food transfers are a daily occurrence when chimpanzees feed on prized (e.g., bananas, sugar cane), or not easily processed (e.g., fruits of *Strychnos, Diplorhyncus*), plant foods. (Details of Gombe chimpanzee diets and food-processing techniques appear in van Lawick-Goodall 1968b and Wrangham 1974b.) The results summarized below are based primarily on records of nonaggressive banana-sharing in camp at Gombe. A standardized food-begging checksheet was used by a variety of observers over 21 months (January 1972 through September 1973).

Banana-sharing occurred between all age-sex classes of chimpanzees (see Figure 2). The vast majority occurred between a mother and her infant, juvenile, or adolescent offspring; food almost always passed from mother to child, but the reverse was also true. If the 37 chimpanzees involved are paired in all possible dyadic combinations, only 5% of these dyads comprise known kinship ties, yet these accounted for 86% of the 457 observed banana distributions. The remaining nonkinship sharings were far from random: 78% were by adult males giving bananas to adult females or their infants. Individual differences in the generosity of adult males emerged that were unrelated to known kinship ties. All age-sex classes avidly consumed bananas, yet adult males rarely begged bananas at all and adult females almost never begged from each other. Overall, distribution of bananas was not random by age: in 88% of cases, recipients were younger than donors, a finding generally in reverse of dominance order. Many of these nonrandom age-sex class food distribution patterns resemble those recorded by Nishida (1970) in a geographically separated chimpanzee population 170 km to the south.

Incidental observations outside camp of the sharing of naturally occurring fruits confirmed these impressions. The hard-shelled, orange-sized fruit of the *Strychnos* requires strength and technique to process it for eating. No Gombe chimpanzee infant was seen to be capable of this, but 2-5-year-old infants almost always cadged fragments from their mothers. The leathery pod of *Diplorhyncus* trees presented an even more interesting case. Each pod contains small amounts of edible material in sticky sap, and adults may process hundreds of these in prolonged feeding sessions. The pod is neatly split in two, and in some chimpanzee mother-infant pairs, the 2-3-year-old infant regularly takes one half (the apparently less desirable one) while the mother eats from the other. Nishida (1970) termed this "halving behavior," with other plant foods. Infants of this age are capable of opening the pods for themselves but are messy and inefficient about it.

Numerous authors (Etkin 1954; Le Gros Clark 1967; Tiger & Fox 1971; Washburn & Lancaster 1968) have suggested that food-sharing arose in hominid evolution in connection with hunting. Because higher-calorie animal foods are further up the food chain than lower-calorie plants, they are an especially useful dietary commodity; at the same time, meat must be used quickly under tropical conditions before other organisms usurp it. Food-sharing in the context of division of labor is

FIGURE 2
The infant eats the banana peel obtained from her mother. Photograph by W. McGrew.

sometimes cited as a differentiating factor between the pongids and hominids (e.g., Sahlins 1965; Tiger & Fox 1971). Reciprocity is implied: it may only be feasible for a man to gamble on hunting if he is mutually bonded to a woman who consistently engages in surplus gathering. That such sharing would be powerfully adaptive in terms of kin selection seems obvious (Hamilton 1972). Those males that provide animal protein for pregnant and lactating females sharing or nurturing their genes would almost certainly enjoy enhanced reproductive success. But individual selection for such altruism need not require shared genes between interactants or their offspring to be adaptive (Trivers 1971). Among the Hadza, special provision in meat distribution is made for pregnant women (Woodburn 1968). Lactation requires an extra 1000 calories daily (Gunther 1971) at a time when females are least suited for an additional foraging workload.

This pattern seems to apply to living foraging peoples, who may have complex and stereotyped procedures for sharing of food items (Coon 1971). Meat-sharing especially may be strictly regulated by ritual or custom (Gould 1967) and administered by an individual whose qualities or position ensure a fair division (e.g., Turnbull 1965). Although all group members will individually eat on the spot many types of foods, especially plant parts (e.g., Woodburn 1968), much food is transported for distribution elsewhere. In gathering, transport virtually requires the use of containers to make the energy expenditure worthwhile, as such food items are usually small, numerous and inefficiently packaged or constituted (Washburn & Lancaster 1968). In hunting, ease of transport depends on such factors as prey size. In some cases it is obviously easier to move the group to the carcass rather than vice versa.

Contrary to some opinions, wild chimpanzees can show a significant degree of food-sharing. Both similarities and differences seem to exist between human and chimpanzee distribution of plant and animal foods. It seems especially notable that chimpanzee meat distribution does not follow straightforward dominance lines. The fact that older (and more experienced and confident?) males are more successful in obtaining meat after predations could constitute a preadaptation for later evolutionary development of distributive roles. Such male sharing would seem to be adaptive in serving to increase group cohesion among adults, perhaps by facilitating cooperation in defense of group range and resources. Perhaps the increased efficiency of cooperative over solitary hunting or defense of prey is involved (see Schaller & Lowther 1969). It seems likely that selection would favor those individuals capable of participating in mutually advantageous collective actions at least in certain appropriate circumstances.

Chimpanzees may transport mammalian prey pieces for more than short distances in a variety of ways, apparently according to its size: by mouth, draped around the neck, slung over the shoulder with one hand supporting. None of these methods seems efficient; Hewes (1961) points out the importance of bipedalism in food transport. All chimpanzee distribution of plant foods takes place at the source, the exceptions being when an individual carries an armful of bananas to cover at the edge of camp or detaches a fruit-laden branch and retires to a more comfortable spot to eat it. Wild chimpanzees have not been seen to use containers. The closest thing to food-sharing occurs in feeding on insects when offspring eat from the mother's tool or pick up single insects overlooked or rejected by her.

If implications for hominid evolution can be made from pongid food distribution, the most important ones may be that food-sharing, and the sex differences in it, are likely to antedate hunting (see also McGrew 1974a). Kavanagh (1972) suggested that it is a general primate characteristic not necessarily associated with hunting-and-gathering, or mammalian predation, or even a limited supply of food. The patterns of plant food-sharing seen in Gombe chimpanzees would seem to support all three points. The predominant type of mother-to-offspring sharing is directly advantageous in terms of kin selection. The sharing by adult males with females and offspring is more marginal, since we assume that chimpanzees are ignorant of paternity and reciprocity is unlikely or long delayed. Recent findings by McGinnis (this volume) and Tutin (1974), however, on selective, nonpromiscuous mating mean that paternal effects may occur, albeit in ignorance. Given the known individual differences in adult male generosity, indirect kin selection could be operating through preferential treatment of favored females and their offspring. Only with the distribution of meat does one find interactions that might function in even more indirect selection for adult, especially male-male cohesion and for prolonged female sexual receptivity.

EVOLUTIONARY IMPLICATIONS: ORIGINS OF TOOL USE

Most recent thinking in anthropology links the origin of tools in hominid evolution to the use of stone implements in the hunting of mammals by cooperating groups of males. For example: "The first tools on earth were butchering tools" (Tiger & Fox 1971, p. 121). Tool use is only one of many aspects of hominization that is meant to be related to hunting (Le Gros Clark 1967; Washburn & Lancaster 1968). Laughlin (1968) termed hunting as "a master integrating pattern" which ". . . played the dominant role in transforming a bipedal ape into a tool-using and tool-making man who communicated by means of speech and expressed a complex culture" (p. 318). Steward (1968) stated that hunting "was presumably the principal factor that created the nuclear family" (p. 331).

With focal interest on the hunting half of the hunter-gatherer

complex, gathering seems to have been disproportionately underempha-
sized. Many workers ignore the probable evolutionary importance of
gathering or dismiss its results as "casually collected foods" (Laughlin
1968, p. 319). Coon (1971, p. 73) refers to "the primacy of hunting"
and states categorically that "it has more impact on social structure
than gathering." Lee (1968) and others have criticized this one-sided
viewpoint, which, he says, gives the impression that early hominid
cultures could be characterized by a "precarious hunting subsistence
base." Careful empirical studies of extant human foragers demonstrate
the importance of gathering in day-to-day life. Among the Kung San of
Botswana, 60-80% by weight and 67% by calorie of the diet derive from
vegetable sources, primarily collected by females (Lee 1968). A similar
situation obtains among the Hadza of Tanzania, who rely mainly on
wild vegetable matter for their food, with no more than 20% by weight
accounted for by meat and honey (Woodburn 1968). Australian
aborigines are basically vegetarian in diet; meat and other fleshy foods
are preferred when available (Gould 1967). In all cases, the male-hunter
and female-gatherer division of labor obtains, with gathering predominating.

As has been discussed, male chimpanzees are proficient hunters of
mammalian prey. Many parallels exist between their patterns of hunting
and those used in various human foraging societies. Several chimpanzees
may work cooperatively to bring down a single prey animal, often by
anticipating and blocking its escape routes (Teleki 1973; Wrangham
1974b). Hunting chimpanzees exchange information nonverbally, not
necessarily with conscious intent, e.g., by gaze direction, hair erection,
silent locomotion (Teleki 1973). Laughlin (1968) has stressed the
importance of silent, inconspicuous signaling in human coordinated
hunting. Alternatively, single chimpanzees may successfully hunt and
kill prey alone. Like the Hadza (Woodburn 1968), the same individuals
may hunt solitarily or cooperatively, depending on the context. Chimpanzees concentrate on immature prey animals, although this varies
among prey species (Teleki 1973; Wrangham 1974b). According to
Laughlin (1968, p. 313) "The archeological and ethnographic record is
unambiguous in the fact that the vast majority of mammals killed are
immature or subadult." Chimpanzees monitor and respond to the
behavior patterns of the prey species as well as intra- and inter-specific
competing predators, apparently to maximize rates of animal protein
intake. At Gombe, 75% of the bush bucks eaten by chimpanzees were
taken from baboons. The chimpanzees arrived at the scene after the kill
had been made by baboons and dispossessed them (Wrangham 1974b).
Such success is based on selective discrimination of vocal cues from
either prey species or fellow conspecific predators; other vocalizations in
nonpredation contexts are largely ignored. Laughlin (1968) has stressed

the considerable investment in practical knowledge of animal behavior made by human hunters, even to the extent of giving this priority over tool improvements. To summarize, it seems that chimpanzees, like the Dobe Bushmen, ". . . eat as much vegetable food as they need, and as much meat as they can." (Lee 1968, p. 41).

One major difference between chimpanzee and human hunting is obvious. Extant human hunters, even at the simplest level, always use tools in the hunting and processing of mammalian prey (for detailed review, see Oswalt 1973). Although chimpanzee predation on mammals has been recorded in at least five different locations in Africa (e.g., Kawabe 1966; Suzuki 1971), and over 100 attempted or successful predations have been seen at Gombe alone, tool use has been seen only once in hunting. Simply put: chimpanzees are successful hunters without tools. The exceptional case is an instructive one, however. F. X. Plooij (personal communication) saw a large group of Gombe chimpanzees surround a group of four bush pigs. The adults kept a piglet between them, and the chimpanzees were effectively kept at bay by the formidable tusks and size of the pigs. Finally, an old male chimpanzee picked up and threw a melon-sized rock, hitting one of the adult pigs. Shortly thereafter, the pigs fled and the chimpanzees chased, caught, killed, and ate the piglet.

In contrast, consider the sophisticated tool-use activities for feeding on social insects specialized in by Gombe female chimpanzees. Connolly (1974) mentioned the greater skill involved in tool-use tasks when "two hands are used in complementary roles to achieve certain ends" (p. 540). The only example of this so far seen in wild chimpanzee tool-using is in ant-dipping: one hand (or foot) holds the wand steady while the other hand sweeps the length of the wand, catching the ants. Although both termite-fishing and ant-dipping are techniques for exploiting social insect foods, the raw materials, manufacturing, and final products of each show important differences (McGrew 1975). A future paper will present a detailed contrasting of the two activities in the Gombe chimpanzee population and cross-cultural comparisons with other chimpanzee populations (e.g., Nishida 1973). During the two-month peak season, termite-fishing occurs on a daily basis, and females obtain termites in all months of the year. A typical bout of ant-dipping may result in about 20 g of ants being ingested. Insects are more than a trivial nutritional source, as has been recognized by some previous commentators (e.g., Lancaster 1968), but the sex difference adds a new dimension to this.

Oswalt (1973) recently distinguished between the basic *implement* and the more advanced *facility* in his taxonomy of food-getting technology. Implements apply energy directly to alter other masses (e.g., a digging stick), whereas facilities apply energy indirectly to influence the conditions of living things (e.g., a pitfall). Facilities are more conceptually complex than instruments and are especially useful when used in conjunction with implements. At Gombe the sole use thus far seen of facility is part of the chimpanzee tactics in driver ant-dipping (McGrew 1974b). Chimpanzees prefer to dip from elevated sites, apparently to avoid the massed defense of painful bites from the ants. In situations where natural elevations, e.g., fallen logs, overhanging tree limbs, etc., are absent or overrun, chimpanzees may bend over a previously upright sapling to a horizontal position, then perch on it to dip. They thus remain within arm's reach of the ants' nest but out of direct contact with the dispersed ants' defense. Four of the five Gombe chimpanzees seen to use this technique were females.

Given the technological contrasts between methods for obtaining meat vs. insects, and the sex differences in their deployment, Gombe chimpanzee behavior may imply new interpretations for supposed protohominid behavior. If any parallels existed between the hominid and pongid lines in the development of implementation, it seems more likely that tool use originated in solitary female-foraging activities rather than cooperative male hunting. This interpretation is not new (see similar speculative account in Etkin 1954; Morgan 1972), but perhaps these new data give it more plausibility.

Isaac (1968, p. 254) has pointed out that "Pleistocene prehistory . . . has come to be expressed almost exclusively in terms of stone tool morphology." Lithic artifacts obviously preserved better in the early archaeological record than did nonlithic ones, e.g., bone, horn, shell, leather, wood, bark, etc. Yet the probability that nonlithic tools preceded lithic ones in cultural evolution seems to have been under-emphasized until recently. For example, while recognizing the use by protohominids of nonlithic materials such as bone and wood, Oakley (1965) implied that the first tools were of stone and that nonlithic materials came in only when stone tools were available to shape them. The controversy over and resistance to Dart's proposals of an "osteo-dontokeratic culture" are well known (e.g., Dart 1964). Even the functional assignments made to early lithic tools seem to be subject to considerable presupposition based on hunting, e.g., the simple Oldowan hammering stones are thought of as tools for smashing open long bones to permit access to marrow (Leakey 1966). It may be just as plausible (vide infra) to assign these to the task of breaking open hard-shelled fruits and nuts, or insect foods. Certain Australian aborigines use stones to smash open woody galls, then eat the insects inside (Cleland & Johnston 1933). Similarly, Gould (1967) has pointed out that stone hand axes, although an important part of the material culture of certain Australian aborigine groups, are *not* used by them in butchering. This fact contrasts with the usual function assigned to Paleolithic stone hand axes, which is the dismantling of large game. Gorman (1971) has noted the conspicuous absence of lithic hunting tools at Mesolithic sites in southeast Asia and has suggested that the ungulate and primate prey found there were obtained with tools of "perishable material."

Happily, this fixation on the hunting and processing of meat seems to be waning. Oswalt (1973) provides detailed comprehensive examples of the simplest tool kits used by a representative worldwide sample of foraging peoples. One is struck by the importance on nonlithic tools, especially the archetypal removal stick, in a variety of successful subsistence activities. Washburn (1972) recently stressed the importance of wood as a raw tool material, stating that it was likely that the principal tools of *Australopithecus* were wooden.

Chimpanzees at Gombe have not been seen to use stones as tools in feeding activities; instead they use a variety of vegetational tools to obtain and process solid and liquid foods. For many of these tasks, stone would be patently unsuitable, e.g., as long, narrow, pliant probes (see Oswalt's 1973 discussion of "flexibles" as a raw material type). This probably applies to the apparent optimal utility of wood as a tool-use raw material in ant-dipping (McGrew 1974b). Gombe chimpanzees, however, smash hard-shelled fruits (e.g., *Strychnos*) against solid objects, including large embedded rocks (van Lawick-Goodall 1968b). Many repeated blows may be struck against the anvil until the rind cracks open. At other African sites, chimpanzees have been recorded as using stones as hammers to smash open nuts or fruits on anvils (Beatty 1951; Rahm 1972; Struhsaker & Hunkeler 1971). Oswalt (1973) has noted this as an example of the *multiplication process,* a technological advancement in food-getting efficiency. All the tool uses cited are in chimpanzee gathering, rather than hunting, activities. (Another, perhaps borderline case has been noted at Gombe in the *processing* of meat: an old male used leaves on one occasion to catch and hold his feces the day after a predation. He extracted from the dung bits of undigested flesh and reingested them (Bauer & Halperin, personal communication). Similar recycling of fecal material has been recorded for human hunter-gatherers (Coon 1971, p. 169).

The validity of the implications drawn from these observations depends on the usefulness of a living great ape, the chimpanzee, as a model for early man. No one would deny that the African great apes are the extant animal group with which man last shared common ancestry. In recent years, however, several simultaneous developments have complicated the problem. Field studies of mammalian ethology in general and a variety of nonhuman primate species in particular have made us aware of the complexity of life types of nonanthropoid species (e.g., Eisenberg 1973). From this has come a number of models for hominization based on species subject to the selection pressures of primarily open African habitats: gelada (Jolly 1970); savanna baboon (DeVore & Washburn 1963); large social carnivores (Schaller & Lowther 1969). With the emphasis on the savanna in hominization, the living African pongids become less relevant by implication, as they are primarily forest-dwelling at the present time.

At the same time, captivity studies of chimpanzees, and other apes, done in the last 10 years have confirmed repeatedly that we previously underestimated the mental capacities of our nearest relations, as shown in studies of language (Hewes 1973), instrumentation (Dohl 1970), communication (Menzel 1974), complex cognitive abilities (Davenport & Rogers 1970), self-concept (Gallup 1970), etc. Categorical statements about qualitative differences between man and the great apes are increasingly muted. Selection pressures capable of bringing about such advanced abilities are more and more difficult to imagine, given the apes' current niches in nature, but the presence of the abilities is undeniable.

Finally, extensive long-term field studies of separate chimpanzee populations in Tanzania by a variety of fieldworkers have revealed the existence of complex nonhuman primate cultures (van Lawick-Goodall 1973; Nishida 1973). The use of terminology is deliberate, for capacities which only 10 years ago would have been reserved to hominids cannot now be denied to pongids (c.f. Weiss 1973). Findings on tool use, carnivorosity, food-sharing, lifelong familial ties, etc., can no longer be dismissed as anecdotal or idiosyncratic to one population. They have been systematically confirmed by field studies involving scores of research workers and thousands of observation hours. The results may be somewhat embarrassing both to scientific theories and to anthropocentric assumptions, but wild chimpanzees continue to demonstrate these activities. In conclusion, it appears that the chimpanzee, increasingly, represents a better evolutionary model for early man.

Acknowledgment

The author thanks the Republic of Tanzania and the Tanzania National Parks for permission to study in the Gombe National Park; J. Goodall for providing advice and the use of the facilities of the Gombe Stream Research Centre; D. Hamburg for providing encouragement and finance through the Department of Psychiatry, Stanford University, and the W. T. Grant Foundation. The author thanks too many Gombe research workers and field assistants to be mentioned by name who contributed observational data, but J. Crocker, C. Tutin, and R. Wrangham were especially helpful in providing stimulating comments and criticisms at various stages, and D. Bygott, J. Moore, and C. Tutin generously provided photographs. Finally, the paper is dedicated to G. G. Rushby, who recommended in 1946 that a reserve for chimpanzees be established at Gombe.

Anne Pusey

Intercommunity Transfer of Chimpanzees in Gombe National Park

In group-living primate species on which long-term studies have been conducted, it is found that outbreeding is generally promoted by movement of individuals between groups. In most species the individuals that move are males, whereas the females remain in the natal group, e.g., rhesus macaques (Koford 1966; Boelkins & Wilson 1972), Japanese macaques (Sugiyama 1976), olive baboons (Ransom 1972; Packer 1975), purple-faced langurs (Rudran 1973), vervet monkeys, (Struhsaker 1967) and howler monkeys (Scott et al., in press).

In chimpanzees we find evidence for the opposite situation. Nishida (Nishida & Kawanaka 1972; Nishida, this volume) describes the many instances of female transfer that have been observed in two communities (which he calls unit groups) of chimpanzees in the Mahale mountains. There, females may transfer temporarily or permanently. Only young nulliparous females have been observed to transfer permanently to a new community. Temporary transfer for varying lengths of time was shown by both young nulliparous females and by some older females, mostly when they resumed estrous cycles after the birth of their previous infant. In a 7-year study period, no male was observed to transfer to another community.

The purpose of this paper is to review the evidence that at Gombe also, females, but not males, transfer between communities. The first section of the paper describes methods of data collection, and discusses the possible relation of the structure and composition of a provisioned group of chimpanzees visiting an artificial feeding area *camp* to the structure of chimpanzee communities as it is now understood. In the second section, evidence mainly from camp attendance data is presented to show that although male membership in the habituated community (and both communities after the original group split in 1972) is very stable, female membership is less stable, and that many young females have joined the camp-visiting community for varying periods, probably from adjacent communities. Finally in the third section, evidence is given that some habituated females have transferred to other communities.

CAMP ATTENDANCE AND COMMUNITY STRUCTURE

Chimpanzees have been provisioned regularly with bananas since 1963 at an artificial feeding area, *camp* (Goodall 1968b; Wrangham 1974). Attendance of each individual, and the estrous state of each female, has been recorded daily from 1963 to the present, with the exception of a few months in 1963 and 1964. Observations were largely restricted to this area until 1968 when individual chimpanzees began to be followed away from camp. From 1972 onward, individual chimpanzees were sometimes followed for several days and a great deal more was learned about out-of-camp ranging, associations, and community structure.

The structure of chimpanzee communities at Gombe, as elucidated by out-of-camp following, has been discussed by Goodall (1975; this volume) and in detail by Bygott (1974; this volume) and Wrangham (1975; this volume). Very briefly, each community appears to consist of a core group of adult males who associate together frequently, and share a common range. The range of these males includes the smaller and distinct but overlapping ranges of a similar or slightly larger number of mature lactating females with offspring. Estrous females and females without offspring tend to range more widely over the whole area with the males. The unit groups of chimpanzees in the Mahale mountains described by Nishida (1968; 1970; this volume) appear to be similar in structure and composition to the Gombe communities.

The usual view of chimpanzee communities (or unit groups) has been that they are bisexual units with both males and females associating much more with each other than with members of other communities, but Wrangham (1975; this volume) has put forward an alternative view. He suggests that although males are grouped into separate communities, lactating females may be distributed geographically in relation to other females and independently of the males. Thus there may be females with a range at the edges of two male communities, who associate with males from each community and with females who have ranges within either of the two male group ranges. There are not yet enough data on the behavior and associations of females with ranges at the edge of male group ranges to say which view is closest to the real situation. In either case, however, there will be many lactating females whose ranges are completely within the male group's range. These will associate with only one group of males and in this sense will *belong* to one community. Also, unlike lactating females, cycling females associate frequently with males and travel with them over the male group's range. A change from associating regularly with one male group to associating regularly with another, is truly intercommunity transfer whether the community is a completely bisexual unit or not.

During one study period, all the chimpanzees that had visited camp since provisioning began in 1963, were observed. Because of the peaceful interactions of all these chimpanzees from the beginning, it is likely that most or all of them had been associating together before provisioning began, and that they probably belonged to one community. From 1963 onward, there was always a subgroup of chimpanzees who arrived from the south and visited less frequently than the others. In 1972 the community finally split into two units. The southern subgroup stopped visiting camp entirely and became the Kahama community with a range in the south. The remainder of the group continued to visit camp and became known as the Kasekela community with a range centered in Kakombe valley.

The females visiting camp by the end of 1964 almost certainly included most of those lactating females who had ranges near camp and frequently associated with the males. Some of the cycling females possibly came from farther away. Probably there were other lactating females in 1963 and 1964 who had ranges toward the edges of the male range, who also associated frequently with the males, but did not come to camp. The effects of provisioning, which reached a peak in 1966-1967 (Wrangham 1974), may have drawn such females to camp after the others.

SEX DIFFERENCE IN CAMP ATTENDANCE, AND EVIDENCE FOR FEMALE TRANSFER INTO THE HABITUATED COMMUNITY[1]

Males

First Visits to Camp Through 1965. Most of the mature males visiting camp during the study period had already started visiting by the end of 1963 (Table 1). A few, however, started visiting after this. Most of the early visitors were full adults but some were young males whose mothers also visited at this time. The later visitors were almost all young males (e.g., MacDee, Hornby, Satan, Jomeo, and Godi). These may have had mothers with more peripheral ranges, and as they reached adolescence, started to follow the adult males away from their mothers and travelled with them as far as camp (Pusey, in preparation). The

TABLE 1
Year of first visit to camp of all individuals born before 1963

	1963	1964	1965	1966	1967	1968	1969	1970	1971	1972	1973	1974
Males												
Mature	18	2	2									
	(4)											
Immature	1	4			1[a]							1[b]
	(1)	(2)			(1)							
Females												
Parous	5	4			2							
Late Adolescent	3	1	4	1			1	1		1		
		(1)										
Early Adolescent		1					1			2		1
		(1)										
Immature	6	2	1									
	(4)	(2)										

Individuals are classified by their age and reproductive state on their first visit to camp.
In each cell the total number of new visitors of each age class each year is written.
In brackets ( ) underneath is written the number of these whose mothers visited for the first time in the same year.

[a] This individual and his mother were seen out of camp in 1963.
[b] This individual was an orphan (age approx. 4-5 years) and arrived with an adolescent female suspected to be his sister.

mothers of two such males (MacDee and Jomeo) arrived in camp for the first time a couple of years after their sons. (See Table 1).

New Visitors to Camp After 1965. After 1965, the only new males to visit camp were those born into the group, and two juveniles, Sherry and Beethoven. Sherry first visited camp in 1966 with his mother, Vodka. His brother Jomeo had been visiting since 1964, and he and his mother had been seen near camp with Jomeo in 1963, although they did not enter camp. Beethoven is a young juvenile who came to camp for the first time in 1974 with a young nulliparous female, Harmony, who apparently transferred into the habituated community at this time (see discussion following). It is suspected that they are siblings and that Harmony adopted Beethoven after their mother's death (Anderson, in preparation).

Since out-of-camp following was begun, no unhabituated males have been observed associating regularly with habituated males of either community.

Disappearances. Having visited camp once, each male continued to do so regularly (although in some cases infrequently) until:

  1. They died and their body was found.
  2. They disappeared and there was good evidence to assume that they had died (for example if they showed extreme old age, illness, or a bad injury or disability when last seen, or disappeared during an epidemic of illness). For details see Goodall (1975).
  3. They stopped visiting after the community split, and continued to be seen in the Kahama community.

Table 2 shows the number of males in each of these categories who stopped visiting.

From these data it can be seen that no males joined or permanently left the habituated community between 1965 and 1972 when the split occurred. Also, after the split, with the exception of Beethoven, no new male joined either community, and there was no transfer of males between the two communities.

Females

First Visits to Camp Through 1965. By the end of 1964, nine mature females and their associated offspring, three unrelated nulliparous adults, and two unrelated immature females were visiting. In 1965, five more females started visiting. One of these (Lita) was a juvenile whose brother (MacDee) was already visiting, and whose mother (Jessica) started visiting in 1966. The others (Athena, Pallas, Nova, and Nope) were young nulliparous females having full sexual swellings, one of

TABLE 2
Disappearances from camp, between January 1963 and March 1975, of individuals born before 1963

	DEATH KNOWN OR SUSPECTED	MEMBER OF KAHAMA COMMUNITY	STILL SEEN OUT OF CAMP (NOT KAHAMA)	LAST SEEN IN GOOD HEALTH
Males				
Mature	13	7		
Immature	1			
Females				
Mature Parous	9	2	1	1
Late Adolescent	2	2 (1)		3
Immature	1 (1)		1 (1)	1

Individuals classified by age and reproductive state on their last visit to camp.
( ) denotes individuals who stopped visiting at the same time as their mothers.

whom (Nope) was newly pregnant when she first came to camp. These females may have been around in the area before provisioning began, and just slow to visit camp. It is also possible, however, that some of them, like the nulliparous females arriving after 1965, came from other communities.

New Visitors to Camp After 1965. Of ten females who visited camp for the first time after 1965, two were lactating mothers and the rest were young nulliparous females (Table 1).

1. Lactating mothers. Two mature females with infants (Jessica and Vodka) first came to camp in 1966. Both were suspected to be the mothers of young mature males who had been visiting since before 1965 (MacDee and Jomeo respectively) and Vodka had actually been seen out of camp with Jomeo in 1963. Another lactating mother, Sprout, who had possibly visited camp once in 1964, started visiting more frequently at this time. She is the suspected mother of Satan, a young male who started visiting more regularly in 1965.

These three mothers were never frequent visitors and they usually came with their sons. Jessica and her dependent female offspring disappeared in 1967 at the end of a polio epidemic and were assumed to have died (Goodall 1968b). Sprout and her dependent offspring never visited camp after 1968 but were still seen out of camp at the northern edge of the males' range in 1975 (Table 2). Vodka and her infant also visited camp for the last time in 1968 and were not seen again. Sherry, her juvenile son, continued to visit camp after she stopped and became a frequent visitor in 1969.

It is probable that these females had ranges relatively far from camp and did not visit early even though their sons joined the core group of adult males. Only during the height of the banana provisioning (Wrangham 1974) did they extend their ranges to include camp more regularly.

2. Young nulliparous females. Eight young nulliparous females started visiting after 1965. Two of these, Walnut and Winkle, arrived before extensive out-of-camp following had started. All the others arrived after this and were actually seen out-of-camp with groups of habituated chimpanzees several times before they visited camp. In some cases they were first seen in peripheral areas of the male range. The early sightings of Harmony, for example, were all in the north. Dove, on the other hand, first joined groups toward the southern part of the range.

The ages of these new females varied. Some (Patti, Winkle, Harmony, and probably Sparrow and Wanda) were still early adolescent when first seen, and not yet being mated by adult males, while others (e.g., Dove) were already getting full sexual swellings. A feature that most had in common was that they had estrous swellings (whether adolescent or full) on the first few occasions they were seen with habituated chimpanzees. Some, for example Dove and Sparrow, only associated with males of the habituated community while in estrus, and then were not seen in the intervening periods of anestrus, although the habituated chimpanzees often remained feeding in the same area.

In addition to eight young females, other young nulliparous females have visited camp once or twice, or have been seen (usually in estrus) associating with habituated chimpanzees outside camp. It is hard to know the exact number of such females since it is very difficult to describe and recognize an unhabituated female. Unless she is seen a number of times, one individual is unlikely to be recognized consistently. This problem is confounded by the fact that observers change over the years and if a rare but named visitor reappears after a long absence, she may not be recognized.

It is very likely that most or all of these young females did not regularly associate with members of the habituated community long before they were first observed and they probably came from adjacent communities. There are two main lines of evidence for this.

First, if they had come from peripheral areas within the male range, and had always associated regularly with the habituated community, they should have been encountered on out-of-camp follows more often than was the case.

Second, if they did not come from other communities, at least some of them should have been seen, while still immature, in the company of their mothers and siblings. Old unhabituated females who could be the mothers of such females have occasionally been encountered in peripheral areas, and judging by the violent aggression that they often receive from individuals of the habituated community (Bygott 1972; Goodall, this volume), most are probably not regular associates of the habituated community. Moreover, if many such peripheral families did exist within the male range, it would be expected that a similar number of young males would have appeared in camp over the years.

Of the eight young nulliparous females arriving since the beginning of 1966, two (Walnut and Female X), like the many unidentified visitors mentioned above, were temporary visitors, and only visited for a few months and then were not seen or at least recognized again (Table 2). The other six are still seen associating with habituated chimpanzees. Four (Wanda, Dove, Winkle, and Sparrow) have had infants. Wanda now associates with the Kahama community, while Dove, Winkle, and Sparrow associate with the Kasekela community. Winkle and Sparrow now have ranges in the center of the range of the Kasekela community (Wrangham 1975). Patti, after making very infrequent visits to camp for 2½ years, became a frequent visitor in 1974 and appears to spend all her time with the Kasekela community. Harmony, in April 1975, was still a new visitor, and it remains to be seen if she will stay permanently.

MOVEMENTS OF HABITUATED FEMALES AND EVIDENCE FOR THEIR TRANSFER BETWEEN COMMUNITIES

Young Nulliparous Females

There were eight females of the habituated community who either reached adolescence during the study period and were already visiting camp as immatures, or were already adolescent in the first two years of the study but associated with older females presumed to be their mothers. It is probable that these eight females were born into the habituated community. One of them (Pooch) sustained a bad injury 1½ years after starting full estrous cycles, then disappeared during a 'flu epidemic and almost certainly died. During her period of cycling, she spent all her time with individuals of the habituated community. The other seven females fall into three groups; two disappeared permanently after reaching sexual maturity, three showed temporary absences from camp and may have visited other communities, and two were actually seen associating with males of another community.[2] The cases of each are described as follows.

Females Who Disappeared Permanently. The two females in this category are Sally and Bumble. Sally was already showing full sexual cycles when she started visiting camp in 1964 with her mother. In 1966 at the height of the banana feeding, she disappeared for 4 months, then visited for 10 months, then disappeared altogether in 1967 at the end of the polio epidemic. Bumble was early adolescent when she visited camp with her old mother Bessie and younger sister Beattle in 1964. They all disappeared for a year, then Bumble and Beattle returned together without their mother, who must have died. Bumble visited infrequently until the middle of 1968, usually coming when she was in estrus. Beattle was not seen in 1968, and Bumble was not seen again after 1968. These two females, Sally and Bumble, either died or transferred permanently to another community.

Females Showing Temporary Absences. Three females, Gigi, Fifi, and Miff, who visited camp regularly from the juvenile stage onward, are still frequent visitors, and two (Fifi and Miff) now have infants. All of them, however, showed periodic absences from camp after they started having adult estrous cycles.

It is common for cycling females at Gombe to consort with a single male for several days, during which time they do not visit camp. McGinnis (1973; this volume) has shown that on many of the occasions when a female is absent from camp, a male from the community is absent for a similar or identical period, and there is good evidence that they are in consort. Such consortships are more common when the female is in estrus, and are often restricted to this time; however, they may occasionally last for several weeks and include long periods when the female is anestrous (McGinnis 1973; this volume; Tutin, in preparation).

All three females often showed absences from camp during their estrous periods, which were coincident with the absences of a male of the habituated community. They also copulated freely with males in camp. On some occasions, however, they were absent for many days during which they had a full estrous period when there was no coincident absence of a habituated male.

Gigi, after 3½ years of full estrous cycles, had several such long absences in 1969 and 1970, including one of 33 days and another of 81 days. On some occasions habituated males were absent for part of these periods, and she may have consorted with them and simply stayed away from camp while they returned. It is also possible, however, that she sometimes joined another community. Some observations support this second hypothesis. On one occasion McGinnis (personal communication) followed Gigi and Humphrey, a male of the habituated community, north while Gigi was in estrus. They went toward the northern edge of the habituated community's range. On hearing the calls of members of the habituated community to the south, Humphrey turned back and Gigi continued by herself. On two other occasions McGinnis followed Gigi north by herself when she was in estrus, but lost her before it could be determined if she joined another community of chimpanzees. A third observation was made by Wrangham (1975) in 1973. Gigi was in a party of Kasekela chimpanzees between Linda and Rutanga valleys (toward the north edge of the range) when they heard calls from a party of strange chimpanzees farther north (Goodall, this volume). Although all but one of the males of her party turned back, Gigi, Miff, and Athena walked toward the calling group.

Miff began full estrous cycles in 1967. In 1967 and 1968, she had absences from camp during which she was in estrous for 7 to 26 days, when no male showed a coincident absence. Like Gigi she may nevertheless have been consorting with habituated males. During some of her absences, however, all the males made regular visits to camp and it is unlikely that any were with her. She generally went north from camp before such absences (see Wrangham's observation described previously). In June 1968, however, she was consorting with Figan, a male of the habituated community, when she almost certainly must have conceived (Tutin, in preparation).

At the end of 1969, Fifi, after 2½ years of adult cycles during which she copulated and consorted with habituated males, started having long absences. In 1970 she was away for periods of up to 58 and 72 consecutive days; no male was away as long during these periods. In September 1970, during one of these long absences, she was seen far to the south in Nyasanga valley with males of the southern subgroup (later to become the Kahama community). None of these males, however, stayed away from camp as long. It is possible that she also joined males of the southern Kalande community at this time. She returned to the Kasekela community pregnant and remained there after the birth of her infant.

Females Seen with Males of Another Community. Only two females at Gombe, Gilka and Little Bee, have actually been observed to transfer to another community of males and associate with them.[3] For this reason their movements will be described in detail.

Gilka started visiting camp as an infant in 1963 with her mother Olly (who died in 1969), and was a frequent visitor until 1971 when she was first mated by adult males. At this time the community was beginning to split, but the two subgroups of males were still associating together. She was mated by males of both subgroups during 1971. During 1972 when the Kahama community became more separated and remained in southern valleys, she went south and associated with the Kahama males for successive estrous periods, returning to the Kasekela community between these periods. She always seemed to go of her own accord, although in the first few months she joined members of the Kahama community when they were in or near camp, and followed them back to the south.

In the middle of 1972 she transferred entirely to the Kahama community for 5 months. She then returned to the Kasekela community and remained there through her pregnancy and the birth of her infant. After losing the infant, however, she remained in the Kasekela community, copulated with these males when she resumed cycling, and in early 1975 became pregnant again.

Little Bee is almost the same age as Gilka and also has visited camp with her mother Madam Bee since 1963. Madam Bee was always a southern female and an infrequent visitor, however, and she and her family stopped visiting camp when the community split in 1972, becoming members of the Kahama community. Little Bee began to have full estrous swellings in 1971. In early 1972 she associated with males of both communities, always accompanied by her mother and younger sister.

After September 1972, her family ceased visiting camp, but for the next 2½ years, Little Bee made infrequent visits by herself to the Kasekela community. She usually appeared in this community while in estrus and copulated with the males. She spent most of the time in her own community, however, and also associated with the males there. From September 1974 onward, she began to move frequently between the two communities, spending most of her estrous periods with the Kasekela males.

Although her movements between the two communities sometimes may have involved traveling by herself, on two occasions she accompanied parties of Kasekela males back north after they had come south and attacked members of the Kahama community with whom she was traveling. On one of these occasions the individual attacked was a male (De); on another, Little Bee was alone with her mother and sister and they were all attacked (Goodall, this volume). Since December 1974 she remained with the Kasekela community most of the time, and was still there in April 1975.

Various points should be made about the movements of these two females. At first, each made short visits to their *new* community while in estrus and returned to their own community in between estrous periods. It was only later that they transferred entirely for several months at a time. This pattern accords well with our observations, described previously, of new unhabituated females appearing for the first time in habituated groups when in estrus. Also, Little Bee's pattern of association with the Kasekela community by very infrequent visits for the first 2 years was very like Patti's, one of the newcomers.

We must not assume that the cases of Gilka and Little Bee are typical, because the communities were still in the process of separating at the time that both females began to copulate with males of these new communities who were not strangers. Also the relationship between the two communities was probably not typical.

Temporary Absences of Mature Females with Offspring

Similar to the chimpanzees of the Mahale mountains, some mature females, after giving birth and resuming cycling, have shown absences of up to 2, and in one case, 4 months from camp. Others, however, were never away so long and on all their absences could be shown to have consorted with males of the habituated community (McGinnis 1973; this volume).

During many of these longer absences, it can be shown that a male was absent for the same period, and even when this is not the case, a known male may have been consorting with the female for part of the time. Olly was away for 4 months in 1966 and no male was absent for so long. In the third month of this absence she was seen in the northern valley Busindi with David (a male of the habituated community); therefore, it is possible that she had been consorting with a series of habituated males but not coming to camp. Since she was far to the north, she could have been with males of the Mitumba community.

Other primiparous or multiparous females (e.g., Athena, Miff, and Melissa) have shown shorter absences of about a month or so when no male was away, and one, Nova, shifted her range further south in 1973, and thereafter rarely came to camp. She was still occasionally seen in the south until her death in 1975. This shift in range coincided with a resumption of cycling, but other factors may also have been involved. We do not know with which males Nova consorted.

In summary, females upon resumption of estrous cycles after giving birth, are sometimes absent for long periods and, in some cases, do not appear to be with a male of their community. Although they may go to other communities, as yet there is no positive evidence for this.

DISCUSSION

These data indicate that the patterns of intercommunity transfer observed in the chimpanzees at Gombe are similar to those of the chimpanzees of the Mahale mountains. The population of habituated males during the study period has been very stable except for seven males who left to form a separate community. There is no evidence for transfer of males into or out of the habituated community before the split, or either of the communities after the split, except for an orphaned juvenile who accompanied an adolescent female, probably his sister, when she joined one of these communities.

The associations of females with the Gombe communities are more unstable. Many young nulliparous females have joined the communities either temporarily or permanently, and others have left and transferred between communities temporarily or possibly permanently. Mature females with offspring have also temporarily changed the frequency of their associations with the males of the habituated community—some apparently peripheral females appearing more often, and others disappearing for many weeks at a time. In none of these cases has it actually been established that transfer between communities is involved.

Some features of the habituated community may not be typical. First, the average age at which young females joined the community tended to be younger than the age at which those growing up in the community left it for the first time. Although some of the newcomers were still early adolescents when first observed, few if any of the females growing up in the habituated community left before developing full estrous swellings; some did not leave until 2 or 3 years after this.

Second, of eight females growing up in the habituated community during the study period, one almost certainly died and only two left permanently (and either or both of these may have died). Four stayed permanently in their natal community although one and possibly all made temporary visits to other communities.[4] In contrast, at least five females permanently joined the habituated community during this time. This imbalance in the number of females leaving and joining the habituated community may simply be due to chance variation in the demographies of the communities in the area. The difference in age, however, cannot be explained in this way.

It is also of interest that of five conceptions of three females who grew up in the habituated community during the study period, three (two of Miff's and one of Gilka's) occurred in the natal community, and only two (Gilka's first and Fifi's first) occurred while the females were possibly in other communities (Tutin, in preparation). Moreover, Fifi resumed cycling again in February 1975 and spent her first two estrous periods copulating with males of her natal community, including her two adult brothers. It remains to be seen if she will go to another community during later cycles.

This degree of inbreeding due to females remaining in their natal community may be unusual. It seems likely that this tendency for females of the habituated community not to leave early, or stay away permanently, may be due in part to provisioning. For several years all the chimpanzees in the habituated community visited camp almost every day (Wrangham 1974). Such frequent association with every member of the community would probably never occur in unprovisioned communities. Perhaps females of the habituated community formed atypically strong social bonds at this time with other members of their natal community. Because they were in camp every day, they may have had unusually little contact with adjacent communities. Such contact may be essential before permanent transfer can take place. In chimpanzees, like many cultures of man (Murdock 1957), and unlike most other primate species (see introductory paragraphs), females and not males move between groups. The sex of the individuals that move between groups must depend on the social structure of the group and the factors involved in maintaining the structure. The relation of chimpanzee social structure to their ecology has been discussed in detail by Wrangham (1975; this volume). In brief he shows that individual chimpanzees occupy rather large core areas relative to their body size (compared to gibbons and orangutans), probably because of the patchy distribution of food sources. The core areas of individuals overlap extensively. He suggests that lactating females have to maximize their feeding efficiency at all times to enhance their reproductive rate. The reproductive success of each male, on the other hand, depends on the number of females to which he has access, and he may range further than females, thus encountering more females, at the expense of his feeding efficiency. Since the individual ranges of all individuals overlap extensively, a group of males can defend a group territory without incurring great costs in competition for food and increased locomotion. Such a territory would be larger in area than the total area of individual territories defended by the same number of males. The group territory would thus encompass the ranges of a larger number of females, and each male would have access to more females than he would by defending an individual territory.

To defend a group territory, males must act cooperatively. Cooperation will be greatest if bonds between males are strong. Bonds should be strongest, if males are incorporated into the male group at a relatively young age and if males are closely related. Both conditions are fulfilled if males remain in their natal group. Further, aggressive competition for females, which could disrupt these male-male bonds, should be minimized if the males are all closely related (Popp & DeVore, this volume). Females, in contrast, are rather solitary in adult life. If there is a selective advantage for any individual in outbreeding, therefore, females are the more likely candidates for moving between groups.

Given that females move, it would be to the advantage of males to show affiliative behavior toward reproductively active female newcomers. There is some evidence that males may try to recruit females from neighboring communities (Goodall, this volume). Resident females, on the other hand, would be expected to show aggression toward incoming females since they would be in competition for food resources. Observations of such behavior have been made (Pusey, in preparation).

The selective advantage of the long period of adolescent sterility observed in female chimpanzees, during which they have regular estrous cycles but do not conceive, may be that it allows them time to investigate new communities, select suitable mates, and form new relationships (Tutin, in preparation). The conspicuous sexual swellings exhibited regularly during this time may serve in part as a long-distance signal to males of neighboring communities and possibly ensure that they receive affiliative behavior from these males.

Although male social relations have been studied extensively (Bygott 1974), a similar study of female social relations still remains to be done. From the existing data, however, it is clear that males associate together more and show more affiliative behavior to each other, such as social grooming, than females do with each other. Furthermore, intracommunity infanticide resulting from female-female aggression has been seen (Goodall, in preparation). These observations are consistent with a model of chimpanzee social structure where males form a kinship group and most females, as a result of intercommunity transfer, are probably only distantly related.

Acknowledgment

This work was supported by grants from the Guggenheim Foundation and the Grant Foundation of New York. I am very grateful to the Tanzanian government for permission to work in Gombe National Park, and to Dr. J. Goodall, the director of the Research Centre. Many students and research assistants have contributed painstakingly to the long-term records without which this work would not have been possible. I am indebted to Craig Packer for numerous discussions during which the ideas presented in this paper were developed. I also had valuable discussions with David Bygott, Caroline Tutin, Barbara Smuts, and others. I thank Professor R. A. Hinde for useful criticism. I remain extremely grateful to my advisor, Dr. D. A. Hamburg, for his constant support and encouragement.

Notes

1. The term *habituated community* will be used to include both the community visiting camp before the split, and the Kasekela community that continued to visit camp after the split. The term *habituated chimpanzees* refers to all chimpanzees who visited camp and became habituated there.

2. Since the time of writing, a third female, Honey Bee, has been observed with males of another community. She is the early adolescent sister of Little Bee and she had been associating with her mother and males of the Kahama community since the community split in 1972. Her mother died in August 1975, and soon after this Honey Bee was observed associating with males of the unhabituated southern Kalande community.

3. See note 2.

4. The eighth female (Little Bee) appears to be a special case. She grew up in the camp-visiting habituated community and then transferred back into it after the community split. Her movements were described in a previous section.

Journeys between food sources are often made in single file.

Richard W. Wrangham

Sex Differences in Chimpanzee Dispersion

Recent field studies (Bygott 1974; Japanese work reviewed by Sugi-yama 1973) have supported the proposition that chimpanzee popula-tions tend to be divided into sets of individuals sharing a single home range. Such sets—here called *communities* but variously referred to as unit-groups (Nishida 1968; 1970; Nishida & Kawanaka 1972; Izawa 1970; Kano 1971b), large-sized groups (Itani & Suzuki 1967) or regional populations (Sugiyama 1973)—consist of 15 to 80 individuals of all age-sex classes. Males have not been seen to transfer between communities, and interactions between males of different communities are normally aggressive.

This paper summarizes results from a study of individuals in two communities (Wrangham 1975). The results suggest that insufficient attention has been paid to sex differences in ranging, and argue for a reappraisal of the community concept.

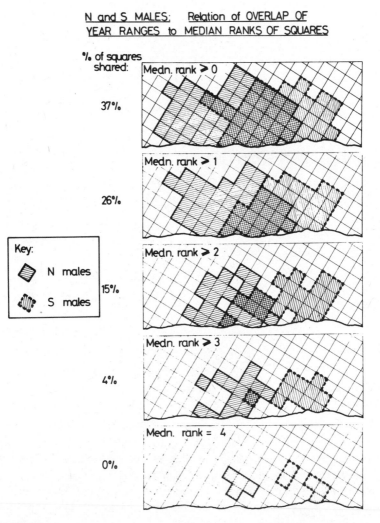

FIGURE 1

Community ranges. Year ranges of all males within a community contribute to the community range. For each male, squares were ranked according to the percent of observation points falling there: rank 1, 0.01-0.99%; 2, 1-2.49%; 3, 2.50-4.99%; 4, 5%. Median ranks were taken across individuals. Data are from all observations (ca. 700 hours) beginning outside the artificial feeding area, February 1972 to January 1973: locations were scored at 30-minute intervals. Squares have sides of 0.5 km.

METHODS

Methods were described in detail elsewhere (Wrangham 1975). The data were collected in Gombe National Park, Tanzania in 1972 and 1973 and are from three main sources.

    1. 1000 observation hours collected by the author. Adult males were the target animals.

    2. 6000 observation hours collected by more than 30 observers on standard checksheets, and contributed by them to a general data pool.

    3. Daily records of attendance by chimpanzees at an artificial feeding area visited regularly by members of one community.

The study population consisted of all chimpanzees habituated to observers. Bygott (1974) analyzed their association patterns up to 1972. A dendrogram of links at the highest level of association revealed two separate clusters *(subcommunities)* of males: most females were linked outside these clusters, but were assigned to the subcommunity that included their most frequently observed companion. By 1972 the level of association between males of the two clusters was sufficiently low for them to be called separate communities. During the present study, the northern community contained nine males and 13 females, and the southern seven males and five females (independent animals only).

RESULTS

Community Ranges

Communities were identified by the association patterns of independent males: the range of a community was therefore defined as the area occupied by all its males. Data were analyzed over 1 year. Northern and southern males were found to occupy partially separate community ranges, covering 13 and 10 sq km respectively (Figure 1). The extent of overlap was inversely related to the degree of use of the range.

Generally, individual males used the community range equally. Differences in size of the year range were attributable to differences in observation time; and over a 4-day period, prime males sometimes approached all four main boundaries. Differences in range use nevertheless existed. Northern males entered the artificial feeding area (near the center of the community range) from different mean directions, which were stable over at least a year: and these directions corresponded to the areas where they were most often seen.

TABLE 1

Distance traveled in one day

CLASS	DISTANCE TRAVELED BETWEEN NIGHT-NESTS (km)										NR	NI	MEDIAN (km)
	0-1	1-2	2-3	3-4	4-5	5-6	6-7	7-8	8-9	9-10			
Male	1	2	19	30	15	17	8	8	4	2	106	15	4.1
Anestrous female	4	13	12	13	8	1	1	0	0	0	52	7	2.8
Estrous female	0	1	3	0	1	2	2	0	0	0	9	4	4.5
Consort pair	0	4	3	5	3	0	0	0	0	0	15	3	3.2

The target individual's path was drawn on a map (1:11,560) and measured subsequently.
NR = number of records, NI = number of individuals (or consort pairs) in sample.

FEMALE RANGES

Females varied more than males in the size of their year ranges: variation in observation time did not explain the differences. Those who were anestrous throughout the year had smaller ranges in relation to observation time than sometimes estrous females (Mann-Whitney $U = 0$, $n_1 = 7$, $n_2 = 11$, $P < 0.001$) and males. Females thus appeared to use the community range less fully than males: this was tested by analysis of ranging behavior over shorter periods.

The best sample was from the northern community. Path length in complete days of observation was scored. Males traveled farther per day than anestrous females (mean scores of eight males and six females compared: $U = 0$, $P < 0.001$). To discount possible bias incurred by sampling male and female 1-day ranges in different seasons, records were matched by date (within 14 days): males again traveled further (Wilcoxon $z = 3.62$, $N = 29$, $P < 0.001$). The data are summarized in Table 1.

The size of 4-day ranges was scored as the number of squares (0.5 km x 0.5 km) occupied. Males covered more squares than anestrous females (mean scores of five males and four females compared: $U = 0$, $P < 0.001$). The data are summarized in Table 2. Figure 2 gives examples of 4-day ranges of a male and a female at one season: their location and size were consistent.

Their small 1-day, four-day, and year ranges suggest that anestrous females occupied smaller core areas than males. (Core area is used to indicate the smallest area within which an individual spends P percent of his time, where P is an arbitrary constant: a useful value of P would be 80%.) Year range data suggested that different females had core areas in different places, but sampling problems made these unreliable (Wrangham 1975): however, observed core areas corresponded to mean directions of entry into the artificial feeding area. This correspondence indicates real differences in the location of female core areas. Their direction from the feeding area is shown in Figure 3. Figure 3 also shows that female entries were more highly concentrated into particular directions than those of males $(U = 18, n_1 = 9, n_2 = 11, P < 0.02)$. Their high directedness confirms that females tended to occupy smaller core areas than males.

Three forms of travel were typical of an estrous female (Tutin, in preparation). She traveled more widely than anestrous females within

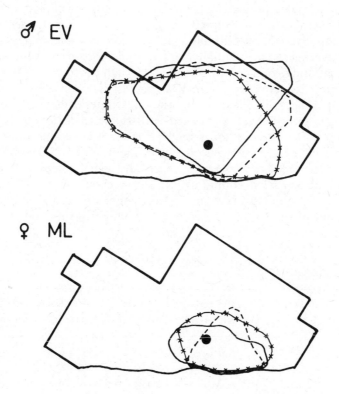

♂ EV

♀ ML

FIGURE 2

Four-day ranges of a northern male and female. For both individuals, three separate 4-day ranges are shown: the lines enclose all paths. The data were collected in July and August 1973, and are shown in relation to the northern community range from February 1972 to January 1973. The black dot marks the artificial feeding area. (I am grateful to M. Thorndahl for providing the data on ML.

the community range; she formed a consortship with an adult male and traveled in a small area (Table 2) on the edge of his community range; or she migrated to a different community range (see also Nishida, this volume).

If an anestrous mother tends to occupy only a small part of the community range and may visit another community range when estrous, we may ask what justification there is for regarding the community as a bisexual social unit (see following discussion).

MODELS OF CHIMPANZEE DISPERSION

Three possible models of chimpanzee dispersion are shown in Figure 4. Figure 4(i) is the "classic" picture (since Itani & Suzuki 1967), with community ranges being shared equally by both sexes.

The data given previously show that females occupied core areas dispersed within the community range and smaller than those of males, as represented in Figure 4(ii) and (iii). For convenience, males are again shown as sharing the area equally. In fact male core areas were dispersed within community ranges, although less so than those of females.

In Figure 4(ii) anestrous females are dispersed in relation to male community ranges: in Figure 4(iii) they are independent of them. This difference is important; however, it is difficult to test which model is more correct. In both, some females overlap two community ranges, but the frequency of overlap is greater in Figure 4(iii). Three anestrous mothers occasionally associated peacefully with males of both the northern and southern communities: this fact shows that females do not defend the community range in the same way as males, but sampling problems prevent using these data to distinguish the models.

TABLE 2
Area covered in four days

NORTHERN MALES			NORTHERN FEMALES			CONSORT PAIRS		
INITIALS	MEAN	N	INITIALS	MEAN	N	INITIALS	MEAN	N
HG	15.0	2	FL	10.0	1	FG / GG	7.5	2
MK	25.0	2	PS	12.0	2	FB / PL	7.3	3
FB	16.7	4	ML	10.2	5			
EV	23.7	3	WK	10.3	1			
FG	18.0	3						

Scores show mean number of 500 m x 500 m squares occupied in 4 days (found by superimposing grid on drawn path). On some occasions more than one individual was present throughout: the record was then assigned to the main target individual.
N = number of 4-day observations.

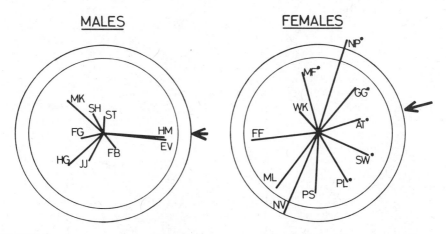

MALES   FEMALES

FIGURE 3

Direction of entry into the artificial feeding area. Relative angles were calculated within sexes. The number of entries in each of eight directions were "normalized" to yield a matrix without a mean angle: individual angles were then calculated. The length of each line is proportional to directedness, or the concentration of scores in that direction: females were more highly directed than males. Overall mean angles (i.e., prior to normalization) are shown by the large black arrows, and were similar for males and females. The circles are visual guides. Females who were estrous at some time are marked with a dot. The data were collected in the following months: July, August, and December 1972, and January, July, and August 1973. Number of entries: males 1001, females 991.

The major testing point is what happens when a male community range is observed to shift.

1. In 1973 the northern males expanded their community range to the north. They then regularly associated with a mother whom they rarely met at other times. There was no evidence that she had moved.

2. Kano (1971b) gave evidence of a community migrating between separate areas of seasonally abundant food. From observations such as changes in the frequency of long-distance calls, he concluded that "except for a few chimpanzees who parted from the main bands, the whole band (community) migrated out of the area." Kano referred to the few chimpanzees remaining as "comprising females and juveniles." It seems possible that some or all anestrous females had remained resident and were thus dispersed independently of the males.

3. Nishida and Kawanaka (1972) studied cases of females *transferring* between communities at a time when one community (M-group) appeared to supplant a smaller one (K-group). Although the term transfer implies movement by the female, their evidence suggests that female transfers occurred when males moved and females stayed where they were. Some females behaved exactly as expected from Figure 4(iii): when M-group supplanted K-group they joined the former, only to return to K-group when it regained its normal range.

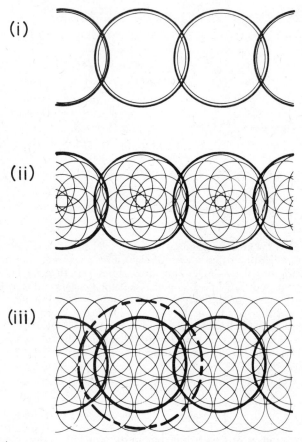

FIGURE 4

Models of chimpanzee dispersion. Circles with thick lines are male community ranges, in which a number of males cover the area fairly equally. The large circles with thin lines (i) are female community ranges. Small circles with thin lines (ii & iii) are core areas of individual females. The diagrams represent ranging patterns of anestrous females. In Figure (iii) the dashed line shows expansion of a male community range: its males consequently associate with more females. This would not occur if community ranges expanded in (i) or (ii): females within the communities would increase the size of their core areas.

These observations suggest that anestrous mothers may reside in small core areas even when males move, and argue against the concept of the community as a bisexual unit.

DISCUSSION

Analysis of party composition showed that males associate with each other more than females associate with either sex (Bygott 1974), and that agonistic interactions between communities mostly involve males. Evidence for the existence of chimpanzee communities has thus been based largely on male behavior. The assumption that the community concept applies equally well to females now appears unjustified. There are two important theoretical consequences.

First, the adaptive significance of chimpanzee social structure can be explained in a new way. The even dispersion of females is attributed to their need to maximize feeding efficiency and hence reproductive rate. If several males can collectively defend a territory larger than the total area of individually defended territories, each would gain reproductive advantage (access to more females) through cooperation: this would be favored especially when locomotion costs are low, i.e., when food is abundantly and widely distributed. Competition within the group territorial system (community) would lead to differential mating success. If mating success is age-related and young males can expect to increase their reproductive opportunities in the future, however, their best strategy may be to aid in community defense whenever possible. This view accords with the emphasis of Goss-Custard et al. (1972) and Crook (1972) of the importance of reproductive competition, and follows Rodman's (1973a) suggestion that the social system of male orangutans is adapted to female dispersion.

Second, the differences between chimpanzees and orangutans, which have been a puzzle to socioecologists, now seem less than previously thought. In both species females are evenly dispersed in separate core areas. The only apparent structural difference is that male competition for females is made by groups of chimpanzees but by solitary orangutans. Future studies may reveal conditions under which even this difference does not exist, as we would expect if a major determinant of the capacity for group territoriality is the cost of locomotion.

These propositions were discussed in detail by Wrangham (1975). Further evidence is needed to distinguish firmly between the concepts of the bisexual and the male-only community. Sex differences in ranging behavior clearly exist; however, the current data favor applying the community concept only to males.

Acknowledgment

I am grateful to the Tanzanian government for permission to work in Gombe National Park, and to Dr. J. Goodall for providing facilities. Many observers contributed to the data: I am especially grateful to Mitzi Thorndahl. The work was financed through the Grant Foundation of New York. I am indebted to Professor R. A. Hinde, F.R.S. for criticism of the manuscript, and to Caroline Tutin and others for valuable discussions.

Stewart D. Halperin

Temporary Association Patterns in Free Ranging Chimpanzees:
An Assessment of Individual Grouping Preferences

It has become apparent from recent field studies (Sugiyama 1969; Nishida and Kawanaka 1972; Goodall 1973; Bygott 1974) that the chimpanzee possesses a much more stable and definable social unit than originally reported (Reynolds & Reynolds 1965; Goodall 1965). Chimpanzee populations appear to be divided into communities (Goodall 1973) or unit groups (Nishida 1968), which have been shown to be geographically and behaviorally distinct social entities. Although the chimpanzee community is similar to other primate groups in many respects, it is unique in the openness and fluidity of the grouping patterns among community members. The chimpanzee exhibits a continually changing group structure in which the individuals in the community are found in temporary associations of varied composition (Goodall 1968b; Reynolds & Reynolds 1965). Previous reports by Goodall (1965), Reynolds & Reynolds (1965), and Izawa (1970) have presented data on the types and frequency distribution of temporary parties within chimpanzee populations. These authors reported that certain types of temporary parties were more frequently observed than others, and the findings were consistent for each of the authors (Table 1). The conclusions reached in all three of these studies were based on a small sample of unsystematically selected temporary parties. The purpose of this paper is to reexamine the temporary grouping structure of free-ranging chimpanzees in light of the more extensive data now available. Special emphasis has been placed on assessing the grouping patterns of individuals from a known community of chimpanzees *(Pan troglodytes schweinfurthi)*.

Over 1200 hours of observation on nine adult males and five anestrous females with dependent offspring are included in an analysis on temporary grouping patterns. The results suggest that there are significant differences in the grouping patterns among different age-sex classes, and that even individuals within the same age-sex class show distinct differences in their grouping patterns. Many differences found in the grouping patterns of each of the subject animals could be related to other factors including social status, kinship relations, and number of offspring.

STUDY AREA, SUBJECTS, AND METHODS

The data presented in this paper were obtained from a more extensive study on the social relationships among adult male chimpanzees, and the long-term behavioral records compiled by the staff of the Gombe Stream Research Centre in Tanzania (Halperin, in preparation; Goodall 1974). They were collected between February 1972 and March 1973 and were restricted to observations on individuals known to be members of the northern (or Kakombe) community (Bygott 1974; Goodall 1974, Wrangham 1975).

In a preliminary analysis on the grouping patterns of five adult males, considerable variation was found in the type of temporary parties in which each male was most frequently observed. The classification system used by previous workers (Reynolds & Reynolds 1965; Goodall 1968b; Izawa 1970) was insufficient for a complete cataloguing of the types of temporary groups in which the males were seen. I therefore decided to do a more comprehensive assessment of the grouping patterns for all the adult males in the Kakombe community (nine animals). In order to compare the grouping patterns of the adult males with another age-sex class in the same community, data were also included for five anestrous females with dependent offspring.

The data were collected on standardized *group travel* charts. Each time a target (focal animal) chimpanzee was followed and observed outside the artificial feeding area (Goodall 1968b), the party composition was continually recorded. The record consisted of noting the arrival and departures of all other chimpanzees during the observation period.

Over 800 hours of data from the group travel charts were used to tabulate the temporary grouping patterns for the nine adult males in the Kakombe community. The males ranged in age from subadult to old male (Goodall 1968). For the five females with dependent offspring, approximately 400 hours of data were used. Two of the five females had two offspring each, and the other three females had one infant each.

Although group travel charts were used every time a chimpanzee was followed, for this analysis only records that were at least 7 continuous hours of observation were included. From these records a sample point was taken on each successive hour of the observation period. On each sample point the type of temporary party in which the focal or target chimpanzee was observed was noted. There were approximately an equal number of sample points for each of the 14 subjects. Observations that began in the artificial feeding area were not included in the analysis, although visits to the area during the course of the observation were not excluded.

Reynolds and Reynolds (1965), Goodall (1968b), and Izawa (1970) used similar classification schemes for assessing temporary grouping patterns. According to their findings, there were four divisions into which they classified the most commonly observed temporary parties:

1. *Nursery Group:* adult females with their respective dependent offspring.
2. *All-Male Group:* two or more adult males.
3. *Bisexual Group:* adults of both sexes, but no females with dependent offspring (occasional adolescents).
4. *Mixed Group:* females and dependent offspring, females without offspring, adolescents, and adult males.

It should be mentioned that Goodall (1968b) also has a separate division for lone males, and although Izawa (1970) did not include lone individuals in his data on temporary parties, he noted their existence in a separate analysis.

For the present analysis, six separate divisions or *party types* were used to classify the sample points for the adult males. Three of the divisions were comparable to those used by the three authors mentioned, and three are additional divisions. It would have been possible to split the divisions even further but only *party types* that accounted for at least 2% of the animals' sample points were given the status of a separate division:

1. *All-Male Party:* two or more adult males.
2. *Bisexual Party:* adults of both sexes, but no females with dependent offspring (occasional adolescents).
3. *Mixed Party:* females with and without offspring, adolescents and adult males.
4. *Lone Adult Males.*
5. *Lone Adult Male and Mature Female without Offspring.*
6. *Adult Male(s) and Female(s) with Dependent Offspring.*

For the five females with dependent offspring, there were four divisions or *party types:*

1. *Mixed Party:* females with and without offspring, adolescents, and adult males.
2. *Female(s) and Dependent Offspring with Adult Male(s).*
3. *Female and Dependent Offspring Alone.*
4. *Nursery Party:* Adult females and their respective offspring and occasional other females.

RESULTS

Adult Males

As can be seen from Table 1, each of the males displayed fairly distinct grouping patterns. The males as a specific class appeared to have no uniform distribution of their sample points over the six party divisions. Some of the males (MK, SH) could be considered loners with a high concentration of their sample points in the *lone-male* divisions. Other males were more likely to have their sample points confined to more social parties, although the type of social party was specific to the individual male.

Mike (MK), the second oldest male in the community, had previously been the dominant male in the community (Goodall 1968; Bygott 1974). He was displaced as the alpha male by a younger male, Humphrey (HM) prior to the onset of this study. Mike's (MK) grouping profile reflects his current social status in relation to the other community members. On three measures that I used to assess social status (display rate, attack frequency, and pant grunts received) Mike (MK) was always ranked in the bottom quarter among the other males. On a separate analysis (Halperin, in preparation) in which dyadic association indices were calculated for all possible pairs of adult males, Mike (MK) was consistently ranked as the least-associated-with male by all the other eight animals. These findings fit well with the high concentration

TABLE 1

Temporary grouping preferences in nine adult male chimpanzees (number in cell indicates percentage of sample points for each group type.)

CHIMP	LONE MALE	ADULT MALES	LONE MALE AND FEMALE WITHOUT YOUNG	ADULT MALES AND FEMALE(S) WITHOUT YOUNG	MIXED GROUP	ADULT MALE(S) AND FEMALE(S) WITH YOUNG
MK	54	12	0	1	4	28
SH	52	9	2	12	17	8
EV	48	14	3	9	9	16
JJ	31	25	0	13	3	28
HG	29	28	4	16	10	14
HM	24	28	0	11	22	16
ST	21	7	10	16	36	10
FB	6	45	11	11	10	17
FG	2	41	10	4	19	24

of Mike's sample points in the lone-male division. His pattern of party preferences is characteristic of an old male who avoids large parties, especially ones with a core of adult males.

Sherry (SH), the youngest of the males, showed a high concentration of his sample points in the lone male division, similar to the older male Mike (MK). His grouping pattern could be considered typical of a young peripheral male who has yet to be integrated into the adult male social hierarchy. During the 14 months that data was collected, Sherry (SH) was the lowest-ranking male on all three of the status measures (display rate, attack frequency, and pant grunts received). Over 50% of Sherry's sample points were in the lone-male division, and only 9% were in the all-male division. Almost the entire 9% in the all male division was attributable to Sherry's association with his older sibling Jomeo (JJ). Thus Sherry (SH) was rarely found in all-male parties, and when not alone, was most likely to be in larger mixed groups.

A very different pattern from these two males was that shown by a prime adult male Humphrey (HM). Humphrey was the dominant male in the community during the first 8 months of the study. His status as the alpha male was evident, as he was consistently ranked the highest of any community member on the three status measures. Less than a quarter of his sample points were in the lone-male division, and the rest of his sample points were fairly well distributed among the other party divisions (excluding the third division). Having the advantage of being the recipient of submissive behavior from all other community members, Humphrey appeared not to be restricted to any particular party type.

Figan (FG), another prime male, presents even a more striking contrast to either the old male Mike (MK) or the young male Sherry (SH). Figan (FG) was a very high-ranking male, probably next highest to Humphrey (HM). Figan (FG) eventually displaced Humphrey (HM) as the dominant male during the course of the study (the details of the social relationships and the relative social status among the males are discussed by Halperin, this volume). Figan showed the lowest lone-male score of any of the other males. The highest concentration of his sample points were in the all-male division (almost 50%). It is interesting to note that Figan (FG) avoided lone travel and tended toward more social parties, and in particular all-male parties. It was in these parties that he could demonstrate his prowess as a high-ranking male (Bygott 1974). Figan's high all-male score is partially attributable to his association with his older sibling Faban (FB). Even with this close association taken into account, Figan's (FG) score in this category was still considerably higher than the other males.

Of all the six party types, the third division (lone male and female without young) comprised the fewest sample points for each of the males. Although this division could include an adult male with any female without an offspring, it was almost totally based on a male's association with an estrous female. This division can be considered similar to what McGinnis (1972) called consortships. Because of the dependence of consortships on the presence of an estrous female, and the fact that a consort pair often move off to the periphery of their normal range (Wrangham 1975), it is a difficult party type on which to obtain an adequate sampling. Even with this difficulty, the results as

TABLE 2

Temporary grouping preferences in five females with young (Numbers in cells are percentages of sample points in each group type.)

CHIMP	FEMALE AND YOUNG ALONE	NURSERY GROUP	MIXED GROUP	FEMALE(S) AND YOUNG AND MALE(S)
PL	55	40	2	3
FF	60	26	9	5
NV	65	32	3	0
ML	68	16	5	11
PS	77	13	3	7

seen in Table 2 (division three) correspond quite well with the rank order Tutin (personal communication) obtained for the frequency of consortships among the males in a more extensive study on sexual behavior.

The size of the parties over the study period revealed a trend toward larger social parties during the rainy season as compared to the dry season. Although this affected the absolute size of the parties in which a male was observed, his grouping preferences appeared to remain characteristic over the entire 14 months.

Anestrous Females with Dependent Offspring

The females present a much more uniform grouping pattern than the adult males (Table 2). Whereas each adult male showed a distinctive combination of the six party types, the females restricted their party preferences almost exclusively to the first two party divisions (female and offspring alone and nursery party). These two divisions accounted for over 80% of the sample points for all five females. The most frequent party type for the females was the first division, in which the female was alone with her dependent offspring.

The two females with the highest scores in the first division (lone female with offspring) were mothers with *two* dependent offspring. The other three females, each with one infant, spent almost twice as much time in the nursery party as the other two females. One possible explanation for the difference between the females could be that a female with two offspring is a social unit in itself; thus, the female seeks other individuals less often. It must be stressed that more data on mothers with dependent offspring are needed in order to test this assumption.

DISCUSSION

Understanding the structure and grouping patterns within a chimpanzee community is complicated by the fact that individuals above the juvenile stage are capable of independent travel and are not restrained by a rigid group structure. Whereas most primate groups are characterized by some degree of constant visual and spatial proximity among group members, the chimpanzee can be found in almost any combination of temporary associations. To obtain an accurate picture of the overall grouping structure (both in terms of variety and frequency) within a chimpanzee community, therefore, one must have a detailed grouping profile on every independent animal.

Only two age-sex classes were sampled in the present study, but very obvious differences were found between them. The adult males were more social in their grouping patterns than the females. The anestrous females (with dependent offspring) spent most of their sampled time alone with their offspring or in nursery parties. The adult males, although showing more diversity in party types and more individual variation in grouping patterns, were as a class more apt to be sampled in social parties than even the most social of the five females.

Because detailed behavioral data was collected on all the adult males, it was possible to compare their grouping data with data on their social interactions. The analysis indicated that an individual's grouping pattern reflects his social relationships within the community. Since factors such as age, social status, and kinship relations are important determinants in establishing social relationships, the same factors were found to affect an individual's grouping patterns. The type of temporary party in which an adult male was observed was seen as another aspect of the male's total *social profile* in the community.

Recent studies by Bygott (1974) and Wrangham (1975) confirm earlier reports that adult males are much more frequently involved in intercommunity agonistic interactions than any other age-sex class. These studies also suggest that, when aggressive encounters did occur, the males were usually in social parties, most often all-male parties. Lone males were rarely seen to travel far into the ranges occupied by adjacent communities. Parties of males, however, were observed at times to travel large distances into the areas most frequented by these adjacent communities. Some males were rarely involved in these intercommunity encounters, but some, when accompanied by other *specific* males, were most often involved. The data from grouping patterns confirms these findings that certain individuals who had higher scores in the all-male division were the same animals involved in intercommunity interactions. Some of the males with very low all-male scores (usually with high lone scores) were not observed to be part of the intercommunity encounters very often.

In the earlier findings on temporary parties reported by Reynolds and Reynolds (1965), Goodall (1968b), and Izawa (1970), the mixed party was the most frequently observed and thought to be the most common party type in chimpanzee populations (Table 3). At the time of these studies, the respective authors did not have the benefit of total habituation and full recognition of all individuals in the populations; they had to rely on unsystematic and opportunistic sampling of temporary parties. The high mixed-party frequency was possibly attributable to the observer being able to locate and sample larger, more vocal mixed parties. The difficulty in locating and recognizing lone individuals, especially timid mothers with young offspring, probably tended to increase the bias towards sampling the larger parties (mixed and bisexual). Consortships, which are normally rare occurrences, were not even considered in these early reports. Males were more often included in all the social parties they sampled, but this could have been related to the

TABLE 3

Type of temporary group (% of sample)

YEAR	REYNOLDS 1965	GOODALL 1968	IZAWA 1970
Number of groups sampled	215	350	78
Mixed group	37	30	52
Bisexual adults	30	18	30
Mothers and young	17	24	10
Males	16	10	8
Lone males	a	18	b

[a] Lone individuals not sampled.   [b] Considered in a separate analysis.

fact that adult males were more habituated and less timid to human observers than other age and sex classes.

The present study was not without its biases. The analysis included data from observations of animals visiting the artificial feeding area during an observation period. This data provided an increased frequency for certain types of parties. For the adult males the increase was most significant in the party type consisting of males with females and their offspring (division 6), whereas for the females, the third and fourth divisions were inflated by the data sampled from the feeding area. The females with dependent offspring probably spent even more time alone with their offspring or in nursery parties than the sample indicated.

The findings from the present analysis suggest that, although seasonality of food availability affects chimpanzee grouping patterns (Reynolds & Reynolds 1965; Wrangham 1975), the most important element in understanding the overall grouping pattern (size, frequency, and variety) is the specific age and sex composition of that chimpanzee community. All independent individuals within a community should be sampled, including estrous females, and adolescents, as well as adult males and females with dependent offspring.

Acknowledgment

I am especially grateful to Dr. Jane Goodall for allowing me the opportunity to study and use the research facility at the Gombe Stream. I am also indebted to the Tanzanian government for allowing me to study within the national parks system. I would also like to thank all the staff of the Gombe Stream for aiding in the collection of the data used in the present analysis. Special thanks goes to Professor David Hamburg for providing both financial and moral support in preparing the manuscript.

Bibliography

Albrecht, H. and Dunnett, S. C.
    1971   *Chimpanzees in western Africa.* Verlag München: R. Piper & Co.

Aldrich-Blake, F. P. G.; Dunn, T. K.; Dunbar, R. I. M.; and Headley, P.M.
    1971   Observations on baboons, *Papio anubis,* in an arid region in Ethiopia. *Folia Primatol.* 15:1-35.

Alexander, B. K.
    1970   Parental care of adult male Japanese monkeys. *Behaviour* 36:270-85.

Altmann, S. A.
    1960   A field study of the sociobiology of rhesus monkeys, *Macaca mulatta.* Unpublished doctoral dissertation, Harvard University.
    1962   A field study of the sociobiology of rhesus monkeys, *Macaca mulatta. Ann. N. Y. Acad. Sci.* 102:338.
    1967   The structure of primate social communication. In S. A. Altmann, ed., *Social communication among primates,* pp. 325-62. Chicago: University of Chicago Press.

Altmann, S. A. and Altmann, J.
    1970   *Baboon ecology.* Chicago: University of Chicago Press.

Andrew, R.
    1963   The origin and evolution of the calls and facial expressions of the primates. *Behaviour* 20:1-109.

Archer, J.
    1970   Effects of population density on behaviour in rodents. In J. H. Crook, ed. *Social behaviour in birds and mammals,* pp. 169-210. New York: Academic Press.

Ardrey, R.
    1966   *The territorial imperative.* New York: Atheneum.

Armstrong, E. A.
    1947   *Bird display and behaviour.* London: Lindsay Drummond.
    1965   *Bird display and behaviour.* Rev. ed. New York: Dover.

Asano, M.
 1967   A note on the birth and rearing of an orangutan, *Pongo pygmaeus* at Tama Zoo, Tokyo. *Int. Zoo Yearbook* 7:95-96.

Asdell, S. A.
 1946   *Patterns of mammalian reproduction.* New York: Comstock.

Ashby, W. R.
 1970   Information flows within coordinated systems. In Rose, ed., *Progress of cybernetics,* pp. 57-64. New York: Gordon & Breach.

Baenninger, L. P.
 1966   The reliability of dominance orders in rats. *Animal Behaviour* 14: 367-71.

Bakwin, H.
 1943   Loneliness in infants. *Amer. J. Dis. Child* 63:30.
 1949   Psychologic aspects of pediatrics. *J. Pediat.* 45:512.

Bartholomew, G. A.
 1970   A model for the evolution of pinniped polygyny. *Evolution* 24:546-59.

Bastian, J., and Bermant, G.
 1973   Animal communication: An overview and conceptual analysis. In G. Bermant, ed., *Perspectives on animal behavior,* pp. 307-57. Glenview, Il.: Scott, Foresman.

Bauer, H. R.
 1975   Behavioral changes about the time of reunion in parties of chimpanzees in Gombe National Park. In *Proceedings of the 5th Congress of the International Primatological Society, Nagoya, Japan, 1974.* Basel: Karger.

Beach, F. A. ed.
 1965   *Sex and behavior.* New York: John Wiley and Sons, Inc.

Beatty, H.
 1951   A note on the behavior of the chimpanzee. *J. Mammal.* 32:118.

Beccari, O.
 1904   *Wanderings in the Great Forests of Borneo.* London: Constable.

Bergerud, A. J.
 1974   Rutting behaviour of Newfoundland caribou. In V. Geist and F. Walther, eds., *The behaviour of ungulates and its relation to management.* Vol. 1, pp. 395-435. IUCN Publications. Morges, Switzerland.

Berkson, G., and Davenport, R. K. Jr.
 1962   Stereotyped movements of mental defectives. I. Initial survey. *Amer. J. Ment. Defic.* 66: 849-52.

Bernstein, I. S.
 1964a   Role of the dominant male rhesus monkey in response to external challenges to the group. *J. Comp. Physiol.* 57:404-6.

1964b The integration of rhesus monkeys introduced to a group. *Folia Primat.* 2:50-63.

1966 Analysis of a key role in a capuchin *(Cebus albifions)* group. *Tulane Studies in Zool.* 13(2): 49-54.

1969 Stability of the status hierarchy in a pigtail monkey group *(Macaca nemestrina). Animal Behav.* 17:452-57.

1970 Primate status hierarchies. In L. A. Rosenblum, ed., *Primate Behavior.* Vol. 1. New York and London: Academic Press.

Bernstein, I. S., and Gordon, T. P.
1974 The function of aggression in primate studies. *American Scientist* 62(3): 304-11.

Bertram, B. C. R.
1973 Lion population regulation. *East African Wildlife Journal.* Vol. 11 (3,4): 215-25.

Bertalanffy, L. von
1960 *Problems of life.* New York: Harper and Row.

Bingham, H. C.
1932 Gorillas in a native habitat. Carnegie Institute Publication.

Bingham, L. R., and Hahn, T. C.
1973 Observations on the birth of a lowland gorilla in captivity. *Breeding* pp. 113-14.

Bischof, Norbert
1975 Comparative ethology of incest avoidance. In Fox, R., ed., *Biosocial Anthropology,* pp. 37-67. London: Malaby Press.

Blaffer Hrdy, S.
1977 *The Langurs of Abu.* Cambridge: Harvard University Press.

Blurton Jones, N. G., and Trollope, J.
1968 Social behaviour of stump-tailed macaques in captivity. *Primates* 9: 365-94.

Boelkins, R. C., and Wilson, A. P.
1972 Intergroup social dynamics of the Cayo Santiago Rhesus *(Macaca mulatta)* with special reference to changes in group membership by males. *Primates* 13: 125-40.

Bouissou, M. Z.
1972 Influence of body weight and presence of horns on social rank in domestic cattle. *Animal Behav.* 20(3): 474-77.

Brandt, E. M., and Mitchell, G.
1972 *Parturition in primates: Behavior related to birth,* pp. 177-223. Davis: National Center for Primate Biology; Univ. of Calif.

Brody, S.
1945 *Bioenergetics and growth.* New York: Hafner.

Bronowski, J., and Bellugi, U.
1970 Language, name, and concept. *Science* 168: 669-73.

Bronson, F. H. and Marsden, H. M.
    1973   The preputial gland as an indicator of social dominance in male mice. *Behavioral Biology* 9:625-28.

Brown, J. K.
    1970   A note on the division of labour by sex. *Amer. Anthropol.* 72: 1073-78.

Brown, J. L.
    1963   Aggressiveness, dominance and social organization in the stellar jay. *The Condor* 65: 460-84.
    1964   The evolution of diversity in avian territorial systems. *Wilson Bulletin* 76:160-69.
    1972   Communal feeding of nestlings in the Mexican jay *(Alphelocoma ultramarina):* interflock comparisons. *Animal Behaviour* 20:395-402.

Bruce, R. H.
    1941   An experimental analysis of social factors affecting the performance of white rats. *J. Comp. Psychol.* 31:395-412.

Buckley, W.
    1967   *Sociology and modern systems theory.* Englewood Cliffs, NJ: Prentice-Hall.

Burt, W. H.
    1943   Territorial and home range concepts as applied to mammals. *Journal of Mammalogy* 24:346-52.

Bygott, J. D.
    1972   Cannibalism among wild chimpanzees. *Nature* 238:410-11.
    1974   Agonistic behaviour and dominance in wild chimpanzees. Ph.D. thesis, University of Cambridge.

Carpenter, C. R.
    1934   A field study of the behavior and social relations of the howling monkeys *(Alouatta palliata). Comp. Psychol. Monogr.* 10:1-168.
    1940   A field study in Siam of the behavior and social relations of the gibbon *(Hylobates lar). Comp. Psychol. Monogr.* 16:1-212.
    1942   Sexual behavior of free-ranging rhesus monkeys, *Macaca mulatta. J. Comp. Psychol.* 33:113-42.
    1965   The howlers of Barro Colorado Island. In I. DeVore, ed., *Primate behavior: Field studies of monkeys and apes,* pp. 250-91. New York: Holt, Rinehart and Winston.
    1969   Approaches to the study of naturalistic communication in nonhuman primates. In Sebeok, T. A. and Ramsay, eds. *Approaches to animal communication,* pp. 40-70. The Hague: Mouton.

Carpenter, B., and Carpenter, J. T.
    1958   The perception of movement by young chimpanzees and human children. *J. comp. physiol. Psychol.* 51, 782-84.

Cassirer, E.
    1944   *An essay on man.* New Haven: Yale University Press.

Chalmers, N.
   1967   The ethology and social organization of the black mangabey *(cercoce-bus albigena)*. Ph.D. thesis, Cambridge University.
   1968   The social behavior of free-living mangabeys in Uganda. *Folia primat.* 8:263-81.

Chance, M. R. A.
   1961   The nature and special features of the instinctive social bond of primates. In S. L. Washburn, ed., *Social life of early man.* New York: Viking Fund Publications in Anthropology, Vol. 31, pp. 17-33.

Chase, I. D.
   1974   Models of hierarchy formation in animal societies. *Behavioral Science.*

Chivers, D. J.
   1972   The Siamang and the gibbon in the Malay Peninsula. In D. Rumbaugh, ed., *Gibbon and Siamang.* 1:103-35. Basel: Karger.

Cleland, J. B., and Johnston, T. H.
   1933   The ecology of the aborigines of central Australia: Botanical notes. *Trans. Roy. Soc. South. Austral.* 57:113-24.

Clutton-Brock, T. H.
   1972   Feeding and ranging behaviour of the red colobus monkey. Ph.D. thesis, Cambridge University.
   1973   Feeding levels and feeding sites of red colobus *(Colobus badius tephrosceles)* in the Gombe National Park. *Folia Primatol.* 19:368-79.

Cody, M. L.
   1971   Finch flocks in the Mohave Desert. *Theoret. Pop. Biol.* 2:142-58.

Collins, K. D.
   1972   The implications of the biological life cycle for the social organization of baboons and other animals. Unpublished manuscript.

Connolly, K.
   1974   The development of skill. *New Scient.* 62:537-40.

Conoway, C. R., and Koford, C. B.
   1964   Estrous cycles and mating behavior in a free-ranging band of rhesus monkeys. *J. Mammal.* 45:577-88.

Coolidge, H. J.
   1933   *Pan paniscus,* pygmy chimpanzee from south of the Congo River. *Am. J. Phys. Anthrop.* 18:1-59.

Coon, C. S.
   1971   *The hunting peoples.* Boston: Little, Brown.

Craig, W.
   1921   Why do animals fight? *International Journal of Ethics* 31:264-78.

Crandall, L. S.
   1964   *The management of wild mammals in captivity.* Chicago: University of Chicago Press.

Crawford, M. P.
   1942   Dominance and the behavior of pairs of female chimpanzees when they meet after varying intervals of separation. *J. Comp. Psychol.* 33:259-65.

Crook, J. H.
    1966   Gelada baboon herd structure and movement. *Symp. Zool. Soc. Lond.*
            18:237-58.
    1968   The nature and function of territorial aggression. In A. Montagu, ed.,
            Man and aggression, pp. 141-78. New York: Oxford Univ. Press.
    1970   The socio-ecology of primates. In J. H. Crook, ed., *Social behavior of
            birds and mammals,* pp. 103-66. New York: Academic Press.
    1971   Sources of cooperation in animals and man. In J. F. Eisenberg and W. S.
            Dillon, eds., *Man and beast: Comparative social behaviour,* pp. 237-72.
    1972   Sexual selection, dimorphism and social organisation in the primates. In
            B. Campbell, ed., *Sexual selection and the descent of man,* pp. 231-81.
            London: Heinemann.
    1974   Social organization and the developmental environment. In N. F. White,
            ed., *Ethology and psychiatry.* University of Toronto Press.

Crook, J. H., and Aldrich-Blake, P.
    1968   Ecological and behavioural contrasts between sympatric ground-dwell-
            ing primates in Ethiopia. *Folia Primatol.* 8:192-227.

Crook, J. H., and Butterfield, P. A.
    1970   Gender role in the social system of quelea. In J. H. Crook, ed., *Social be-
            haviour in birds and mammals,* pp. 211-48. New York: Academic Press.

Crook, J. H., and Gartlan, J. S.
    1966   Evolution of primate societies. *Nature* 210:1200-03.

Dart, R. A.
    1964   The ecology of the south African man-apes. In D. H. S. Davis, ed., *Eco-
            logical studies of southern Africa,* pp. 49-66. The Hague: Dr. W. Junk.

Darwin, C.
    1859   *The origin of species.* London: John Murray.
    1872   *The expression of the emotions in man and animals.* London: John
            Murray.

Dare, R.
    1974   Food-sharing in free-ranging *Ateles geoffreyi* (red spider monkeys). *Lab.
            Prim. Newslet.* 13:19-21.

Davenport, R.
    1967   The orang-utan in Sabah. *Folia Primat.* 5:247-63.

Davenport, R. K.; Menzel, E. W. Jr.; and Rogers, C. M.
    1961   Maternal care during infancy. Its effect on weight gain and mortality in
            the chimpanzee. *Amer. J. Orthopsychiat.* 31:No. 4.

Davenport, R. K., and Berkson, G.
    1963   Stereotyped movements of mental defectives. II. Effects of novel
            objects. *Amer. J. ment. Defic.* 67:879-82.

Davenport, R. K.; Menzel, E. W. Jr.; and Rogers, C. M.
    1966   Effects of severe isolation on normal chimpanzees. *Arch. Gen. Psychiat.*
            14:134-38.

Davenport, R. K., and Rogers, C. M.
1968   Intellectual performance of differentially reared chimpanzees. I. Delayed response. *Amer. J. ment. Defic.* 72:674-80.

Davenport, R. K.; Rogers, C. M.; and Menzel, E. W. Jr.
1969   Intellectual performance of differentially reared chimpanzees. II. Discrimination-learning set. *Amer. J. ment. Defic.* v. 73.

Davenport, R. K., and Rogers, C. M.
1970a   Inter-modal equivalence of stimuli in apes. *Science* 168:279-80.
1970b   Differential rearing of the chimpanzee. A project survey. In G. H. Bourne, ed., *The chimpanzee.* Vol. 3. Basel: Karger.
1971   Perception of photographs by apes. *Behaviour* 39:318-20.

Davenport, R. K.; Rogers, C. M.; and Rumbaugh, Duane M.
1973   Long-term cognitive deficits in chimpanzees associated with early impoverished rearing. *Dev. Psych.* v. 9.

Davenport, R. K.; Rogers, C. M.; and Russell, I. S.
1973   Cross modal perception in apes. *Neuropsychologia* 11:21-28.

Davis, J.
1973   Forms and norms: the economy of social relations. *Man* 8:159-76.

Deag, J. M.
1973   Intergroup encounters in the wild barbary macaque *Macaca sylvanus L.* In R. P. Michael and J. H. Crook, eds., *Comparative ecology and behaviour of primates,* pp. 315-73. New York: Academic Press.
1974   A study of the social behaviour and ecology of the wild Barbary macaque, *Macaca sylvanus.* Ph.D. thesis, Bristol University.

Deag, J. M., and Crook, J. H.
1971   Social behaviour and 'agonistic buffering' in the wild barbary macaque *Macaca sylvana L. Folia primat.* 15:183-200.

DeVore, I.
1962   The social behavior and organization of baboon troops. Ph.D. thesis, University of Chicago.
1965a   *Primate behavior.* New York: Holt, Rinehart and Winston.
1965b   Male dominance and mating behavior in baboons. In Frank Beach, ed., *Sex and behavior,* pp. 266-89. New York: Wiley.
1966   Dynamics of male dominance in a baboon troop. (A 16mm, narrated, color film). New York: Modern Learning Aids.

DeVore, I., and Washburn, S. L.
1963   Baboon ecology and human evolution. In F. C. Howell and F. Bourlière, eds., *African ecology and human evolution,* pp. 335-67. Chicago: Aldine.

Döhl, V. J.
1968   Uber die Fähigkeit einer Schimpansin, Umwege mit selbständigen Zwischenzielen zu überblicken. *Z. Tierpsychol.,* 25:89-103.
1969   Versuche mit einer Schimpansin über Abkürzungen vei Umwegen mit selbständigen Zwischenzielen. *Z. Tierpsychol.* 26:200-7.

1970   Zielorientiertes Verhalten beim Schimpansen. *Naturwissenschaft und Medizin* 7:43-57.

Dollard, J.; Doob, L. W.; Miller, N. E.; Mowrer, O. H.; and Sears, R. R.
1939   *Frustration and aggression.* New Haven: Yale University Press.

Durham, N. M.
1971   Effects of altitude differences on group organization of wild black spider monkey *(Ateles paniscus). Proc. 3rd Int. Congr. Primatol.,* Zurich, 1970. Basel: Karger.

Eibl-Eibesfeldt, I.
1961   The fighting behavior of animals. *Scientific American* 205:112-22.
1975   *Ethology: the biology of behavior.* 2nd ed., New York: Holt, Rinehart and Winston.

Eisenberg, J. F.
1973   Mammalian social systems: Are primate social systems unique? In E. W. Menzel, ed., *Precultural primate behaviour,* pp. 232-49. Basel: Karger.

Eisenberg, J. F.; Muckenhiren, N. A.; and Rudran, R.
1972   The relation between ecology and social structure in primates. *Science* 176: 863-74.

Ellefson, J. O.
1968   Territorial behaviour in the common white-handed gibbon, *Hylobates lar* Linn. In P. C. Jay, ed., *Primates: Studies in adaptation and variability,* pp. 180-99. New York: Holt, Rinehart and Winston.

Emlen, J. T., and Lorenz, F. W.
1942   Pairing responses of free-living valley quail to sex hormone pellet implants. *Auk* 59:369-78.

Etkin, W.
1954   Social behavior and the evolution of man's mental faculties. *Amer. Natural.* 88:129-42.

Fantz, R. L.
1965   Ontogeny of perception. In A. M. Schrier, H. F. Harlow, and F. Stollnitz, eds., *Behavior of nonhuman primates.* Vol. 2, pp. 365-403. New York: Academic Press.

Farentinos, R. C.
1972   Social dominance and mating activity in the tassel-eared squirrel *(Sciurus aberti ferreus). Behav.* 20(2):316-26.

Fisher, R. A.
1958   *The genetical theory of natural selection.* New York: Dover.

Fortes, M.
1949   Time and social structure: an Ashanti case study. In M. Fortes, ed., *Social Structure.* New York: Russell and Russell.

Fossey, D.
1970   Making friends with mountain gorillas. *Nat. Geogr.* 137:48-68.
1972   Vocalizations of the mountain gorilla *(Gorilla gorilla beringei). Anim. Behav.* 20:36-53.
1974   Observations on the home range of one group of mountain gorillas *(Gorilla gorilla beringei). Anim. Behav.* 22:568-81.

Fossey, D., and Harcourt, A. H.
1977   Feeding ecology of free-ranging mountain gorilla *(Gorilla g. beringei).* In T. H. Clutton-Brock, ed., *Primate Ecology: Studies in feeding and ranging behavior of lemurs, monkeys, and apes.* London and New York: Academic Press.

Fouts, R.
1972   The use of guidance in teaching sign language to a chimpanzee. *Journal of Comparative and Physiological Psychology* 80:515-22.
1973   Acquisition and testing of gestural signs in four young chimpanzees. *Science* 180:978-80.
1975   Communication with chimpanzees. In G. Kurth and I. Eibl-Eibesfeldt, eds., *Hominisation and behavior.* Stuttgart: Gustav Fischer Verlag.

Fouts, R.; Chown, W.; and Goodin, L.
1973   The use of vocal English to teach American sign language (ASL) to a chimpanzee: Translation from English to ASL. Paper presented at the Midwestern Psychological Association meeting in Chicago, Illinois, May, 1973.

Fouts, R.; Chown, W.; Kimball, G.; and Couch, J.
1976   Comprehension and production of American sign language by a chimpanzee (Pan). Paper presented at the XXI International Congress of Psychology, Paris, July 18-25, 1976.

Fouts, R.; Mellgren, R.; and Lemmon, W.
1973   American sign language in the chimpanzee: Chimpanzee-to-chimpanzee communication. Paper presented at the Midwestern Psychological Association meeting, Chicago, May, 1973.

Fox, R.
1972   Alliance and constraint. In B. A. Campbell, ed., *Sexual selection and the descent of man.* Chicago: Aldine.

Free, J.
1958   The defense of bumble bee colonies. *Behaviour* 12:233-42.

Frisch, J. E.
1973   The hylobatid dentition. In Duane Rumbaugh, ed., *Gibbon and Siamang.* Vol. 2. Basel: Karger.

Furaya, Y.
1969   On the fission of troops of Japanese monkeys. II. General view of troop fission of Japanese monkeys. *Primates* 10:47-69.

Furness, W.
1916   Observations on the mentality of chimpanzees and orangutans. *Proceedings of the American Philosophical Society* 45:281-90.

Gadgil, M.
1972   Male dimorphism as a consequence of sexual selection. *American Naturalist* 106:574-80.

Galdikas, B. M. F.
In   Orangutan adaptations at Tanjung Puting Reserve: mating and ecology.
press

Gallup, G. G.
1970   Chimpanzees: Self-recognition. *Science* 167:86-87.

Gardner, B. T., and Gardner, R. A.
1971   Two-way communication with an infant chimpanzee. In A. M. Schrier and F. Stollnitz, eds., *Behavior of nonhuman primates,* 4. New York: Academic Press.

Gardner, R. A., and Gardner, B. T.
1969   Teaching sign language to a chimpanzee. *Science* 165:664-72.

Gartlan, J. S.
1968   Structure and function in primate society. *Folia primat.* 8:89-120.

Gartlan, J. S., and Brain, C. K.
1968   Ecology and social variability in *Cercopithecus aethiops* and *C. mitis.* In P. C. Jay, ed., *Primates: Studies in adaptation and variability,* pp. 253-92. New York: Holt, Rinehart and Winston.

Gavan, J. A.
1953   Growth and development of the chimpanzee: A longitudinal and comparative study. *Human Biol.* 25:93-143.

Geist, V.
1966   The evolution of horn-like organs. *Behaviour* 27:175-214.
1971   *Mountain sheep.* Chicago: University of Chicago Press.

Gibson, J. J.
1966   *The senses considered as perceptual systems.* New York: Houghton-Mifflin.

Gijzen, A., and Tijskens, J.
1971   Growth in weight of the lowland gorilla *(Gorilla g. gorilla)* and of the mountain gorilla. London: Int. Zool. Yearbook.

Goffman, E.
1969   *The presentation of self in everyday life.* Middlesex: Penguin.

Goodall, J.
1963a   Feeding behaviour of wild chimpanzees. *Symp. Zool. Soc. London* 10:39-48.
1963   My life among wild chimpanzees. *National Geographic Magazine* 124:272-308.
1965   Chimpanzees of the Gombe Stream Reserve. In I. DeVore, ed., *Primate behavior.* New York: Holt, Rinehart and Winston.
1967   Mother-offspring relationships in chimpanzees. In D. Morris, ed., *Primate ethology,* pp. 287-346. London: Weidenfeld & Nicolson.
1968a   A preliminary report on expressive movements and communication in the Gombe Stream Chimpanzees. In P. C. Jay, ed., *Primates.* New York: Holt, Rinehart and Winston.

1968b The behaviour of free-living chimpanzees in the Gombe stream area. *Animal Behaviour Monographs* 1:161-311.

1962 Nest building behavior in the free ranging chimpanzee. *Annals New York Academy of Sciences.* 102:455-67.

1971 *In the shadow of man.* Boston: Houghton Mifflin.

1973a Cultural elements in a chimpanzee community. In E. W. Menzel, ed., *Precultural primate behavior,* pp. 144-84. Basel: Karger.

1973b The behaviour of the chimpanzee in their natural habitat. *Amer. J. of Psychiatry* 130:1-12.

1974 Personal communication (July 20-28, 1974).

1975 Chimpanzees of Gombe National Park: Thirteen Years of Research. In I. Eibesfeldt, ed., *Hominisation und Verhalten.* Stuttgart: Gustave Fischer Verlag.

Goodman, M., and Tashian, R., eds.
1976 *Molecular Anthropology.* New York: Plenum Press.

Gorman, C.
1971 The Hoabinhian and after: Subsistence patterns in southeast Asia during the late pleistocene and early recent periods. *World Archaeol.* 2:300-20.

Goss-Custard, J. D.; Dunbar, R. I. M.; and Aldrich-Blake, F. P. G.
1972 Survival, mating and rearing strategies in the evolution of primate social structure. *Folia primat.* 17:1-19.

Gould, R. A.
1967 Notes on hunting, butchering and sharing of game among the Ngatatjara and their neighbours in the west Australian desert. *Kroeber Anthropol. Soc. Pap.* 36: 41-66.

Graham, C. E.
1970 Reproductive physiology of the chimpanzee. In G. H. Bourne, ed., *The chimpanzee.* Vol. 3. Basel: Karger.

Groom, A. F. G.
1973 Squeezing out the mountain gorilla. *Oryx* 12:207-15.

Groves, C. P.
1970 *Gorillas.* London: Arthur Barker.

Grubb, P.
1974 Mating activity and the social significance of rams in a feral sheep community. In V. Geist and F. Walther, eds., *The behavior of ungulates and its relation to management,* Morges, Switzerland: IUCN Publications.

Gunther, M.
1971 *Infant feeding.* Harmondsworth: Penguin.

Guthrie, R. D.
1970 Evolution of human threat display organs. In T. Dobzhansky, Max K. Hecht, W. C. Steele, eds., *Evolutionary biology.* Vol. 4, pp. 257-300. New York: Appleton-Century-Crofts.

Guthrie, R. D., and Petocz, R. G.
    1970   Weapon automimicry among mammals. *American Naturalist* 104: 585-88.
Guttman, R.; Naftali, G.; and Nevo, E.
    1975   Aggression patterns in three chromosome forms of the mole rat, *Spalax ehrenbergi. Animal Behavior* 23:485-93.

Haber, R. N., and Hershenson, M.
    1973   *The psychology of visual perception.* New York: Holt, Rinehart & Winston.
Hall, K. R. L.
    1962   The sexual, agonistic and derived social behaviour patterns of the wild chacma baboon, *Papio ursinus. Proc. Zool. Soc. Lond.* 139:283-327.
    1965   Behaviour and ecology of the wild patas monkey, *Erythrocebus patas,* in Uganda. *Journal of Zoology* 148:15-87.
Hall, K. R. L., and DeVore, I.
    1965   Baboon social behavior. In I. DeVore, ed., *Primate behavior: Field studies of monkeys and apes.* New York: Holt, Rinehart & Winston.

Hall, K. R. L., and Gartlan, J. S.
    1965   Ecology and behaviour of the vervet monkey, *Cercopithecus aethiops,* Lolui Island, Lake Victoria. *Proceedings of the Zoological Society of London* 145:37-56.

Hall, K. R. L., and Mayer, B.
    1967   Social interactions in a group of captive patas monkeys *(Erythrocebus patas). Folia primat.* 5:213-36.
Hallett, G.
    1969   *Wittengenstein's definition of meaning as use.* New York: Fordham University Press.

Halperin, S. D.
    In   Social relationships: a quantitative assessment of social interactions
    prepa- among adult male chimpanzees. Ph.D. thesis. Washington University.
    ration
Hamburg, B. A.
    1978   The biosocial basis of sex difference. In S. L. Washburn and E. R. McCown, eds., *Human evolution: biosocial perspectives.* Vol. IV. Menlo Park, Ca.: Benjamin/Cummings.

Hamburg, D. A.
    1971a Crowding, stranger contact and aggressive behavior. In L. Levi, ed., *Society, stress and disease.* New York: Oxford University Press.
    1971b Recent research on hormonal factors relevant to human aggressiveness. *International Social Science Journal.* 23:36-47.
    1971c Psychological studies of aggressive behavior. *Nature.* 230:19-23.

Hamilton, C. L.
    1972   Long term control of food intake in the monkey. *Physiol. and Behav.* 9:1-6.

Hamilton, W. D.
    1964   The genetical evolution of social behavior. *J. Theoret. Biol.* 7:1-52.

1972 Altruism and related phenomena, mainly in social insects. *Ann. Rev. Ecol. Systemat.* 3:193-232.

Hanby, J. P.
1975 The social nexus: problems and solutions in the portrayal of primate social structures. *Proceedings of 5th Congress of Internat. Primatol. Soc., Nagoya, Japan.* Basel: Karger.

Harcourt, A. H.
In prepa- ration Social relations and determinants of ranging patterns of mountain gorilla. Ph.D. thesis, Cambridge University.

Harcourt, A. H., and Groom, A. F. G.
1972 Gorilla census. *Oryx* 11:355-63.

Harding, R. S. O.
1973 Predation by a troop of olive baboons *(Papio anubis). Amer. J. Phys. Anthropol.* 38:587-92.

Harlow, H. F.
1950 Analysis of discrimination learning by monkeys. *J. exp. Psychol.* 40: 26-38.
1965 Sexual behavior in the rhesus monkey. In F. A. Beach, ed., *Sex and Behavior.* New York: Wiley.

Harlow, H. F.; Dodsworth, R. O.; and Harlow, M. K.
1965 Total social isolation in monkeys. *Proc. Nat. Acad. Sci.* 54:90-97.

Harlow, H. F., and Harlow, M. K.
1968 Effects of social isolation on learning by rhesus monkeys. *Proc. 2nd Int. Congr. Primatol., Atlanta, Ga.* Basel/New York: Karger.
1969 Effects of various mother-infant relationships on rhesus monkey behaviors. In B. M. Foss, ed., *Determinants of infant behaviour.* Vol. 4. London: Methuen.

Harlow, H. F., and Yudin, H. C.
1933 Social behavior of primates 1. Social facilitation of feeding in the monkey and its relation to attitudes of ascendance and submission. *J. Comp. Psych.* 16:171-85.

Harrisson, B.
1962 *Orangutan.* London: Collins.

Hatch, J. J.
1970 Predation and piracy by gulls at a ternery in Maine. *Auk* 87:244-54.

Hayes, C.
1951 *The ape in our house.* New York: Harper and Row.

Hayes, K., and Hayes, C.
1951 The intellectual development of a home-raised chimpanzee. *Proceedings of the American Philosophical Society* 95:105-9.
1952 Imitation in a home-raised chimpanzee. *J. of Comp. and Physiol. Psychol.* 45:450-59.
1953 Picture perception in a home-raised chimpanzee. *J. comp. physiol. Psychol.* 46:470-74.

1954   The cultural capacity of chimpanzee. *Human Biol.* 26:288-303.

Hayes, K. J., and Nissen, C. H.
1971   Higher mental functions of a home-raised chimpanzee. In A. M. Schrier & F. Stollnitz, eds., *Behavior of nonhuman primates,* pp. 60-115. New York: Academic Press.

Hebb, D. O.
1947   Spontaneous neurosis in chimpanzee: Theoretical relations with clinical and experimental phenomena. *Psychosom. Med.* 9:3-19.
1949   *The organization of behavior.* New York: Wiley.

Heider, F.
1958   *Interpersonal relations.* New York: Wiley.

Heiligenberg, W.
1965   A quantitative analysis of digging movements and their relationships to aggressive behavior in cichlids. *Behav.* 13:163-70.

Hess, J. P.
1973   Some observations on the sexual behaviour of captive lowland gorillas. In R. P. and J. H. Crook, eds., *Comparative ecology and behaviour of primates,* pp. 507-81. London: Academic Press.

Hewes, G. W.
1961   Food transport and the origin of hominid bipedalism. *Amer. Anthropol.* 63:687-710.
1973a   Pongid capacity for language acquisition. In W. Montagna and E. Menzel Jr., eds., *Symposia of the Fourth International Congress of Primatology, Nagoya, Japan.* Basel: Karger.
1973b   Primate communication and the gestural origin of language. *Current Anthropology* 14:5-24.

Hinde, R. A.
1952   The behaviour of the great tit *(Parus major)* and other related species. *Behaviour,* Supplement, II.
1955   The courtship and copulation of the greenfinch *(Chloris chloris). Behaviour* 7:207-32.
1967   The nature of aggression. *New Society,* March 2.
1970   *Animal behaviour: A synthesis of ethology and comparative psychology.* 2nd ed. New York: McGraw-Hill.
1971   Development of social behavior. In A. M. Schrier and F. Stollnitz, eds., *Behavior of nonhuman primates: modern research trends.* 3, pp. 1-68. New York: Academic Press.
1972a   *Non-verbal communication.* London and New York: Cambridge Univ. Press.
1972b   Social behavior and its development in subhuman primates. *Condon Lectures.* Oregon State System Higher Education, Eugene.
1974   *The biological bases of human social behaviour.* New York: McGraw-Hill.
In   Interactions, relationships, and group structure in non-human primates.
press   *Proc. 4th Int. Congr. Primatol., Nagoya, Japan.* Basel: Karger.

Hinde, R. A., and Simpson, M. J. A.
  1975 Qualities of mother-infant relationships in monkeys. In *The Parent-Infant Relationship,* Ciba Foundation Symposium 1974. Amsterdam: Associated Scientific Publishers.

Hinde, R. A., and Spencer-Booth, Y.
  1967 The effect of social companions on mother-infant relations in rhesus monkeys. In D. Morris, ed., *Primate Ethology.* London: Weidenfeld and Nicolson.

Hockett, C. F.
  1960 Logical considerations in the study of animal communication. In Lanyon and Tavolga, *Animal sounds and communication,* pp. 392-430. Washington: American Institute of Biological Sciences.

Holloway, R. L., Jr.
  1969 Culture: A human domain. *Curr. Anthropol.* 10:395-412.

Holttum, R. E.
  1969 *Plant life in Malaya.* London: Longman Group.

Holzberg, S., and Schroder, J. Horst
  1975 The inheritance of aggressiveness in the convict cichlid fish, *Cichlasoma nigrofasciatum* (Pisces: Cichlidae). *Animal Behavior* 23:625-31.

Homans, G. C.
  1961 *Social behaviour, its elementary forms.* London: Routledge and Kegan Paul.

Horn, H. S.
  1968 The adaptive significance of colonial nesting in the brewer's blackbird *(Euphagus cyanocephalus).* Ecology 49:682-94.

Hornaday, W. T.
  1885 *Two years in the jungle.* London.

Horr, D. A.
  1972 The Bornean orang-utan. Population structure and dynamics in relationship to ecology and reproductive strategy. Unpublished paper presented at meeting of *Am. Ass. Phys. Anthrop.*
  In Orang-utan social structure: A computer simulation strategy.
  press

Howard, H. E.
  1935 *The nature of a bird's world.* Cambridge: Cambridge University Press.

Hrdy, S. B.
  1974 Male-male competition and infanticide among the langurs *(Presbytis entellus)* of Abu, Rajastan. *Folia Primatol.* 22:19-58.
  1976 Care and exploitation of nonhuman primate infants by conspecifics other than the mother. *Advances in the Study of Behavior* 6:101-58.

Huxley, T. H.
  1863 *Evidence as to man's place in nature.* William and Norgate, London. See also "Introduction to the Ann Arbor Paperbacks edition" of this work, by A. Montagu. University of Michigan Press.

515

Imanishi, K.
 1960   Social organization of subhuman primates in their natural habitat. *Curr. Anthropol.* 1:393.

Itani, I.
 1954   *The monkeys of Takasakiyama.* Tokyo: Kobunsha. (In Japanese).
 1959   Paternal care in the wild Japanese monkey. *Primates* 2:61-93.
 1963   Vocal communication of the wild Japanese monkey, *Primates* 4:11.
 1963   Paternal care in the wild Japanese monkey. In C. H. Southwick, ed., *Primate Social Behavior,* pp. 91-97. New York: Van Nostrand.

Itani, J., and Suzuki, A.
 1967   The social unit of wild chimpanzees. *Primates* 8:355-81.

Isaac, G. L.
 1968   Traces of Pleistocene hunters: An East African example. In R. B. Lee and I. DeVore, eds., *Man the hunter,* pp. 253-261. Chicago: Aldine-Atherton.

Izawa, K.
 1970   Unit-groups of chimpanzees and their nomadism in the savannah woodland. *Primates* 11:1-46.

Jackson, G., and Gartlan, J. S.
 1965   The flora and fauna of Lolui Island, Lake Victoria. *Journal of Ecology* 53: 573-97.

Jantschke, F.
 1972   *Orang-utans in Zoologischen Gärten.* Munich: Piper.

Jay, P. C.
 1963   Mother-infant relations in langurs. In H. L. Rheingold, ed., *Maternal behavior in mammals,* pp. 282-304. New York: Wiley.
 1963   The Indian langur monkey *(Presbytis entellus).* In C. H. Southwick, ed., *Primate social behavior,* Princeton: Van Nostrand.
 1965   The common langur of North India. In I. DeVore, ed., *Primate behavior: Field studies of monkeys and apes,* pp. 197-249. New York: Holt, Rinehart and Winston.

Johnson, R. N.
 1972   *Aggression in man and animals.* Philadelphia: W. B. Saunders.

Jolly, A.
 1967   *Lemur behavior, a Madagascar field study.* Chicago: University of Chicago Press.
 1972   *The evolution of primate behavior.* New York: MacMillan.

Jolly, C. J.
 1970   The seed-eaters: A new model of hominid differentiation based on a baboon analogy. *Man* 5:5-26.

Kano, T.
 1971a Distribution of the primates on the eastern shore of Lake Tanganyika. *Primates* 12:281-304.
 1971b The chimpanzee of Filabanga, Western Tanzania. *Primates* 12:229-46.
 1972 Distribution and adaptation of the chimpanzee on the eastern shore of Lake Tanganyika. *Kyoto Univ. Afr. Studies* 7:37-129.
 1974 The pygmy chimpanzee. *Shizen* 29:28-37.

Katchadourian, H. A., and Lunde, D. T.
 1972 *Fundamentals of human sexuality.* New York: Holt, Rinehart and Winston.

Kaufman, I. C., and Rosenblum, L. A.
 1966 A behavioral taxonomy for *Macaca nemestrina* and *Macaca radiata* based on longitudinal observation of family groups in the laboratory. *Primates* 7:205-58.

Kaufmann, J. H.
 1962 Ecology and social behavior of the coati, *Nasua narica,* on Barro Colorado Island, Panama. *University of California Publications in Zoology* 60:95-222.
 1965 A three-year study of mating behavior in a free-ranging band of rhesus monkeys. *Ecology* 46:500-12.
 1967 Social relations of adult males in a free-ranging band of rhesus monkeys. In S. A. Altmann, ed., *Social communication among primates.* Chicago: University of Chicago Press.

Kavanagh, M.
 1972 Food-sharing behaviour within a group of douc monkeys, *(Pygathrix nemaeus nemaeus). Nature* 239:406-7.

Kawabe, M.
 1966 One observed case of hunting behaviour among wild chimpanzees living in the savanna woodland of western Tanzania. *Primates* 7:393-96.

Kawai, M.
 1958 On the system of social ranks in a natural group of Japanese monkeys. *Primates* 1:111-48.
 1965 On the system of social ranks in a natural troop of Japanese monkeys. Basic rank and dependent rank. In K. Imanishi and S. A. Altmann, eds., *Japanese monkeys: a collection of translations.* Atlanta: Yerkes Regional Primate Center.

Kawai, M., and Mizuhara, H.
 1962 An ecological study of the wild mountain gorilla *(Gorilla gorilla beringei). Primates* 2:1-42.

Kawamura, S.
 1958 Matriarchal social ranks in the Minoo-B troop: A study of the rank system of Japanese monkeys. *Primates* 2:181-252.

Kawanaka, K., and Nishida, T.
   1974   Recent advances in the study of inter-unit-group relationships and social structure among chimpanzees of the Mahali Mountains. Paper presented at the *5th Congress of the International Primatological Society, Nagoya, Japan.* Basel: Karger.

Kellog, W. N.
   1968   Communication and language in the home-raised chimpanzee. *Science* 162:423-27.

Kellogg, W. N., and Kellogg, L. A.
   1967   *The ape and the child.* New York: Hafner. (Original published 1933.)

Kimura, D., and Archibald, Y.
   1974   Motor functions of the left hemisphere. *Brain* 97:337-50.

Kimura, D.; Battison, R.; and Lubert, B.
   1976   Impairment of nonlinguistic hand movements in a deaf aphasic. *Brain and language* 3:566-71.

Koford, C. B.
   1963   Group relations in an island colony of rhesus monkeys. In C. H. Southwick, ed., *Primate social behavior,* pp. 136-52. Princeton: Van Nostrand.
   1966   Population changes in rhesus monkeys, 1960-1965. *Tulane Studies in Zool.* 13:1-7.

Köhler, W.
   1951   *The mentality of apes.* New York: Humanities Press. (Original published 1925.)
   1971   Die Methoden der psychologischen Forschung an Affen. In Emil Abderhalden, (Hrsg.), *Handbuch der biologischen Arbeitsmethoden,* 1921 (Abt. 6, Teil D), 69-120. Translated in English in M. Henle, ed., *The selected papers of Wolfgang Köhler,* pp. 197-223. New York: Liveright.

Kortlandt, A.
   1940   Eine Übersicht über die angeborenen verhaltensweisen des mitteleuropaischen Kormorans. *Arch. Neerl. Zool.* 4:401-42.
   1962   Chimpanzees in the wild. *Scientific American* 206:128-38.
   1972   *New perspectives on ape and human evolution.* Amsterdam: Stichting voor Psycholbiologie.
   1974   Ecology and paleoecology of ape locomotion. *Proc. 5th International Congress of Primatology, Nagoya, Japan.* Basel: Karger.

Kravitz, H.; Rosen, V.; Teplitz, Z.; Murphy, J. B.; and Lesser, R. E.
   1960   A study of head banging in infants and children. *Dis. nerv. Syst.* 21:203-8.

Krushinskii, L. V.
   1962   *Animal behavior: Its normal and abnormal development.* New York: Consultants Bureau.

Kruuk, H.
   1972   *The spotted hyena.* Chicago: University of Chicago Press.

Kummer, H.
 1957 Soziales verhalten einer mantelpavian-gruppe. *Beiheft zur Schweizeris-chen Zeitschrift fur Psychologie und ihre anwendungen* 33:1-91.
 1967 Tripartite relations in hamadryas baboons. In S. A. Altmann, ed., *Social communication among primates*. Chicago: Univ. of Chicago Press.
 1968 *Social organization of hamadryas baboons*. Chicago: Univ. of Chicago Press.
 1970 Immediate causes of primate social structures. In H. Kummer, ed., *Proceedings of the 3rd International Congress of Primatology, Zurich*. Basel: Karger.

Kummer, H., and Kurt, F.
 1963 Social units of a free-living population of hamadryas baboons. *Folia primatologica*. Vol. 1, 1:4-19.
 1971 *Primate Societies*. Chicago: Aldine.
 In Rules of bond formation among captive gelada baboons. *Proc. 5th Int.*
 press *Congr. Primatol. Soc., Nagoya, Japan*. Basel: Karger.

Kummer, H.; Gotz, W.; and Angst, W.
 1974 Triadic differentiation: An inhibiting process protecting pair bonds in baboons. *Behaviour* 49:62-87. 87.

Kurtén, B.
 1972 *Not from the apes*. New York: Random House. (Pantheon Books)

Lack, D.
 1954 *The natural regulation of animal numbers*. Oxford University Press.
 1966 *Population studies of birds*. Oxford: Clarendon Press.

Lancaster, J. B.
 1968 On the evolution of tool-using behavior. *Amer. Anthropol.* 70:56-66.
 1971 Play mothering: the relations between juvenile females and young infants among free-ranging vervet monkeys *(Cercopithecus aethiops)*. *Folia Primat.* 15:161-82.
 1968 Primate communication systems and the emergence of human language. In P. C. Jay, ed., *Primates: Studies in adaptation and variability*, pp. 439-57. New York: Holt, Rinehart & Winston.

Langer, S. K.
 1942 *Philosophy in a new key*. Cambridge, Mass.: Harvard University Press.

Lashley, K. S.
 1951 The problem of serial order in behavior. In L. A. Jeffress, ed., *Cerebral mechanisms in behavior*, pp. 112-36. New York: Wiley.

Laughlin, W. S.
 1968 Hunting: An integrating biobehavior system and its evolutionary importance. In R. B. Lee and I. DeVore, eds., *Man the hunter*, pp. 304-20. Chicago: Aldine-Atherton.

Leach, E. R.
 1966 Rethinking anthropology. *L.S.E. Monogr. Social Anthrop.* 22. London: Athlone.

Leakey, M. D.
    1966   A review of the Oldowan culture from Olduvai Gorge, Tanzania. *Nature* 210:462-66.

LeBoeuf, B. J.; Whiting, R. J.; and Grant, R. F.
    1972   Perinatal behavior of northern elephant seal females and their young. *Behaviour* 43:121-56.

LeBoeuf, B. J., and Peterson, R. S.
    1969   Social status and mating activity in elephant seals. *Science* 163:91-93.

Lee, R. B.
    1968   What hunters do for a living, or, How to make out on scarce resources. In R. B. Lee and I. DeVore, eds., *Man the hunter*, pp. 30-48. Chicago: Aldine-Atherton.

Lee, R. B., and DeVore, I., eds.
    1968   *Man the hunter.* Chicago: Aldine-Atherton.

Le Gros Clark, W. E.
    1967   Human food habits as determining the basic patterns of economic and social life. In J. Kuhnau, ed., *Proceedings of the seventh international congress of nutrition.* Vol. 4, pp. 18-24. Braunschweig: Vieweg and Sohm.

Lenneberg, E. H.
    1967   *Biological foundations of language.* New York: Wiley.

Leskes, A., and Acheson, N. H.
    1971   Social organisation of a free-ranging troop of black and white colobus monkeys *(Colobus abyssinicus). Proc. 3rd Int. Congr. Primatol., Zurich.* Basel: Karger.

Lévi-Strauss, C.
    1953   Social structure. In A. L. Kroeber, ed., *Anthropology Today.* Internat. Symp. on Anthropology, Chicago.

Lieberman, P.; Crelin, E.; and Klatt, D.
    1972   Phonetic ability and related anatomy of the newborn and adult human, neanderthal man, and the chimpanzee. *American Anthropologist* 74: 287-307.

Lindburg, D. G.
    1969   Rhesus monkeys: Mating season mobility of adult males. *Science* 166: 1176-78.
    1971   The rhesus monkey in north India: An ecological and behavioral study. In L. A. Rosenblum, ed., *Primate behavior: Developments in field and laboratory research.* Vol. 2, pp. 1-106. New York: Academic Press.

Lockard, J. S.; McDonald, L. L.; Clifford, D. A.; and Martinez, R.
    1976   Panhandling: Sharing of resources. *Science* 191:406-8.

Lockie, J. D.
    1956   Winter fighting in feeding flocks of rooks, jackdaws, and carrion crows. *Bird Study* 3:180-90.

Lorenz, K.
  1950 The comparative method in studying innate behaviour patterns. *Sym. Soc. exp. Biol.* 4:221-68.
  1966 *On Aggression.* London: Methuen.
  1971 *Studies in animal and human behavior.* Vol. 2. Cambridge: Harvard University Press.

Lott, D.
  1974 Sexual and aggressive behavior of adult male American bison *(Bison bison).* In V. Geist and F. Walther, eds., *The behavior of ungulates and its relation to management.* Morges, Switzerland: IUCN Publications.

Loy, J.
  1970 Behavioral response of free-ranging rhesus monkeys to food shortage. *American J. Physical Anthropology* 33:263-71.

MacKay, D. M.
  1969 *Information, mechanism and meaning.* Cambridge: M.I.T. Press.

MacKinnon, J.
  1969 The Oxford University Expedition to Sabah 1968. *Oxford Univ. Explor. Club Bull.* 17,4:53-70.
  1971 The orang-utan in Sabah today. *Oryx* 11,2-3:141-91.
  1973a The behavior and ecology of the orang-utan *(Pongo pygmaeus).* Ph.D. thesis, Oxford University.
  1973b Orang-utans in Sumatra. *Oryx* 12,2:234-42.
  1974 The behaviour and ecology of wild orang-utans. *Pongo pygmaeus. An. Behav.* 22:3-74.
  1975 Distinguishing characters of the insular forms of orang-utan. *Int. Zoo Yearbook* 15:95-97.
  In press Reproductive behaviour in wild orang-utan populations.

Mark, V., and Ervin, F. R.
  1970 *Violence and brain.* New York: Harper and Row.

Marler, P.
  1961 The filtering of external stimuli during instinctive behaviour. In W. H. Thorpe and O. L. Zangwill, eds., *Current problems in animal behaviour,* pp. 150-66. Cambridge University Press.
  1965 Communication in monkeys and apes. In DeVore, *Primate behavior,* pp. 544-84. New York: Holt, Rinehart and Winston.
  1968 Aggregation and dispersal: Two functions in primate communication. In P. C. Jay, ed., *Primates: Studies in adaptation and variability,* pp. 429-38. New York: Holt, Rinehart & Winston.

Marsden, H.
  1972 The effect of food deprivation on intergroup relations in rhesus monkeys. *Behavioral Biology* 7:369-74.

Maslow, A. H., and Flanzbaum, S.
  1936 The role of dominance in the social and sexual behavior of infra-human primates. II. An experimental determination of the behavior syndrome of dominance. *J. Genet. Psychol.* 48:278-309.

Mason, W. A.
  1960 The effects of social restriction on the behavior of rhesus monkeys. I. Free social behavior. *J. comp. Physiol. Psychol.* 53:582-89.
  1967 Motivational aspects of social responsiveness in young chimpanzees. In H. W. Stevenson et al., eds., *Early behavior: Comparative and developmental approaches.* New York: Wiley.
  1968 Use of space by Callicebus groups. In P. C. Jay, ed., *Primates: Studies in adaptation and variability,* pp. 200-16. New York: Holt, Rinehart and Winston.
  1968 Early social deprivation in the nonhuman primates: Implications for human behavior. In D. C. Glass, ed., *Environmental Influences.* New York: Rockefeller University Press.
  1970 Chimpanzee social behavior. In G. H. Bourne, ed., *The chimpanzee.* Vol. 2, pp. 265-88. Basel: Karger.
  1973 Field and laboratory studies of social organization in *Saimiri* and *Callicebus.* In L. A. Rosenblum, ed., *Primate Behavior.* Vol. 2. New York: Academic Press.

Mason, W. A.; Davenport, R. K.; and Menzel, E. W. Jr.
  1968 Early experience of social development of rhesus monkeys and chimpanzees. In G. Newton and S. Levine, eds., *Early Experience and Behavior,* pp. 440-80. Springfield: Thomas.

Maynard Smith, J.
  1964 Group selection and kin selection. London: *Nature* 201:1145-7.
  1972 *On Evolution.* Edinburgh: Edinburgh University Press.
  1974 The theory of games and the evolution of animal conflicts. *J. Theor. Biol.* 47:209-21.

Maynard Smith, J., and Price, G. R.
  1973 The logic of animal conflict. *Nature* 246:15-18.

Mayr, E.
  1972 Sexual selection and natural selection. In B. Campbell, ed., *Sexual selection and the descent of man 1871-1971.* Chicago: Aldine.

McClure, H. E.
  1966 Flowering, fruiting and animals in the canopy of a tropical rain forest. *The Malayan Forester* 29:182-203.

McGinnis, P. R.
  1973 Patterns of sexual behaviour in a community of free-living chimpanzees. Ph.D. thesis, Cambridge University.

McGrew, W. C.
    1974a Patterns of plant food sharing by wild chimpanzees. In *Proceedings of the 5th Congress of the International Primatolotical Society, Nagoya, Japan.* Basel: Karger.
    1974b Tool use by wild chimpanzees in feeding upon driver ants. *J. Hum. Evol.* 3.
    1975 Socialization and object manipulation in wild chimpanzees. In S. Chevalier-Skolnikoff and F. E. Poirier, eds., *Primate Socialization.* Chicago: Aldine-Atherton.

Mead, G. H.
    1934 *Mind, self and society.* Chicago: Univ. of Chicago Press.

Mech, L. D.
    1966 *The wolves of isle royale.* Washington, D.C.: U.S. Government Printing Office.

Mellgren, R.; Fouts, R.; and Lemmon, W.
    1973 American sign language in the chimpanzee: Semantic and conceptual functions of signs. Paper presented at the Midwestern Psychological Association Meeting in Chicago, May 1973.

Menzel, E. M., Jr.
    1960 Selection of food by size in the chimpanzee, and comparison with human judgments. *Science* 131:1527-28.
    1966 Responsiveness to objects in free-ranging Japanese monkeys. *Behaviour* 26:130-50.
    1969 Chimpanzee utilization of space and responsiveness to objects: Age differences and comparison with macaques. In C. R. Carpenter, ed., *Proceedings of the 2nd International Congress of Primatology, Atlanta.* Basel: Karger.
    1971a Group behavior in young chimpanzees: Responsiveness to cumulative novel changes in a large outdoor enclosure. *J. comp. physiol. Psychol.* 74:46-51.
    1971b Communication about the environment in a group of young chimpanzees. *Folia primat.* 15:220-32.
    1973a Leadership and communication in a chimpanzee community. In E. W. Menzel, Jr., ed., *Precultural primate behaviour,* pp. 192-225. Basel: Karger.
    1973b Chimpanzee spatial memory organization. *Science* 182:943-45.
    1974 A group of young chimpanzees in a one-acre field. In A. M. Schrier and F. Stollnitz, *Behavior of nonhuman primates.* Vol. 5, pp. 83-153. New York: Academic Press.

Meyburg, B.
    1974 Sibling aggression and mortality among nestling eagles. *Ibis* 116:224-8.

Michotte, A.
    1950 The emotions regarded as functional connections. In Reymert, ed., *Feelings and emotions: The Moosehart Symposium,* pp. 114-26. New York: McGraw-Hill.

Miles, R. C.
 1965   Discrimination-learning sets. In A. M. Schrier, H. Harlow, and F. Stoll-
        nitz, eds., *Behavior of Nonhuman Primates*. Vol. 1, pp. 51-95. New
        York: Academic Press.

Miller, N. E., and Dollard, J.
 1941   *Social learning and imitation*. New Haven: Yale University Press.

Miller, R. E., and Banks, J. H.
 1962   The determination of social dominance in monkeys by a competitive
        avoidance method. *J. Comp. Physiol. Psychol.* 55:137-41.

Mohnot, S. M.
 1971   Some aspects of social changes and infant-killing in the hanuman lan-
        gur, Presbytis entellus (Primates: Cercopithecidae) in Western India.
        Mammalia 35:175-98.

Montagu, A.
 1968   The new litany of "innate depravity" or original sin revisited. In A.
        Montagu, ed., *Man and aggression*. pp. 3-17.
 1976   *The nature of human aggression*. New York: Oxford University Press.

Moorehead, A.
 1960   *The white Nile*. New York: Harper and Row.

Morgan, E.
 1972   *The descent of woman*. New York: Bantam.

Morris, D.
 1952   Homosexuality in the ten-spined stickleback *(Pygosteus sungitus L.)*.
        Behaviour 4:233-61.
 1957   'Typical Intensity' and its relation to the problem of ritualization.
        Behaviour 11:1-12.

Morrison, J. A., and Menzel, E. W., Jr.
 1972   Adaptation of a free-ranging rhesus monkey group to division and trans-
        portation. *Wildlife Monographs* 31:1-78.

Morse, D. H.
 1970   Ecological aspects of some mixed-species foraging flocks of birds.
        Ecological Monographs 40:119-68.

Moyer, K. E.
 1968   Kinds of aggression and their physiological basis. *Communications in
        Behavioral Biology* 2:65-87.

Murdock, G. P.
 1957   World ethnographic sample. *American Anthropologist* 59:664-87.

Murdock, G. P., and Provost, C.
 1973   Factors in the division of labour by sex: A cross-cultural analysis.
        Ethnol. 12:203-25.

Myers, R. E.
 1976   Comparative neurology of vocalization and speech: Proof of a dichot-
        omy. In S. R. Harnad, H. D. Steklis, and J. Lancaster, eds., *Origins and
        Evolution of Language and Speech,* pp. 745-57. New York Academy of
        Sciences.

Mykytowycz, R., and Dudzinski, M.
    1972   Aggressive and protective behaviour of adult rabbits *Oryctolagus cuniculus L.* towards juveniles. *Behaviour* 43:7-120.

Nagel, U.
    1971   Social organization in a baboon hybrid zone. *Proc. 2nd Int. Congr. Primatol., Zurich, 1970.* Basel: Karger.

Napier, J. R., and Napier, P. H.
    1967   *A handbook of living primates.* London: Academic Press.

Nerlove, S. B.
    1974   Women's workload and infant feeding practices: A relationship with demographic implications. *Ethnol.* 13:207-14.

Nishida, T.
    1966   A sociological study of solitary male monkeys. *Primates* 7:141-204.
    1967   The society of wild chimpanzees. *Shizen* 22:31-41.
    1968   The social group of wild chimpanzees in the Mahali Mountains. *Primates* 9:167-224.
    1970   Social behavior and relationship among wild chimpanzees of the Mahali Mountains. *Primates* 11:47-87.
    1972a A note on the ecology of the red colobus monkeys living in the Mahali Mountains. *Primates* 13:57-64.
    1972b Preliminary information on the pygmy chimpanzees *(Pan paniscus)* of the Congo Basin. *Primates* 13(5):415-25.
    1973a *Children of the Mountain Spirits.* Tokyo: Chikuma-Shobo.
    1973b The ant-gathering behaviour by the use of tools among wild chimpanzees of the Mahali Mountains. *J. Hum. Evol.* 2:357-70.
    1974   Ecology of wild chimpanzees. In R. Ohtsuka et al., eds., *Human Ecology,* pp. 15-60. Tokyo: Kyoritsu-Shuppan.
    In   Chimpanzees of the Mahali Mountains. (I) Ecology and the structure of
    press the unit-group.

Nishida, T., and Kawanaka, K.
    1972   Inter-unit-group relationships among wild chimpanzees of the Mahali mountains. *Kyoto Univ. Afr. Stud.* 7:131-69.

Nishida, T.; Uehara, S.; and Nyundo, R.
    In   Predatory behavior of wild chimpanzees of the Mahale Mountains.
    press

Nishimura, A.
    1973   The third fission of a Japanese monkey group at Takasakiyama. In C. R. Carpenter, ed., *Behavioral regulators of behavior in primates.* Lewisburg: Bucknell Univ. Press.

Nissen, H. W.
  1931 A field study of the chimpanzee. *Comp. Psychol. Monogr.* 8:1-122.
  1946 Primate psychology. In P. L. Harriman, ed., *Encyclopedia of psychology,* pp. 546-70. New York: Philosophical Library.
  1954 Development of sexual behavior in chimpanzees. Amherst Symposium: Genetic, psychological and hormonal factors in the establishment and maintenance of patterns of sexual behavior in mammals.
  1956 Individuality in the behavior of chimpanzees. *Amer. Anthropol.* 58: 407-13.

Nissen, H. W., and Crawford, M. P.
  1936 A preliminary study of food-sharing behavior in young chimpanzees. *J. Comp. Psychol.* 22:383-419.

Nowlis, V.
  1941 The relation of degree of hunger to competitive interaction in chimpanzees. *J. Comp. Psychol.* 32:91-115.
  1941 Companionship preference and dominance in the social interaction of young chimpanzees. *Comparative Psychology Monograph* 17:1-57.

Oakley, K. P.
  1965 *Man the tool-maker.* London: British Museum (Natural History).

Oki, J., and Maeda, Y.
  1973 Grooming as a regulator of behavior in Japanese macaques. In C. R. Carpenter, ed., *Behavioral regulators of behavior in primates.* Lewisburg: Bucknell University Press.

Osborn, R. M.
  1963 Behaviour of the mountain gorilla, *Symp. Zool. Soc. Lond.* 10:29-37.

Oswalt, W. H.
  1973 *Habitat and technology.* New York: Holt, Rinehart and Winston.

Owen-Smith, R. N.
  1974 The social system of the white rhinoceros. In V. Geist and F. Walther, eds., *The Behavior of Ungulates and its Relation to Management.* Vol. 1, Morges, Switzerland: IUCN Publications.

Packer, C.
  1975 Male transfer in olive baboons. *Nature* 255:219-20.
  In   Male transfer and intertroop relationships. Ph.D. thesis, Sussex University.
  press

Parker, G. A.
  1974a Assessment strategy and the evolution of fighting behaviour. *Journal Theor. Biol.* 47:223-43.
  1974b Courtship persistence and female-guarding as male time investment strategies. *Behaviour* 48:157-84.

Parker, S.
  1976 The precultural basis of the incest taboo: toward a biosocial theory. *American Anthropologist* 78:285-305.

Parsons, T., and Shils, E. A., eds.
  1951   *Towards a General Theory of Action.* Cambridge: Harvard University Press.

Pavlov, I. P.
  1957   *Experimental psychology and other essays.* New York: Philosophical Library.

Penniman, T. K., ed.
  1965   *A hundred years of anthropology.* 3rd ed. London: Gerald Duckworth.

Pfeiffer, J. E.
  1972   *The emergence of man.* 2nd ed. New York: Harper and Row.

Phillips, R. E.
  1972   Sexual and agonistic behaviour in the killdeer. *Animal Behavior* 20:1-9.

Pianka, E. R.
  1970   On r- and K-selection. *American Naturalist* 104:592-97.
  1972   "r" and K selection or b and d selection. *American Naturalist* 106:581-88.

Pierce, J. R.
  1972   Communication. *Sci. Amer.* 27:31-41.

Pinneau, S. R.
  1955   The infantile disorders of hospitalism and anaclitic depression. *Psychol. Bull.* 52:429-52.

Poirier, F. E.
  1968   Nilgiri langur *(Presbytis johnii)* territorial behavior. *Primates* 9:351-64.
  1972   *Primate Socialization.* New York: Random House.

Polanyi, M.
  1959   *The study of man.* Chicago: University of Chicago Press.

Premack, A. J., and Premack, D.
  1972   Teaching language to an ape. *Scientific American* 227:92-99.

Premack, D.
  1970   A functional analysis of language. *Journal of the Experimental Analysis of Behavior* 14:107-25.
  1971a Language in chimpanzee? *Science* 172:808-22.
  1971b On the assessment of language competence in the chimpanzee. In A. M. Schrier and F. Stollnitz, eds., *Behavior of Nonhuman Primates.* 4, pp. 56-228. New York: Academic Press.
  1972   Concordant preferences as a precondition for affective but not symbolic communication (or How to do experimental anthropology). Paper presented at Conference Behavioral Basis of Mental Health, Galway.

Prestrude, A. M.
  1970   Sensory capacities of the chimpanzee: A review. *Psychol. Bull.* 74:47-67.

Pusey, A.
  1977   Social development of adolescent chimpanzees in Gombe National Park. Ph.D. thesis, Stanford University.

Rahaman, H., and Parthasarathy, M. D.
  1969   Studies on the social behavior of bonnet monkeys. *Primates* 10:149-62.

Rahm, U.
  1972   L'emploie d'outils par les chimpanzées de l'ouest de la Cote-d'Ivoire. *Terre et la Vie* 25:506-9.

Ralls, R.
  1974   Scent marking in captive Maxwell's duikers. In V. Geist and F. Walther, eds., *The behavior of ungulates and its relation to management.* Vol. 1, Morges, Switzerland: IUCN Publications, pp. 114-23.

Ransom, T. W.
  1972   Ecology and social behavior of baboons in the Gombe National Park. Doctoral dissertation, University of California, Berkeley.

Reed, T., and Gallagher, E.
  1962   Gorilla birth at National Zoological Park, Washington. *Zool. Garten* 27: 279-92.

Rensch, B., and Döhl, J.
  1968   Wahlen zwischen zwei überschaubaren Labyrinthwegen durch einen Schimpansen. *Z. Tierpsychol.* 25:216-31.

Reynolds, P.
  1972   Play and the evolution of language. Ph.D. thesis, Yale University.

Reynolds, V.
  1966   Open groups in honinid evolution. *Man* (n.s.) 1:441-52.

Reynolds, V., and Luscombe, G.
  1969   Chimpanzee rank order and the function of displays. In C. R. Carpenter, ed., *Proceedings of the 2nd International Congress of Primatology, Atlanta.* Basel: Karger.

Reynolds, V., and Reynolds, F.
  1965   Chimpanzees in the Budongo Forest. In I. DeVore, ed., *Primate behavior: Field studies of monkeys and apes.* New York: Holt, Rinehart and Winston.

Ribbands, C. R.
  1954   The defence of the honeybee community. *Proc. Roy. Soc. B.* 142: 514-24.

Ribble, M. A.
  1943   *The rights of infants: Early psychological needs and their satisfaction.* New York: Columbia University Press.

Richards, S. M.
    1974   The concept of dominance and methods of assessment. *Animal Behaviour* 22:914-30.

Riesen, A. H.
    1966   Sensory deprivation. In E. Stellar and J. Spraque, eds., *Progress in physiological psychology.* New York: Academic Press.
    1970   Chimpanzee visual perception. In G. H. Bourne, ed., *The chimpanzee,* *Vol. 2.* Basel: Karger.

Rijksen, H. D.
    1974   Social structure in a wild orang-utan population. *Proceedings of the 6th International Congress of Primatology.* London: Academic Press.

Ripley, S.
    1967   Intertroop encounters among Ceylon grey langurs *(Presbytis entellus).* In S. A. Altmann, ed., *Social communication among primates.* Chicago: University of Chicago Press.
    1970   The social organization of foraging in gray langurs. In J. R. Napier and P. H. Napier, eds., *Old World monkeys,* pp. 481-509. New York: Academic Press.

Riss, D. C.
    In   An account of the take-over of the alpha rank in a Gombe Stream chim-
    prepa- panzee community.
    ration

Robinson, B. W.
    1976   Limbic influences on human speech. In S. R. Harnad, H. D. Steklis, and J. Lancaster, eds., *Origins and evolution of language and speech,* pp. 761-71. New York Academy of Sciences.

Rodman, P. S.
    1971   Orang-utans of the Kutai Nature Reserve. Presented to a joint symposium of the Zoological Society of London and the Society for the Study of Animal Behaviour. November 1971.
    1973a   Population composition and adaptive organization among orang-utans of the Kutai Reserve. In R. P. Michael and J. H. Crook, eds., *Ecology and behaviour of primates,* pp. 171-209. London and New York: Academic Press.
    1973b   Synecology of Bornean primates. Ph.D. thesis, Harvard University, Cambridge, Mass.

Rogers, C. M., and Davenport, R. K.
    1969   Effects of restricted rearing on sexual behavior of chimpanzees. *Develop. Psychol.* 1:No. 3.
    1970   Chimpanzee maternal behavior. In G. H. Bourne, ed., *The chimpanzee.* Vol. 3. Basel: Karger.
    1971   Intellectual performance of differentially reared chimpanzees. III. Oddity. *Amer. J. ment. Defic.* 75:No. 4.

Rosenbleuth, A.; Wiener, N.; and Bigelow, J.
    1943   Behavior, purpose, and teleology. *Philosophy of Science* 10:18-24.

Rosenblum, L. A.
    1971   Kinship interaction patterns in pigtail and bonnet macaques. *Proc. 3rd Internat. Congress. Primatol., Zurich, 1970.* Basel: Karger.

Roth, W. E.
    1901   Food: Its search, capture and preparation. *North Queensland Ethnography Bull.* 3:1-31.

Rothenbuhler, W. C.
    1964   Behavior genetics of nest cleaning in honeybees. IV. responses of F_1 and backcross generations to disease killed brood. *American Zoologist* 4:111-23.

Rowell, T. E.
    1964   The habitat of baboons in Uganda. *Proceedings of the East African Academy* 2:121-27.
    1966a Forest living baboons in Uganda. *J. Zool.,* Lond. 149:344-64.
    1966b Hierarchy in the organization of a captive baboon group. *Anim. Behav.* 14:430-33.
    1967   A quantitative comparison of the behaviour of a wild and a caged baboon group. *Animal Behaviour* 15:499-509.
    1968   Grooming by adult baboons in relation to reproductive cycles. *Animal Behaviour* 16:585-88.
    1974   The concept of social dominance. *Behavioral Biology* 11:131-54.

Rowell, T. E.; Hinde, R. A.; and Spencer-Booth, Y.
    1964   "Aunt"-infant interaction in captive rhesus monkeys. *Anim. Behav.* 12:219-26.

Rudran, R.
    1973   Adult male replacement in one-male troops of purple-faced langurs *(Presbytis senex senex)* and its effects on population structure. *Folia primat.* 19:166-92.

Rumbaugh, Duane M.
    1970   Learning skills of anthropoids. In L. A. Rosenblum, ed., *Primate behavior: Developments in field and laboratory research.* Vol. 1. New York: Academic Press.

Rumbaugh, D.; Gill, T. V.; and von Glaserfeld, E. C.
    1973   Reading and sentence completion by a chimpanzee (Pan). *Science* 182:731-33.

Russell, C., and Russell, W. M. S.
    1968   *Violence, monkeys and man.* London: Macmillan.

Saayman, G. S.
    1971a Behaviour of chacma baboons. *Afr. Wild Life* 25:25-29.
    1971b Baboons' response to predators. *Afr. Wild Life* 25:46-49.

Sackett, G. P.
    1968   The persistence of abnormal behaviour in monkeys following isolation rearing. In R. Porter, ed., *The role of learning in psychotherapy.* London: Churchill.

Sade, D. S.

  1967 Determinants of dominance in a group of free-ranging rhesus monkeys. In S. A. Altmann, ed., *Social communication among primates*. Chicago: University of Chicago Press.

  1972a A longitudinal study of social behavior of rhesus monkeys. In R. Tuttle, ed., *The functional and evolutionary biology of rhesus monkeys*. Chicago: Aldine-Atherton.

  1972b Sociometrics of *Macaca mulatta*. I. Linkages and cliques in grooming matrices. *Folia. Primatol.* 18:196-223.

Sahlins, M. D.

  1965 The social life of monkeys, apes and primitive man. In J. N. Spuhler, ed., *The evolution of man's capacity of culture*. Detroit: Wayne State University Press.

Savage, T. S., and Wyman, J.

  1847 Notice of the external characters and habits of troglodytes gorilla, a new species of orang. *Jour. Nat. Hist.* 5:417-43.

Sawin, P. B., and Crary, D. D.

  1953 Genetical and physiological background of reproduction in the rabbit II. Some racial differences in the pattern of behavior. *Behaviour* 6: 128-46.

Schaller, G.

  1961 The orangutan in Sarawak. *Zoologica* 46:73-82.

  1963 *The mountain gorilla: Ecology and behaviour*. Chicago: University of Chicago Press.

  1972 *The Serengeti lion*. Chicago and London: University of Chicago Press.

Schaller, G. B., and Lowther, G. R.

  1969 The relevance of carnivore behavior to the study of early hominids. *Southwest. J. Anthropol.* 25:307-41.

Schiller, P. H.

  1951 Figural preferences in the drawings of a chimpanzee. *J. comp. physiol. psychol.* 44:101-11.

Schnierla, T.

  1972 Problems in the biopsychology of social organization. In L. R. Aronson, E. Tobach, J. Rosenblatt, and D. Lehrman, eds., *Selected writings of T. C. Schnierla*, pp. 417-39. San Francisco: W. H. Freeman.

Schoener, T. W.

  1971 Theory of feeding strategies. *Annual Review of Ecology and Systematics* 2:369-404.

Schouteden, H.

  1929 A propos du chimpanzé de la rive gauche du Congo. *Bull. cercl. Zool. Congol.* 6.

  1930 Les chimpanzés de la rive gauche du Congo. *Bull. cercl. Zool. Congol.* 6:67-69.

  1930 Le chimpanzé de la rive gauche du Congo. *Bull cercl. Zool. Congol.* 7:114-19.

1931　Quelques notes sur le chimpanzé de la rive gauche du Congo, *Pan saty-rus paniscus. Rev. zool. Bot. Afr.* 20:310-14.

1936　Le chimpanzé de la rive gauche du fleuve. *Bull. cercl. Zool. Congol.* 13:15.

Schultz, A. H.
1938　The relative weight of testes in primates. *Anat. Rec.* 72:387-94.
1969a　*The life of primates.* New York: The Universe Natural History Series.
1969b　*The life of primates.* London: Weidenfeld and Nicolson.

Schwarz, E.
1928　Le chimpanzé de la rive gauche du Congo. *Bull. cercl. Zool. Congol.* 5: 70-71.
1934　On the local races of the chimpanzee. *Ann. Mag. Nat. Hist.* 10:576-83.

Scott, J. P.
1956　The analysis of social organization in animals. *Ecology* 37:213-21.
1958　*Aggression.* Chicago: University of Chicago Press.

Scott, J. P., and Fredericson, E.
1951　The causes of fighting in mice and rats. *Physiol. zool.* 24:273-309.

Scott, J. P., and Fuller, J. L.
1965　*Genetics and the social behavior of the dog.* Chicago: University of Chicago Press.

Scott, J. W.
1942　Mating behavior of the sage grouse. *Auk* 59:477-98.

Scott, N. J.; Malmgren, L. A.; and Glander, K. E.
In　Grouping behaviour and sex ratio in mantled howling monkeys *(Alou-press atta palliata).* In D. J. Chivers and C. Harcourt, eds., *Proc. 6th Congr. Int. Primatal. Soc.* Vol. 1. London: Academic Press.

Selander, R. K.
1972　Sexual selection and dimorphism in birds. In B. Campbell, ed., *Sexual selection and the descent of man 1871-1971.* Chicago: Aldine.

Siegel, S.
1956　*Non-parametric statistics for the behavioral sciences.* New York: McGraw-Hill.

Simmons, K.
1970　Ecological determinants of breeding adaptation and social behavior in two fish-eating birds. In J. H. Crook, ed., *Social behavior in birds and mammals,* pp. 37-77. New York: Academic Press.

Simonds, P. E.
1965　The bonnet macaque in south India. In I. DeVore, ed., *Primate behavior: Field studies of monkeys and apes.* New York: Holt, Rinehart and Winston.

Simpson, M. J. A.
1973a　The social grooming of male chimpanzees. In R. P. Michael and J. H. Crook, eds., *Comparative ecology and behaviour of primates.* New York: Academic Press.

1973b Social displays and the recognition of individuals. In P. P. G. Bateson and P. H. Klopfer, eds., *Perspectives in ethology.* New York: Plenum.

Skinner, B. F.
1974 *About behaviorism.* New York: Alfred Knopf.

Smith, D. A.
1973 Systematic study of chimpanzee drawing. *J. comp. physiol. Psychol.* 82:406-14.

Sommerhoff, G.
1950 *Analytical biology.* London: Oxford University Press.

Southwick, C. H.
1967 An experimental study of intragroup agonistic behavior in rhesus monkey *(Macaca mulatta). Behaviour* 28:182-209.

Southwick, C. H.; Beg, M. A.; and Siddiqi, M. R.
1965 Rhesus monkeys in north India. In I. DeVore, ed., *Primate behavior: Field studies of monkeys and apes.* New York: Holt, Rinehart and Winston.

Spencer-Booth, Y.
1968 The behaviour of group companions towards rhesus monkey infants. *Anim. Behav.* 16:541-57.

Spitz, R. A.
1945 Hospitalism: an inquiry into the genesis of psychiatric conditions in early childhood. *The psycho-analytic study of the child.* Vol. 1. International University Press.
1956 The influence of the mother-child relationship, and its disturbances. In Soddy, *Mental health and infant development.* Vol. 1, p. 103. New York: Basic Books.

Spitz, R. A., and Wolfe, K. M.
1946 Anaclitic depression: An inquiry into the genesis of psychiatric conditions in early childhood. *The psycho-analytic study of the child.* Vol. 2. International University Press.

Steward, J. H.
1968 Causal factors and processes in the evolution of prefarming societies. In R. B. Lee and I. DeVore, eds., *Man the hunter,* pp. 321-34. Chicago: Aldine-Atherton.

Stoltz, L. P., and Saayman, G. S.
1970 Ecology and behaviour of baboons in the northern Transvaal. *Ann. Transv. Mus.* 26:99-143.

Struhsaker, T. T.
1967 Social structure among vervet monkeys *(Cercopithecus aethiops). Behaviour* 29:83-121.
1969 Correlates of ecology and social organization among African cercopithecines. *Folia Primatol.* 9:123-34.

Struhsaker, T. T., and Gartlan, J. S.
    1970   Observations on the behavior and ecology of the patas monkey *(Ery-throcebus patas)* in the Waya Reserve, Cameroon. *J. Zool.* 161:49-63.

Struhsaker, T. T., and Hunkeler, P.
    1971   Evidence of tool-using by chimpanzees in the Ivory Coast. *Folia Primatol.* 15:212-19.

Sugiyama, Y.
    1960   On the division of a natural troop of Japanese monkeys at Tagasaki-yama. *Primates* 2:109-48.
    1967   Social organization of hanuman langurs. In S. A. Altmann, ed., *Social communication among primates,* pp. 221-236. Chicago: University of Chicago Press.
    1968 · Social organization of chimpanzees in the Budongo forest, Uganda. *Primates.* 9:225-58.
    1969   Social behavior of chimpanzees in the Budongo forest, Uganda. *Primates* 10:197-225.
    1973   The social structure of wild chimpanzees: a review of field studies. In R. P. Michael and J. H. Crook, eds., *Comparative ecology and behaviour of primates,* pp. 375-410. London: Academic Press.
    1976   Life history of male Japanese monkeys. *Advances in the science of behaviour* 7:255-84.

Suzuki, A.
    1971   Carnivority and cannibalism observed among forest-living chimpanzees. *J. Anthrop. Soc. Nippon* 79:30-48.

Sweeney, G.
    1947   Food supplies of a desert tribe. *Oceania* 17:289-99.

Syme, G. J.
    1974   Competitive orders as measures of social dominance. *Animal Behavior* 22:931-40.

Taylor, C. R.; Schmidt-Nielson, K.; and Robb, J. L.
    1970   Scaling of energetic cost of running to body size in mammals. *Amer. J. Physiol.* 219:1104-07.

Teleki, G.
    1973   *The predatory behavior of wild chimpanzees.* Lewisburg, Pa.: Bucknell University Press.

Thibaut, J. W., and Kelly, H. H.
    1959   *The social psychology of groups.* New York: Wiley.

Tiger, L., and Fox, R.
    1971   *The imperial animal.* New York: Dell.

Tinbergen, N., and van Iersel, J. J.
    1947   Displacement reactions in the three-spined stickleback. *Behaviour* I:56-63.

Tokuda, K.
1961-   A study of the sexual behavior in the Japanese monkey troop. *Primates*
1962   3(2):1-40.

Tolman, E. C.
1932   *Purposive behavior in animals and men.* New York: Century.

Trivers, R. L.
1971   The evolution of reciprocal altruism. *Quart. Rev. Biol.* 46:35-57.
1972   Parental investment and sexual selection. In B. Campbell, ed., *Sexual selection and the descent of man 1871-1971.* Chicago: Aldine.
1974   Parent-offspring conflict. *American Zoologist* 14:249-64.

Turnbull, C. M.
1965   The Mbuti pygmies: An ethnographic survey. *Anthropol. Pap. Amer. Mus. Nat. Hist.* 50:137-282.

Tutin, C. E. G.
1975a   Exceptions to promiscuity in a feral chimpanzee community. *Proceedings of the 5th Congress of the International Primatological Society, Nagoya, Japan, 1974.* Basel: Karger.
1975b   Mating patterns in a community of wild chimpanzees. Ph.D. thesis, University of Edinburgh.

Ulrich, R.
1966   Pain as a cause of aggression. *American Zoologist* 6:643-62.

van Hoof, J. A. R. A. M.
1967   The facial displays of the catarrhine and apes. In D. Morris, *Primate ethology,* pp. 7-68. London: Weidenfeld & Nicholson.
1971   *Aspects of the social behaviour and communication in human and higher non-human primates.* Rotterdam: Bronder-Offset.
1972   A comparative approach to the phylogeny of laughter and smiling. In R. A. Hinde, ed., *Non-verbal communication.* London: Cambridge Univ. Press.

van Lawick, Hugo
1973   *Solo: The story of an African wild dog.* London: Collins.

Virgo, H. B., and Waterhouse, M. J.
1969   The emergence of attention structure amongst rhesus macaques. *Man* (n.s.) 4:85-93.

von Frisch, K.
1955   *The dancing bees.* New York: Harcourt.

Wallace, A. R.
  1869   *The Malay archipelago.* London: Macmillan.

Wallace, B.
  1973   Misinformation, fitness, and selection. *American Naturalist* 107:1-7.

Walther, F.
  1974   Some reflections on expressive behaviour in combats and courtships of certain horned ungulates. In V. Geist and F. Walther, eds., *The behaviour of ungulates and its relation to management.* Vol. 1, pp. 56-106. Morges, Switzerland: ICUN Publications.

Ward, P.
  1965   Feeding ecology of the black-faced dioch *Quelea quelea* in Nigeria. *Ibis* 107:173-214.

Ward, P., and Zahavi, A.
  1973   The importance of certain assemblages of birds as "information-centres" for food-finding. *Ibis* 115:517-34.

Warren, J. M., and Maroney, R. J.
  1958   Competitive social interaction between monkeys. *J. Soc. Psychol.* 48: 223-33.

Washburn, S. L.
  1972   Human evolution. In T. Dobzhansky, M. K. Hecht, and W. C. Steere, eds., *Evolutionary biology.* Vol. 6., pp. 349-60. New York: Appleton-Century-Crofts.

Washburn, S. L., and Hamburg, D. A.
  1968   Aggressive behavior in Old World monkeys and apes. In P. C. Jay, ed., *Primates: Studies in Adaptation and Variability,* pp. 458-78. New York: Holt, Rinehart and Winston.

Washburn, S. L., and Lancaster, C. S.
  1968   The evolution of hunting. In R. B. Lee and I. DeVore, eds., *Man the hunter,* pp. 293-303. Chicago: Aldine-Atherton.

Washburn, S. L., and Harding, R. S. O.
  1975   Evolution and human nature. In D. A. Hamburg and H. K. H. Brodie, eds., *American handbook of psychiatry,* 2nd ed. Vol. BI, pp. 3-13. New York: Basic Books.

Watson, A., and Moss, R.
  1970   Dominance, spacing behaviour and aggression in relation to population limitation in vertebrates. In A. Watson, ed., *Animal populations in relation to their food resources.* Oxford: Blackwell Scientific Publications.

Weiss, G.
  1973   A scientific concept of culture. *Amer. Anthropol.* 75:1376-1413.

Westermarck, E. A.
  1891   *The history of human marriage* (reprinted 1921). New York: Allerton Book Co.

White, L. A.
1942   On the use of tools by primates. *J. comp. Psychol.* 34:369-74.

White, R. W.
1948   *The abnormal personality.* New York: Ronald Press.

Whiten, A., and Rumsey, T. J.
1973   'Agonistic buffering' in the wild barbary macaque, *Macaca sylvana L.* *Primates* 14:421-25.

Whyte, L. L.
1950   *The next development in man.* New York: New American Library of World Literature.

Williams, G. C.
1971   *Group selection.* Chicago: Aldine-Atherton.

Wilson, E. O.
1971   *The insect societies.* Cambridge: Harvard University Press.
1973a Review. *Science* 179:466-67.
1973b Group selection and its significance for ecology. *BioScience* 23(11): 631-38.
1975   *Sociobiology.* Cambridge, Mass.: Belknap Press of Harvard University Press.

Wilson, R. H.
1974   Agonistic postures and latency to the first interaction during initial pair encounters in the red jungle fowl, *Gallus gallus. Animal Behaviour* 22: 75-82.

Wilson, W. L., and Wilson, A. C.
1968   Aggressive interactions of captive chimpanzees living in a semi-free ranging environment. *Report of the 6571st Aeromedical Research Laboratory,* Holloman Air Force Base, New Mexico.

Wise, L. A., and Zimmerman, R. R.
1973   The effects of protein deprivation on dominance measured by shock avoidance competition and food competition. *Behavioral Biology* 9: 317-29.

Wislocki, G. B.
1942   Size, weight and histology of the testes in the gorilla. *Journal of Mammalogy* 23:281-87.

Woodburn, J.
1968   An introduction of Hadza ecology. In R. B. Lee and I. DeVore, eds., *Man the hunter,* pp. 49-55. Chicago: Aldine-Atherton.

Wrangham, R. W.
1974   Artificial feeding of chimpanzees and baboons in their natural habitat. *Animal Behaviour* 22:83-93.
1975   The behavioural ecology of chimpanzees in Gombe National Park, Tanzania. Ph.D. thesis. University of Cambridge.

Wynne-Edwards, V. C.
    1962   *Animal dispersion in relation to social behavior.* New York: Hafner.

Yamada, M.
    1963   A study of blood-relationship in the natural society of the Japanese macaque. *Primates* 4:43-65.
    1971   Five natural troops of Japanese monkeys on Stodoshima Island. II. A comparison of social structure. *Primates* 12:125-50.

Yerkes, R. M.
    1925   Traits of young chimpanzees. In R. M. Yerkes and B. W. Learned, eds., *Chimpanzee intelligence and its vocal expression.* Baltimore: Williams and Wilkins.
    1927   *Almost human.* New York: Century.
    1940   Social behavior of chimpanzees: dominance between mates, in relation to sexual status. *J. Comp. Psychol.* 30:147-86.
    1943   *Chimpanzees. A laboratory colony.* New Haven: Yale University Press.

Yerkes, R. M., and Yerkes, A. W.
    1929   *The great apes.* New Haven: Yale University Press.

Yoshiba, K.
    1968   Local and intertroop variability in ecology and social behavior of common Indian langurs. In P. C. Jay, ed., *Primates: Studies in adaptation and variability.* New York: Holt, Rinehart and Winston.

Zuckerman, S.
    1932   *The social life of monkeys and apes.* London: Routledge and Kegan Paul.
    1933   Functional affinities of man, monkeys and apes. New York: Harcourt, Brace.

Name Index

Subject Index